MEASUREMENT AND MODELLING IN ECONOMICS

CONTRIBUTIONS
TO
ECONOMIC ANALYSIS

195

NORTH-HOLLAND
AMSTERDAM • NEW YORK • OXFORD • TOKYO

MEASUREMENT AND MODELLING IN ECONOMICS

Edited by

Gareth D. MYLES
Department of Economics
University of Warwick
Coventry CV4 7AL, U.K.

1990

NORTH-HOLLAND
AMSTERDAM • NEW YORK • OXFORD • TOKYO

ELSEVIER SCIENCE PUBLISHERS B.V.
Sara Burgerhartstraat 25
P.O. Box 211, 1000 AE Amsterdam, The Netherlands

Distributors for the United States and Canada:

ELSEVIER SCIENCE PUBLISHING COMPANY INC.
655 Avenue of the Americas
New York, N.Y. 10010, U.S.A.

ISBN: 0 444 88515 3

PRINTED IN THE NETHERLANDS

INTRODUCTION TO THE SERIES

This series consists of a number of hitherto unpublished studies, which are introduced by the editors in the belief that they represent fresh contributions to economic science.

The term "economic analysis" as used in the title of the series has been adopted because it covers both the activities of the theoretical economist and the research worker.

Although the analytical methods used by the various contributors are not the same, they are nevertheless conditioned by the common origin of their studies, namely theoretical problems encountered in practical research. Since for this reason, business cycle research and national accounting, research work on behalf of economic policy, and problems of planning are the main sources of the subjects dealt with, they necessarily determine the manner of approach adopted by the authors. Their methods tend to be "practical" in the sense of not being too far remote from application to actual economic conditions. In additon they are quantitative.

It is the hope of the editors that the publication of these studies will help to stimulate the exchange of scientific information and to reinforce international cooperation in the field of economics.

The Editors

DEDICATION

This volume is dedicated to Terence Gorman on the occasion of his 67th birthday. Terence has made a unique and important contribution to the discipline of economics through both his research and his teaching. He has been a highly respected teacher or colleague to almost all the contributors to this volume and to many of today's distinguished economists. His influence can be seen throughout the following pages and in our discipline generally.

PREFACE

The papers contained in this volume represent the proceedings of a conference held at Nuffield College, Oxford in May 1987. Terence Gorman organised the conference and invited the participants. The Economic and Social Research Council provided most of the financial support.

The volume consists of eleven papers which cover a broad area of economic and econometric theory and practice. The papers are arranged in the order in which they were presented to the conference. Each paper is followed by discussion extracted from a recording of the proceedings. It is hoped that this will procedure will capture some of the atmosphere of the conference.

In preparing this volume I have benefitted from the advice of Terence Gorman and from the encouragement of the contributors. I would also like to thank Margaret Kitchen without whose patient and efficient secretarial skills the volume would never have been completed.

Gareth D Myles

CONTENTS

Dedication vii

Preface ix

List of conference participants xiii

1 Introduction 1

2 *Equations and Inequalities in the Theory of Measurement*
Wolfgang Eichhorn 11
Discussion. 26

3 *Nonparametric Tests of Additive Derivative Constraints*
Tom Stoker 43
Discussion 93

4 *Poverty Indices and Decomposability* James E Foster and
Anthony Shorrocks 109
Discussion 118

5 *Optimal Uniform Taxation and the Structure of Consumer*
Preferences Timothy Besley and Ian Jewitt 131
Discussion. 148
A Comment by Terence Gorman 153

6 *Aggregate Production Functions and Productivity Measurement:*
A New Look John Muellbauer 157
Discussion 204

7 *A Model of Cake Division with a Blind Player & Gamblers and*
Liars Christopher Bliss 219
Discussion 262

8 *The Estimation of Engel Curves* Stephen Pudney 267
Discussion 305

9 *Econometric Approaches to the Specification and Estimation of Intertemporal Consumer Behaviour* Richard Blundell 325
Discussion 371

10 *More Measures for Fixed Factors* Terence Gorman 381
Discussion 412

11 *Use and Misuse of Single-Consumer Results in a Multi-Consumer Economy: The Optimality of Proportional Commodity Taxation* Charles Blackorby, Russell Davidson and William Schworm 425
Discussion 446

Author Index 455

Subject Index 459

LIST OF CONFERENCE PARTICIPANTS

Name	Abbreviation
Professor Anindya Banerjee Department of Economics University of Florida	AnB
Dr Alan Beggs Nuffield College Univesrity of Oxford	AB
Dr Timothy Besley Woodrow Wilson School Princeton University	TB
Professor Charles Blackorby Department of Economics University of British Columbia	CB
Dr Christopher Bliss Nuffield College University of Oxford	CJB
Professor Richard Blundell University College London	RB
Professor Wolfgang Eichhorn University of Karlsruhe West Germany	WE
Professor James Foster Department of Economics University of Essex	JF
Mr Terence Gorman Nuffield College University of Oxford	TG
Professor Michael Jerison University of Bonn West Germany	MJ
Mr Ian Jewitt Department of Economics University of Bristol	IJ
Dr John Muellbauer Nuffield College University of Oxford	JM

Name	Abbreviation	Name	Abbreviation
Dr Gareth Myles Department of Economics University of Warwick	GM	Professor Anthony Shorrocks Department of Economics University of Essex	AS
Mr Stephen Pudney Department of Applied Economics University of Cambridge	SP	Professor Tom Stoker Sloan School of Management MIT	TS

INTRODUCTION

This volume presents the proceedings of a conference held at Nuffield College, Oxford, in May 1987. Terence Gorman organised the conference and drew up the list of invited participants. Financial support was provided mainly by the Economic and Social Research Council.

The formal title of the conference was "Measurement and Modelling in Economics". The speakers were chosen first with an eye to variety, secondly to conviviality, and thirdly for their willingness to venture outside their own fields and to discuss each others' work. The result was papers ranging from the econometric theory of Tom Stoker and Steve Pudney through applied econometrics and game theory to the applied mathematics of Wolfgang Eichhorn. Almost all came to all the papers and discussed them freely and agreeably, however far from their own fields.

In organising the conference, one of the prime objectives was the encouragement of discussion between participants, so that the presenters, in particular, could benefit from the ideas of people in quite different fields. With the increasing focus of modern conferences upon precise subject areas, and the sub–division of the larger conferences, such possibilities are becoming fewer. Having created an atmosphere that encouraged active participation, an attempt has been made to convey in these proceedings some of the character of the conference.

It would have been possible to nominate a formal discussant for each paper and to include a prepared comment in this volume. However, to reflect the spontaneity of the conference it was felt preferable to record the informal discussion that occurred as the paper was presented and to include this, in lightly edited form, in the conference volume. I hope that this process, which is something of a break with tradition, will help you to capture the spirit of the conference.

Wolfgang Eichhorn began the conference with a discussion of the "equation of measurement":

$$\phi(Ax + p) = \alpha(A,p)\phi(x) + \beta(A,p),$$

where α and β are known continuous functions and the function ϕ is to be determined by the equation. The name given to this equation reflects how important, and numerous, special cases of it are in economics and the sciences in general. Prior to his talk, Wolfgang had circulated a paper "On a Class of Inequality Measures" that described one special case of the equation of measurement and on the day he presented further applications in the theory of prices indices and aggregation. In addition, Wolfgang also talked about the solution of the general case. Much of the discussion focused on this solution and, in particular, upon the effects of introducing further restrictions.

The paper that Wolfgang has included in this volume describes the equation of measurement and review some of its applications in various areas of economics. The paper proceeds one step further than his conference presentation and introduces the "inequalities of inequality measurement", a set of inequalities that occur whenever two vectors are to be compared; a situation arising in the construction of measures of income inequality, statistical dispersion, risk, and information.

Whilst Wolfgang's paper was at the mathematical extreme of the spectrum, Tom Stoker's on "Nonparametric Tests of Additive Derivative Constraints", which came second, was at the statistical end. That the same people attended, and discussed, both was very agreeable and instrumental in making the conference a success.

The purpose of Tom's paper was to construct nonparametric tests of constraints upon econometric models that are additive in the derivatives of the model. Such additive derivative constraints arise naturally in many economic models, a typical example being the homogeneity properties that are implied by optimising behaviour. In addition, they may be used as prior restrictions on the model being estimated, for example the assumption of constant returns to scale. The tests themselves are based on nonparametric kernel estimators.

Several issues were raised in the discussion of the paper. Amongst these were the possibility of incorporating dummy variables into the model and the range of applications of the tests. Particular interest was shown in the relationship between the use of flexible functional forms and Tom's nonparametric tests and in the results of Monte Carlo tests of the methods. In fact, the Monte Carlo results indicate the tests to work well in realistically finite samples.

Tony Shorrocks and James Foster reported their joint work in the third paper, on "Subgroup Consistent Poverty Measures". The idea for subgroup consistency follows from the natural reasoning that if poverty falls in any subgroup of the population, and does not rise elsewhere, then measured poverty should also fall. After noting that the popular Foster, Greer, and Thorbecke family of indices satisfied this condition, the paper derived the general class of subgroup consistent measures. Following from this, Tony introduced the idea of pairs of relative and absolute measures of poverty which agree with each other in the sense that if either falls, so does the other. To satisfy this condition, the measures must be functions of the headcount ratio and a Foster, Greer, Thorbecke poverty index.

For this volume, Tony has written a detailed summary of the original conference paper. "Poverty Indices and Decomposability" details the motivation for considering subgroup consistent poverty indices and reports the major results. The conference discussion follows the paper closely and focuses on the interpretation and consistency of the standard axioms of poverty indices, the implications of the results and the strengthening of the continuity conditions.

In recent years a number of authors have presented alternative sets of preferences that are sufficient for uniform commodity taxation to be optimal. The next conference paper "Optimal Uniform Taxation and the Structure of Preferences", by Tim Besley and Ian Jewitt, placed previous contributions into context and drew out the interrelations between them. In particular, an implicit representation of the direct utility function was derived that encompassed the preference structures proposed by Angus Deaton and by James Mirrlees. For all utility functions implied by this representation, uniform taxation was shown to be optimal. The extension of the analysis to the taxation of "aggregate" goods was described, uniform taxation within each aggregate was optimal if the implicit representation was separable in each aggregate.

In their analysis Ian and Tim relied on the local separability conditions being extendable to a region and much of the discussion revolved around this issue. Also discussed was the most appropriate choice of modelling strategy; whether to approach the analysis via the primal of one of the dual representations. A further note, by Terence Gorman, follows the discussion and considers the issue of what the local conditions imply for the general form of preference relation.

In the fifth session John Muellbauer took a new look at "Aggregate Production Functions and Productivity Measurement". John considered there to be three major problems in the estimation of aggregate production functions: unobservables, aggregation and simultaneous equations effects. The theoretical and empirical arguments that John presented supported the case that the first of these, unobservables, was the most serious. To overcome the unobservability of undertime working for operatives, a method was employed for estimating undertime from observed data on overtime and a postulated trade–off between the two. A similar procedure was used to determine mean capital utilisation from survey data. The estimated production function indicated that productivity in the UK, after a substantial dip in 1979–80, had reverted back to its old trend line with a rate of growth of approximately 2.5%.

This conclusion generated a good deal of discussion about the nature of the data used, its possible shortcomings and about the modelling of the undertime–overtime trade–off. Also considered were the possible changes in managerial style over the period in question which might have implications for John's estimates of undertime.

Next came Christopher Bliss who covered two papers in his presentation: "A Model of Cake Division with a Blind Player" and "Gamblers and Liars". The inclusion of a blind player in "A Model of Cake Division" was intended to represent the asymmetric information that can characterise many bargaining situations. The concern of the paper was not the determination of the equilibrium division of the cake but in the implementation of the chosen division. Christopher derived an incentive compatible implementation scheme under which it was optimal for the sighted player to reveal the true size of the cake. This scheme possessed an inbuilt inefficiency in that if the announced size was below a certain value, the blind player would refuse to come to an agreement and the cake would be lost.

"Gamblers and Liars" investigated how schemes based on the notion of "random exclusion" should be implemented. Individuals would be invited to reveal private information but would be excluded from benefiting from the social use of this information in a random manner, the randomness allowing incentive compatible incentive schemes to be implemented. The moral of the story, and the reason for the title, was that it is preferable to enter into arbitrated bargaining with highly risk averse people rather than with "gamblers" with low risk aversion since gamblers require a high probability of random exclusion before they will tell the truth. Both the papers are included in this volume together with the discussion.

Stephen Pudney, presenting "The Estimation of Engel Curves" was concerned with statistical models that related long–term average rates of consumption to observed expenditure in short–term surveys. One of the problems of such surveys is the number of observed zero expenditures which, as in the case of clothes, do not accurately reflect consumption. Conversely for durable goods, as defined by the British Family Expenditure Survey (FES), there are fewer zero expenditures than would be expected due to the heterogeneity of this category, including as it does knives, forks, and the like.

To model the alternative reasons generating the zero consumptions, Stephen identified four types of good. Type 1 are those that are consumed by all, so that an observed zero expenditure is purely fortuitous. Type 2 goods have some economic non–consumers; people who would consume if the price were lower. The possibility of conscientious abstention categorised type 3 goods and, finally, type 4 combined abstention and economic non–consumption. For each type, Stephen proposed a statistical model that was designed to capture the behaviour in question.

The relationship between consumption and expenditure was then generated by a purchasing probability function. A simplified form of this model was applied to data from the FES and the results of this preliminary work were given in the paper. An estimation procedure for the general model was also described. The discussion centred on the nature of the data and on alternative models of the purchasing probability.

A considerable quantity of recent research in the microeconometrics of individual life–cycle models was surveyed by Richard Blundell in his paper "Econometric Approaches to the Specification and Estimation of Intertemporal Consumer Behaviour". Richard assumed, for the most part, intertemporal separability of utility so that unobservable life–cycle variables could be captured by a single summary statistic: the marginal utility of wealth. Furthermore, if the model was formulated so that the marginal utility entered the equations as an individual–specific intercept term, it could be eliminated via first differencing panel data. Richard also explored the restrictions placed upon preferences by the functional forms estimated by previous authors.

The paper also presented the results of an empirical application of the methodology to data from the FES. An important conclusion of this study was the fact that the homogeneity restriction was not rejected in the demand system, in marked contrast to the conclusions typical of aggregate studies. In the discussion, the assumption of intertemporal separability was raised as were the restrictions involved in within–period preferences. In addition, both the formation of the cohorts in the empirical application and the modelling of their preferences were extensively discussed.

As background reading for his paper "More Measures for Fixed Factors", Terence Gorman produced as a handout a primer on Cauchy's functional equation and its use in deriving affine equations in many areas of economics. This was illustrated by the derivation of a demand system for which aggregate demand was independent of the distribution of income and the construction of aggregates for intermediate goods. In the event, most of Terence's presentation was devoted to describing the applications in the handout leaving little time for the paper itself.

"More Measures" is concerned with exploring the implications of alternative restrictions on the relationship between net supply functions, output prices and the aggregate measures of fixed factors. In general, if supply can be written as a function of the aggregate measure then subaggregates must exist for each firm which sum to the economy–wide aggregate measure and the profit function for each firm must be of the form

$$\pi^f(p,u_f) = \lambda^f(a(p),u_f) + \mu^f(p)$$

where p represents the vector of prices and u_f the vector of firm f's fixed inputs. The strictness of this condition casts doubts upon the validity of macroeconomic exercises that employ aggregate measures of capital.

Charles Blackorby presented the final paper of the conference. "Use and Misuse of Single Consumer Results in a Multi-consumer Economy: an Application to the Optimality of Second-best Policy" which described the joint work of Charles Blackorby, Russell Davidson and William Schworm. It took as its starting point the results of Ian Jewitt that characterise when piecemeal second–best policy is optimal in a single-consumer model and considered what further restrictions must be placed upon preferences to extend this result to a many-consumer model. The main finding was that considerably stronger requirements must be met and substantial structure is placed upon the permissible forms of household preferences. The discussion of the paper considered the interpretation of the results and the motivation behind the use of single consumer models.

The paper included in this volume "Use and Misuse of Single-consumer Results in a Many-Consumer Economy: the Optimality of Proportional Commodity Taxation" applies the results of the conference paper to derive necessary and sufficient conditions upon community preferences to imply the optimality of proportional commodity taxes. Applying the results of the conference paper then determines the household preferences that will generate the required community preferences. The discussion that follows the paper is best seen as providing the results that form the foundation of the main paper.

In order to capture the discussion, each presentation was recorded on a reel–to–reel tape recorder, and then transferred onto cassette tapes. The next step was to transcribe questions and answers from the cassettes to paper. This, as anyone who has attempted it will be aware, is a time-consuming and awkward process. For the conference, fifteen hours of recordings had to be slowly sifted through. Having transferred the discussion to paper a first sorting was carried out. This eliminated questions that were just seeking points of clarification. Remaining comments were then adjusted into a vaguely grammatical form and the hand–written notes typed.

It is at this stage that the procedure varies for different papers – eventually four alternatives emerged. For the majority, the presentations were accompanied by a handout or by the use of an overhead projector. Where this was the case, the discussion was inserted at appropriate points into the handout or projector transparencies. This invariably involved a re-arrangement of the actual order of discussion. If no handout was provided or if the transparencies had too little structure, the discussion has simply been re–ordered to follow the paper.

The exceptions to the above rules are Richard Blundell's and Terence Gorman's papers. For his, Richard kindly provided a set of notes, written around the discussion, to connect this with both the main text and the handout that accompanied his presentation. In Terence's case, the handout was really just an introduction to the paper circulated. Since all the discussion related to the handout, it has not been keyed into the text of the paper itself.

In all of the papers the discussion may be found after the main text. Lower–case Roman numerals found as superscripts in the main text indicate relevant points of discussion. Finally, for some of the papers, equations in the discussion pages are numbered to correspond with equations in the main text. One complication was that participants were apt to point at equations on the board or screen and comment on them. Often they could be traced; when they could not, the comments had to be dropped.

Terence had urged the speakers to discuss work in progress, so that they and the audience should learn from each other. It worked. Most of the papers were changed in detail in response to the discussion, and several were completely rewritten, Chuck Blackorby's in particular to demonstrate the strength of the assumptions required for Tim Besley and Ian Jewitt's results to hold many consumer economies.

Measurement and Modelling in Economics
G.D. Myles (Editor)

EQUATIONS AND INEQUALITIES IN THE THEORY OF MEASUREMENT

by Wolfgang Eichhorn*

1 Introduction

Essential parts of my paper presented at the Nuffield conference on "Measurement and Modelling in Economics" have appeared elsewhere in the meantime (Eichhorn and Gleissner (1988)). Consequently, this contribution contains only an extended abstract of my paper. The space saved renders it possible to add one further idea. I had only sketched this idea at the end of my Nuffield talk for two reasons. My time had been at its last stage and my further idea at its first.

The paper is organized as follows. Section 2 is the extended abstract. It gives an insight into the various applications of an equation that we called the equation of measurement in Eichhorn and Gleissner (1988). One of the fields of application is the measurement of (economic) inequality, eg, of income inequality.

In this field a system of (mathematical) inequalities plays an even more important role than the equation of measurement. These inequalities will be introduced in Section 3. I propose to call them the inequalities of inequality measurement for the following reason. They turn out to be essential whenever the inequality (spread, diversity, dispersion or, inversely, concentration) of the components of a vector $x \in X^k$ has to be compared with the inequality of the components of a vector $y \in X^\ell$ (X a Banach space, k and ℓ natural numbers).

* Institut für Wirtschaftstheorie und OR, Universität Karlsruhe.

In Section 4 it will be shown that the inequalities mentioned above are essential for defining measures of:

— economic inequality (example: comparison of income distributions),
— statistical dispersion (example: comparison of distributions in statistics),
— risk (example: comparison of lotteries)

and

— information (example: comparison of information structures in information theory).

Section 5 concludes with some remarks on the necessity of characterising particular measures.

2 The Equation of Measurement

Influenced and impressed by certain functional equations and their applications to be found in Aczél (1988), Aczél, Roberts and Rosenbaum (1986) and Paganoni (1987), Gleissner and the author (1988) have proposed calling the functional equation

$$\phi(Ax + p) = \alpha(A,p)\phi(x) + \beta(A,p) \tag{0}$$

the <u>equation of measurement</u>.[ii] Here ϕ: $X \to V$ is an unknown function from a nonempty subset X of a real vector space U into a real vector space V, and the nonempty subsets $Y \ni p$ and $S \ni A$ of L(U) (the algebra of linear operators of U into itself) are such that

$$(Ax + p) \in X \text{ for all } x \in X,\ p \in Y,\ A \in S \subseteq L(U).$$

The functions $\alpha : S \times Y \to \mathbb{R}$ (the real numbers) and $\beta : S \times Y \to V$ are 'given'.[iv,v]

For obvious reasons Paganoni (1987) calls equation (0) a "functional equation concerning affine transformations".[vii,x] We call it the equation of measurement for reasons which are clear by what is written in Aczél (1988) and in Aczél, Roberts, and Rosenbaum (1986) about the 'laws of science', dimensional analysis, scales, and applications to measurement in economics. Furthermore, a wealth of additional contributions to the field of measurement in economics contain equations or systems of equations that are special cases of equation (0).[viii]

In what follows, we sketch some examples from the theories of measurement of prices, aggregation, utility and inequality. For details, see Eichhorn and Gleissner (1988).

2.1 Price Measurement

Let $\phi : \mathbb{R}^n_{++} \to \mathbb{R}_{++}$, where $\mathbb{R}_{++}$ is the set of all positive real numbers. With the real (n,n)–matrices A_1, A_2, A_3, A_4 and the (column) vector x suitably chosen the functional equations

$$\phi(A_1x) = \lambda\phi(x) \tag{1}$$

$$\phi(A_2x) = 1 \tag{2}$$

$$\phi(A_3x) = \phi(x) \tag{3}$$

$$\phi(A_4x) = \phi(x) \tag{4}$$

can be conceived as the axioms of linear homogeneity, identity, price dimensionality and commensurability for a (statistical) price index.[vi] Note that equations (1), (2), (3), and (4) are all special cases of equation (0). Here and in what follows, a vector at the right side of a matrix or a vector is conceived to be a column vector.

2.2 Aggregation

The problem of aggregation in production theory, say, leads, at least in an important special case, to the problem of solving the equation

$$\phi(x + p) = \phi(x) + f(p),$$

ie, equation (0) in the case where

$\phi:\mathbb{R}_+^n \to \mathbb{R}$ ($\mathbb{R}$ the real numbers, $\mathbb{R}_+$ the non–negative reals),
$A = I =$ unit matrix, $\alpha(A,p) \equiv 1$, $\beta(A,p) \equiv f(p) \in \mathbb{R}$.

2.3 Utility

Let $\phi : \mathbb{R}_+^n \to \mathbb{R}$ be a continuous nonconstant cardinal utility function. We call it scale–invariant for equal sacrifices (see Young (1987)) if

$$\phi(Tx) - \phi(Ty) = \phi(Tx') - \phi(Ty') \tag{5}$$

holds for every quadruple of commodity vectors $x, y, x', y' \in \mathbb{R}_+^n$ satisfying

$$\phi(x) - \phi(y) = \phi(x') - \phi(y'). \tag{6}$$

Here

$$T = \begin{bmatrix} t_1 & & & 0 \\ & t_2 & & \\ & & \ddots & \\ 0 & & & t_n \end{bmatrix}, \quad t_1, t_2, \ldots, t_n \in \mathbb{R}_{++}.$$

One can show (see Aczél (1987b)) that implication (6) $\Rightarrow$ (5) yields the functional equation

$$\phi(Tx) = \alpha(T)\phi(x) + \beta(T)$$

which again is a special case of equation (0).

2.4 Inequality Measurement

Let

$$x = (x_1,\dots,x_n) \in \mathbb{R}^n_+ \text{ and } y = (y_1,\dots,y_n) \in \mathbb{R}^n_+$$

be two vectors of incomes satisfying $x_1+\dots+x_n = y_1+\dots+y_n$. The vector x (or, rather, the distribution of incomes $x_1,\dots,x_n$) is called less unequal than the vector y if x can be composed from y by a finite sequence of transfers of the following kind: Subtract from an income, y_i say, the amount z_i ($y_i \geqq z_i \geqq 0$). Add z_i to another income, y_j say ($j \neq i$), but only if $y_j + z_i \leqq y_i$.

This definition is equivalent (see, eg, Marshall and Olkin (1979)) to: There exists a doubly stochastic (n,n)–matrix B, ie, the elements of B are nonnegative real numbers and the sum of the elements in each row as well as in each column of B equals one, such that

$$x = By. \tag{7}$$

Our definition establishes only a partial ordering. For example, two income situations

$$x = (x_1,\dots,x_k) \in \mathbb{R}^k_+ \text{ and } y = (y_1,\dots,y_\ell) \in \mathbb{R}^\ell_+$$

cannot be compared with respect to 'less unequal than' if

(i) $k \neq \ell$,

(ii) $k = \ell$ and $x_1+\dots+x_k \neq y_1+\dots+y_k$,

(iii) $k = \ell$, $x_1+\dots+x_k = y_1+\dots+y_k$, $x_1 \leqq \dots \leqq x_k$, $y_1 \leqq \dots \leqq y_k$ and $\sum_{i=1}^{j} x_i \geqq \sum_{i=1}^{j} y_i$ for not all $j = 1,\dots,k-1$.

In order to get a complete ordering that preserves the partial ordering, one introduces so–called inequality measures. An inequality measure is a sequence of functions

$$\phi_n\colon \mathbb{R}^n_+ \to \mathbb{R} \quad (n = 2,3,\ldots) \tag{8}$$

that fulfills the following condition.

$\phi_n(x) \leqq \phi_n(y)$ for all natural numbers $n \geqq 2$ and all $x \in \mathbb{R}^n_+$, $y \in \mathbb{R}^n_+$ and doubly stochastic (n,n)–matrices B such that $x = By$, ie, $\phi_n(By) \leqq \phi_n(y)$. (9)

$\{\phi_n\}$ is a strict inequality measure, if it also satisfies:

$\phi_n(By) < \phi_n(y)$ if and only if the components of vector By are not a permutation of the components of vector y. (10)

Functions (8) satisfying (9) are called Schur–convex, a function (8) satisfying (9) and (10) is called strictly Schur–convex. Obviously, these functions satisfy the mean value condition

$$\phi_n(a,\ldots,a) \leqq \phi_n(x_1,\ldots,x_n) \leqq \phi_n(0,\ldots,0,\ x_1+\ldots+x_n)$$

for all $x \in \mathbb{R}^n_+$ $(a:=(x_1+\ldots+x_n)/n)$ and are symmetric, ie,

$\phi_n(Px) = \phi_n(x)$ for all $x \in \mathbb{R}^n_+$ and all (n,n)–permutation matrices P. (11)

We notice here

— that the permutation matrices are a subset of the doubly stochastic matrices,
— that the extreme values of the components of By, which are convex combinations of the components of y, are at least as close as the extreme values of the components of y

and

— that equation (11) is another special case of (0), the equation of measurement.

Pfingsten (1986a), (1986b) introduced inequality measures that additionally satisfy, for a fixed $\mu \in [0,1]$, the equation

$$\phi_n(x + \tau(\mu x + (1 - \mu)e)) = \phi_n(x)$$

$$(e := (1,\ldots,1)) \text{ for all } x \in \mathbb{R}^n_+\backslash\{0\} \text{ and all } \tau \in \mathbb{R} \text{ such that } (x + \tau(\mu x + (1 - \mu)e)) \in \mathbb{R}^n_+\backslash\{0\}.^{\text{iii}} \tag{12}$$

Note that for x fixed, $x + \tau(\mu x + (1 - \mu)e)$ is a ray in $\mathbb{R}^n_+$ along which ϕ_n is constant. Note further that this ray is 'the more parallel', intuitively speaking, to the ray ξe (of equal incomes, say) the closer μ is to 0, and that for $\mu = 0$ and $\mu = 1$ we get the equations

$$\phi_n(x + \tau e) = \phi_n(x) \tag{13}$$

and

$$\phi_n(\lambda x) = \phi_n(x) \text{ for all } \lambda \in \mathbb{R}_{++}, \tag{14}$$

respectively, which are called in the context of the measurement of inequality of incomes, leftists' and rightists' equation, respectively. Obviously, equations (13) and (14) represent extreme positions as compared to the intermediate position (12), $0 < \mu < 1$; see, in this connection, Kolm (1976), Pfingsten (1986a), (1986b), Aczél (1987a), (1988), Eichhorn (1978), (1988).

We notice that equations (11), (12), (13), and (14) are all special cases of equation (0).

The general solution of the system of equations (11), (13), and (14) can be found in Aczél (1987a), (1988) or in Eichhorn and Gehrig (1982), whereas that of the system of equations (11) and (12) is contained in

Aczél (1987a) and Eichhorn (1988). In Eichhorn (1988) all Schur–convex functions (see (8), (9)) satisfying (12) are also determined.

In what follows, we generalise the approach to the theory of inequality measurement dealt with in this section.

3 The Inequalities of Inequality Measurement

Let X be a Banach space. We write

$$u = (u_1,\ldots,u_k) \in \mathbb{R}^k_+,\ u_1+\ldots+u_k = 1,\ x = (x_1,\ldots,x_k),\ x_i \in X \qquad (15)$$

and denote by (u,x) the distribution that assigns to $x_i \in X$ the probability $u_i \geqq 0$ $(i = 1,\ldots,k)$. Analogously, let (v,y) be a distribution that assigns to $y_j \in X$ the probability $v_j \geqq 0$ $(j = 1,\ldots,\ell)$. Let $x_1+\ldots+x_k = y_1+\ldots+y_\ell$.

The spread of a distribution (u,x) <u>is less than</u> that of a distribution (v,y) if

$$\sum_{i=1}^{k} u_i f(x_i) \leqq \sum_{j=1}^{\ell} v_j f(y_j) \qquad (16)$$

for all convex functions $f: X \to \mathbb{R}$.

This definition is equivalent (see Bourgin (1983)) to:

The spread of (u,x) <u>is less than</u> that of (v,y) if there exists a (k,ℓ)–matrix B, which is <u>row–stochastic</u>, ie, the elements of B are nonnegative real numbers and the sum of the elements in each row equals 1, and such that

$$v = uB \quad \text{and} \quad x = By. \qquad (17)$$

Because of $ux = uBy = vy$ both distributions have the same mean value.

We point out here that a special case of (17), namely

$k = \ell$, $u = v = (1/k,\ldots,1/k)$ (hence B is doubly stochastic), $X = \mathbb{R}_+$, represents the situation described in Section 2.4.

Thus one is led to the idea to define, as a measure of spread or generalised inequality measure, a sequence of functions

$$\Phi_{2k}\colon \mathbb{R}_+^k \times X^k \to \mathbb{R} \quad (k = 2,3,\ldots) \tag{18}$$

that satisfy the following conditions (19) and (20).

The inequalities of inequality measurement:

$\Phi_{2k}(u,By) \leqq \Phi_{2\ell}(uB,y)$ for all $k \geqq 2$, $\ell \geqq 2$, $u \in \mathbb{R}_+^k$, $y \in X^\ell$
and for all row–stochastic (k,ℓ)–matrices B. (19)

Homogeneity of degree 0 with respect to $u \in \mathbb{R}_+^k$:

$$\Phi_{2k}(\mu u,x) = \Phi_{2k}(u,x) \text{ for all } k \geqq 2,\ \mu = \mathbb{R}_{++},\ u \in \mathbb{R}_+^k,\ x \in X^k. \tag{20}$$

Condition (20) is required, since the components of $u \in \mathbb{R}_+^k$ are supposed to be probabilities:

$$\Phi_{2k}(u,x) = \Phi_{2k}\left[\frac{u}{u_1+\ldots+u_k},x\right].$$

We notice that the (terms of the) measures of spread are Schur–convex with respect to $y \in X^k$ and Schur–concave with respect to $u \in \mathbb{R}_+^k$. This follows from (19) with $k = \ell$. Since the set of the row–stochastic (k,k)–matrices strictly contains the set of the doubly stochastic (k,k)–matrices, the converse is not true: a function (18) that is Schur-convex with respect to $y \in X^k$ and Schur–concave with respect to $u \in \mathbb{R}_+^k$ is not necessarily a (term of some) measure of spread.

4 Applications

As the author has learnt from an unpublished manuscript of Nermuth (1988), the partial order introduced by (16) or, equivalently, by (17) gives rise to a wealth of applications in rather distinct fields of research.

We sketch some of them in terms of our notion of a generalised inequality measure. Such a measure defines a complete order within the structure involved, ie, allows for comparing any two situations.

4.1 Comparison of Income Distributions

In (18), (19) take $\{1/k\}^k$ instead of $\mathbb{R}_+^k$ and let $k = \ell =: n$, $u = (1/n,...,1/n)$, $X = \mathbb{R}_+$, B doubly stochastic. If we write

$$\Phi_{2n}(1/n,...,1/n,y) =: \phi_n(y)$$

(19) becomes $\phi_n(By) \leqq \phi_n(y)$, ie, we obtain the defining inequalities (9) of an inequality measure.

4.2 Comparison of Lotteries

In the special situation, where

$$X = \mathbb{R}_+,\ k = \ell,\ \text{B and y such that } By = y,$$

relations (18), (19) and (20) respectively become

$$\Phi_{2k} : \mathbb{R}_+^{2k} \to \mathbb{R}, \tag{18)*}$$

$\Phi_{2k}(u,By) \leqq \Phi_{2k}(uB,y)$ for all $k \geqq 2$, $u \in \mathbb{R}_+^k$ and for all $y \in \mathbb{R}_+^k$ and row–stochastic matrices B satisfying $By = y$ (19)*

and

$\Phi_{2k}(\mu u,y) = \Phi_{2k}(u,y)$ for all $k \geqq 2$, $\mu \in \mathbb{R}_{++}$, $u \in \mathbb{R}_+^k$, $y \in \mathbb{R}_+^k$. (20)*

The sequence of functions (18)* satisfying (19)* and (20)* can be understood as a measure of the risk of a lottery $(u,y) \in \mathbb{R}_+^{2k}$ where $y = (y_1,\dots,y_k) \in \mathbb{R}_+^k$ are the prizes (including zero) of the lottery and $u = (u_1,\dots,u_k) \in \mathbb{R}_+^k$ satisfying $u_1+\dots+u_k = 1$ are the probabilities of the prizes. Lottery $(u,y) \in \mathbb{R}_+^{2k}$ is called <u>less risky than</u> lottery $(v,y) \in \mathbb{R}_+^{2k}$ if

$$\sum_{i=1}^{k} u_i U(y_i) \geqq \sum_{i=1}^{k} v_i U(y_i) \text{ for all concave functions } U:\mathbb{R}_+ \to \mathbb{R}. \tag{21}$$

Obviously, (21) compares the expected values of the(utility) function U with respect to the(probability) distributions u and v.

This definition is equivalent (see Mosler (1982) and Bourgin (1983)) to the following one. Lottery (u,y) is less risky than lottery (v,y) if there exists a row–stochastic matrix B such that

$$v = uB \quad \text{and} \quad y = By. \tag{22}$$

The partial ordering imposed by this definition is preserved within the complete ordering defined by any function (18)* satisfying (19)* and (20)*. This suggests calling such functions <u>risk measures</u> for lotteries.

4.3 Comparison of Information Structures

Now let $X \subset \mathbb{R}_+^q$ be the set of all probability distributions on a finite state space Ω. The information structure (u,x) gives with probability x_i the signal i, and in this case

$$x_i = (x_{i},\dots,x_{iq}) \in X$$

is the a posteriori probability distribution on Ω. Given the signal i has been observed, x_{ij} is the a posteriori probability of sample j.

Let $\mathcal{A}$ be a finite set of possible actions and $g: \mathcal{A} \times \Omega \to \mathbb{R}$ a von Neumann–Morgenstern utility function. Given the signal i the agent chooses an action $a \in \mathcal{A}$ that maximises his utility. We write this

$$h(x_i) := \max_{a \in \mathcal{A}} \sum_{j=1}^{k} x_{ij} g(a,j). \tag{23}$$

The function $h: X \to \mathbb{R}$ is convex for every g.

The value of the information structure $(u,x) \in \mathbb{R}_+^k \times X^k$ with utility function g is defined by

$$W(u,x,g) := \sum_{i=1}^{k} u_i h(x_i). \tag{24}$$

An information structure $(v,y) \in \mathbb{R}_+^\ell \times X^\ell$ is called more informative than (u,x) if

$$W(u,x,g) \leqq W(v,y,g) \text{ for all utility functions g.} \tag{25}$$

Blackwell's theorem: The following conditions are equivalent:

(i) (v,y) is more informative than (u,x).
(ii) The spread of (u,x) is less than that of (v,y).

In this connection, we refer to Blackwell (1953), Blackwell and Girshick (1954), and Nermuth (1982).

We point out here that the structure whose spread is greater is more informative. If there is no spread, then there is no information (case of entropy maximum).

From Blackwell's theorem and from what we have learnt in Section 3 it suggests itself to propose the following definition of an information measure.

An information measure on a finite information structure $(x,u) \in \mathbb{R}_+^k \times X^\ell$ is a measure of spread or generalised inequality measure as defined in Section 3 (see (18), (19), (20)), supposing that $X \subset \mathbb{R}^q$ is the set of all probability distributions on a finite state space Ω.

5 Characterisations of Particular Measures

For the practical purpose of measuring one has to choose a uniquely specified measure. Which one we choose, ie, which further conditions we add to the defining properties dealt with in Sections 3 and 4, depends on the information that we want to obtain from the particular measure.

Whenever a measure that satisfies a system S of further properties is uniquely determined by (its defining properties and) S, we speak of a characterisation of this measure.

Many interesting characterisations of well–known particular price and inequality measures (eg, the indices of Laspeyres and Fisher in the field of price measurement or the indices of Gini and Theil in the field of inequality measurement) have already been established by several authors.

In a slightly less general setting than that framed in Section 4.3, Aczél and Daróczy (1975) have characterised a wealth of particular information measures.

References

Aczél, J (1987a): *A Short Course on Functional Equations Based Upon Recent Applications to the Social and Behavioral Sciences*; D Reidel Publishing company, Dordrecht.

Aczél, J (1987b), "Scale–Invariant Equal Sacrifice in Taxation and Conditional Functional Equations", *Aequationes Mathematicae* 32, pp 336–349.

Aczél, J (1988): "'Cheaper by the dozen': Twelve Functional Equations and Their Applications to the 'Laws of Science' and to Measurement in Economics"; in W Eichhorn (ed), *Measurement in Economics, Theory and Applications of Economic Indices*, Physica-Verlag, Heidelberg, pp 3–17.

Aczél, J & Z Daróczy (1975): *On Measures of Information and Their Characterisations*; Academic Press, New York.

Aczél, J, F S Roberts & Z Rosenbaum (1986): "On Scientific Laws Without Dimensional Constants", *Journal of Mathematical Analysis and Applications* 119, pp 389–416.

Blackwell, D (1953): "Equivalent Comparisons of Experiments", *Annals of Mathematical Statistics* 24, pp 265–272.

Blackwell, D & M A Girshick (1954): *Theory of Games and Statistical Decisions*, Wiley, New York.

Bourgin, R D (1983): "Geometric Aspects of Convex Sets with the Radon–Nikodym Property", *Springer Lecture Notes in Mathematics* 993, Springer–Verlag, Berlin.

Eichhorn, W (1978): "What is an Economic Index? An Attempt at an Answer"; in W Eichhorn, et al (eds), *Theory and Applications of Economic Indices*, Physica–Verlag, Würzburg, pp 3–42.

Eichhorn, W (1988): "On a Class of Inequality Measures", *Social Choice and Welfare* 5, pp 171–177.

Eichhorn, W & W Gehrig (1982): "Measurement of Inequality in Economics", in B Korte, (ed) *Modern Applied Mathematics, Optimisation and Operations Research*, North–Holland, Amsterdam, pp 657–693.

Eichhorn, W & W Gleissner (1988): "The Equation of Measurement", in W Eichhorn, (ed), *Measurement in Economics, Theory and Applications of Economic Indices*; Physica–Verlag, Heidelberg, pp 19–27.

Krantz, D G, R D Luce, P Suppes and A Tversky (1971): *Foundations of Measurement*, Vol I Academic Press, New York

Kolm, S–C (1976): "Unequal Inequalities I", *Journal of Economic Theory* 12, pp 416–442.

Luce, R D (1959): "On the Possible Psychophysical Laws"; *Psychological Review*, 66, pp 81–95.

Marshall, A W & I Olkin (1979): *Inequalities: Theory of Majorization and Its Applications*; Academic Press, New York.

Mosler, K C (1982): "Entscheidungsregeln bei Risiko: Multivariate stochastische Dominanz"; *Springer Lecture Notes in Economics and Mathematical Systems* 204, Springer–Verlag, Berlin.

Nermuth, M (1982): "Information Structures in Economics"; *Springer Lecture Notes in Economics and Mathematical Systems* 196, Springer–Verlag, Berlin.

Nermuth, M (1988): "Verschiedene ókonomische Theorien mit gleicher formaler Struktur: Risikomessung, Einkommensungleichheit, Informationsstrukturen etc", unpublished manuscript.

Paganoni, L (1987), "On A Functional Equation Concerning Affine Transformations"; *Journal of Mathematical Analysis and Applications* 127, pp 475–491.

Pfingsten, A (1986a): *The Measurement of Tax Progression*; Studies in Contemporary Economics 20, Springer–Verlag, Berlin.

Pfingsten, A (1986b): "Distributionally–Neutral Tax Changes for Different Inequality Concepts"; *Journal of Public Economics* 30, pp 385–393.

Roberts, F S (1979): *Measurement Theory, with Applications to Decision making, Utility and the Social Sciences*, Addison–Wesley, Reading, Massachusetts.

Young, H P (1987): "Progressive Taxation and the Equal Sacrifice Principle"; *Journal of Public Economics*, 32, pp 203–214.

Conference Discussion

I Principle of Theory Construction

The general form of a "scientific law" is greatly restricted by knowledge of the "admissible transformations" of the dependent and independent variables (Luce [1959])

Examples: Transformations from pounds to kilograms, inches to metres, Fahrenheit scales to Celsius scales.

The restrictions are discovered by formulating a functional equation from knowledge of the "admissible transformations".

Formulation of the functional equation: Suppose $x_1,\ldots,x_n,x_{n+1}$ are n + 1 variables, T_i is the set of admissible transformations for the i–th variable, i = 1,...,n, n+1, and x_{n+1} is some unknown function $\phi(x_1,\ldots,x_n)$.

Problem: Find the general form of the function u knowing the sets T_i, ie: find the general form of the "scientific law"

$$x_{n+1} = u(x_1,\ldots,x_n).$$

We assume

$$\phi: R_1\times\ldots\times R_n \to R_{n+1},$$

where $R_1,\ldots,R_n$, R_{n+1} are subsets of $\mathbb{R}$. The knowledge of T_i comes from a theory of measurement.

Examples: if the i–th variable defines a "ratio scale", then T_i consists of all functions

$$\tau_i: R_i \to R_i \text{ , where } \tau_i(x) = a_i x,$$

with some $a_i > 0$. If the i–th variable defines an "interval scale", then T_i consists of all functions

$$\tau_i : R_i \to R_i, \text{ where } \tau_i(x) = a_i x + p_i,$$

with some $a_i > 0$, $p_i \in \mathbb{R}$.
(Theory of scale types: Krantz, Luce, Suppes, and Tversky, [1971], Roberts [1979])

Luce's "principle of theory construction": Assuming that there are no "dimensional constants" (which enter the relation ϕ and cancel out the effects of transformations) admissible transformations in the independent variables should lead to an admissible transformation of the dependent variable:

For all $\tau_1 \in T_1,..,\tau_n \in T_n$ there is a

$$\tau_{n+1} = \tau(\tau_1,..,\tau_n) \in T_{n+1}$$

so that

$$\phi(\tau_1[x_1],...,\tau_n[x_n]) = \tau(\tau_1,...,\tau_n)[\phi(x_1,...,x_n)].$$

i Admissible mappings

TG That does imply a certain immutability in the concept of the transformed ϕ. If it were a utility function, for instance, one should be willing to transform the τ_i in all sorts of ways without having done any transformations of the variables.

WE This is the main thing in the theory. From this it comes out that only a very few ϕ's can have these properties. The aim of Luce was to find all such functions ϕ that have this property: that they are invariant to these sets of transformations.

CB Are there any restrictions on the τ's?

WE Yes. Wait a moment. What you see is a very general formulation. In a moment we restrict the τ's.

Handout Examples (treated and completely solved by Aczel, Roberts, Rosenbaum [1987]):

$$\phi(a_1x_1,\dots,a_nx_n) = \alpha(a_1,\dots,a_n)\phi(x_1,\dots,x_n)$$
$$\phi(a_1x_1,..,a_nx_n) = \alpha(a_1,..,a_n)\phi(x_1,..,x_n) + \beta(a_1,..,a_n)$$
$$\phi(a_1x_1 + p_1,..,a_nx_n + p_n)$$
$$= \alpha(a_1,p_1,..,a_n,p_n)\phi(x_1,..,x_n) + \beta(a_1,..,a_n,p_1,..,p_n)$$
$$\phi(ax_1,\dots,ax_n) = \alpha(a)\phi(x_1,\dots,x_n)$$
$$\phi(ax_1 + p,..,ax_n + p) = \alpha(a,p)\phi(x_1,\dots x_n) + \beta(a,p)$$

ii TG If you look at the equation above i.

WE This is the most general formulation of the problem.

TG So this already seems to me to be saying that the value of the transformed ϕ has to be the value of the original ϕ, ie unless you're transforming the right–hand side, there's no particular meaning about transforming the value of ϕ. For instance, there is a ratio scale, there isn't an interval scale; but now, suddenly, we come to a case where apparently, No 1: the equation above i holds which I am claiming says that there isn't any particular class of transformation of T_{n+1} on its own, but these other equations seem to be saying that if that held there were a class of allowable transformations of ϕ on its own. In which case I'd expect a parameter about that particular transformation allowing for ϕ, ie, for x_{n+1}. So I don't see the status of his dozen compared with his one.

WE One can say that the easiest cases of this equation are these here, easy in the sense of ratio scale and interval scale. When you take other scales here, for instance, the logarithmic scale, which is also considered in the literature, then you get other functional equations and other solutions for ϕ. ϕ is called 'scientific law', so you get other scientific laws.

TG Therefore, if you have a complicated general problem, it is quite sensible to set it up as a number of special cases which are easy to solve. The only question is: I thought Aczel was making two claims, sensible to start because easy and sensible to start because particularly meaningful, and I don't see the meaningful part.

WE One can say much to that. For instance, Luce was not such a mathematician as Aczel. He was firstly interested in some of these 12 equations and could only solve three or four of them, and then with continuity and differentiability assumptions, and when Aczel realised this in a meeting where he spoke with Fred Roberts, both of them turned on this problem and solved it completely. Aczel, who knew me, knew that many of them were interesting in economics. In my book you can find some of them with motivation.

Handout II Measurement in Economics: Examples

Example 1:
Measurement of Inequality

Let $x = (x_1,..,x_n) \in \mathbb{R}^n_+$ be a vector of

- incomes (inequality measurement)
- turnovers (concentration measurement)
- sizes (diversification measurement)
- probabilities (x = ...) (information measurement).

ϕ: $\mathbb{R}^n_+ \to \mathbb{R}$, $x \to \phi(x)$ is called a (statistical) <u>inequality measure</u>, if it satisfies, for all $x \in \mathbb{R}^n_+$,

<u>A.1</u>: (Symmetry):
$\phi(A_5x) = \phi(x)$ for all permutation matrices A_5

A2: Transfer Principle
A3: $\phi(\xi,\ldots,\xi) = 0$

Rightists' Equation: $\phi(\lambda x) = \phi(x)$ for all $x \in \mathbb{R}^n_+$, $\lambda \in \mathbb{R}_{++}$
Leftists' Equation: $\phi(x + \tau e) = \phi(x)$ for all $x \in \mathbb{R}^n_+$, $\tau \in \mathbb{R}$

such that $(x + \tau e) \in \mathbb{R}^n_+$, $e = (1,..,1) \in \mathbb{R}^n$.

Theorem

If $n \geqq 3$, there exist infinitely many functions $\phi = \mathbb{R}^n_+ \rightarrow \mathbb{R}$ which satisfy

$\phi(\lambda x + \tau e) = \phi(x)$ for all...
$\phi(A_5 x) = \phi(x)$ for all ...
$\phi(\xi,\ldots,\xi) = 0$ for all ...

simultaneously.

Centrists' Equation: (Pfingsten [1986b])

$\mu \in (0,1)$, fixed

$$\phi(x + \tau(\mu x + (1-\mu)e)) = \phi(x)$$

for all $x \in \mathbb{R}^n_+$ and $\tau \in \mathbb{R}$, where

$$x + \tau(\mu x + (1-\mu)e) \in \mathbb{R}^n_+$$
$$\phi((1+\tau\mu)x + \tau(1-\mu)e) = \phi(x).$$

iii TG I am about to state something which is bound to make me seem very foolish in a moment but since the argument seems very sensible, it may be useful for you to show why it's wrong. The equation above (iii) is something I used to call additively and multiplicatively homogeneous of degree zero.

Now take any particular x and the one–vector: that will define a plane. Now taking any y in that plane, you can put it in the form $\lambda x + \tau e$. Obviously you're going to have $\lambda \neq 0$ initially because you don't want from the beginning y(x) = u(te) = u(1) = 0, so it must be you're making you y not on that line, but off it. But anything on the plane through the one–vector and the x vector is going to have the same value of u. Now just let ϕ be continuous then, of course, because one of the points in that plane is the one–vector ϕ has to be the same as on the one–vector everywhere in that plane. But that could be any plane, therefore u has to be a constant, and indeed zero, because of your third requirement.

WE Everything that you said is true, but I forgot to say that these infinitely many solutions are not continuous everywhere.

TG So there has to be a discontinuity about the one–vector?

WE Yes, yes. It is so.

TG So it is constant along any plane off the one vector, so as you move from plane to plane you can have discontinuity as well?

WE In the paper we prove that there are discontinuities in the solutions.

TG There are commonly two classes of discontinuity on the plane as you hit the one–vector, and as you move from one plane to another?

WE Yes, that is true.

Handout Example 2:
Price Level Measurement

$\phi(\lambda x) = \lambda\phi(x)$ for all $x \in \mathbb{R}^n_+$, $\lambda \in \mathbb{R}_+$,
$\phi(x + p) = \phi(x) + \beta(p)$ for all $x \in \mathbb{R}^n_+$, $p \in \mathbb{R}^n_+$.

Theorem

Let ϕ: $\mathbb{R}^n_+ \to \mathbb{R}$, $\beta : \mathbb{R}^n_+ \to \mathbb{R}$ be a solution of these equations. Then there exist non–negative real constants $c_1,\ldots,c_n$ such that

$$\phi(x) = c_1x_1+\ldots+c_nx_n,$$

$$\beta(x) = c_1x_1+\ldots+c_nx_n.$$

Conversely, ...

III The "Equation of Measurement"

$$\phi(Ax + p) = \alpha(A,p)\phi(x) + \beta(A,p) \tag{1}$$

the problem
$$\begin{cases} \phi(Ax + p) = \alpha(A,p)\phi(x) + \beta(A,p), \text{ where} \\ \phi: X \to V \text{ unknown function,} \\ X \subseteq U, x \in X, \\ A \in S \subseteq L(U) = \text{algebra of linear operators of} \\ U \text{ into itself,} \\ p \in Y \subseteq U \\ \text{and} \\ (Ax + p) \in X \text{ for all } x \in X, p \in Y, A \in S \subseteq L(U) \\ \left.\begin{array}{l} \alpha: S\times Y \to \mathbb{R} \\ \beta: S\times Y \to V \end{array}\right\} \text{given functions} \end{cases}$$

iv On Solution to Equation

TG I believed myself to have proved that there was only one, and that a very easily understood, solution, and I now see why. I took my Y to be the same as X and that gave the excess power. If you can imagine that Y, the space of p, is the space of Ax, then you could write this A(x + y) and then you see a symmetry between the x and y bits, a symmetry that you might feel like exploiting, whilst there isn't any symmetry here to exploit. So I guess that's why there's such a small class of solutions when you put that constraint, but in general you might be able to generate many more solutions in that more general situation.

v Restricting A

AS If you take that to its logical extreme, A just becomes the identity operator only, and p becomes the origin. Any function will satisfy it?

WE If A is the identity then ϕ (x + p) and the linear functions are solutions, if $\alpha(I,p) \equiv 1$.

AS Well, any function will satisfy $\phi(x) = \phi(x)$. It becomes trivial. I don't quite understand the question you're asking with this. You've got this function, you have a set S and a set Y and you then ask what is the set of solutions, is it ϕ or p that satisfy that?

WE The set of solutions depends on the restrictions you make here, that's clear. For instance, if S contains only the unit matrix, the equation becomes $\phi(x + p) = \alpha(p)\phi(x) + \beta(p)$ and this has solutions.

AS Are you seeking a general solution which would tell you what the function ϕ is as a function of sets S and Y?

WE No, not so. Wait a little. I will give a hint what to do with this equation before I answer this question.

Handout Examples for the Sets S

$S = T_n(\mathbb{R})$, $T_n(\mathbb{R})$ the full matrix ring over the real numbers $\mathbb{R}$.

$$S = \begin{bmatrix} \lambda_{11} \cdots \lambda_{1n} \\ \vdots \quad\quad \vdots \\ \lambda_{1n} \cdots \lambda_{nn} \end{bmatrix}, \; A\hat{A} \in S \text{ for all } A \in S, \hat{A} \in S.$$

$$S = \{A \mid A = \begin{bmatrix} \lambda_1 & & 0 \\ & \ddots & \\ 0 & & \lambda_n \end{bmatrix}, \lambda_1 \in \mathbb{R},\ldots,\lambda_n \in \mathbb{R}\}$$

Then $A\hat{A} \in S$ for all $A \in S$, $\hat{A} \in S$.

$$S = \{A \mid A = \begin{bmatrix} \lambda & & 0 \\ & \ddots & \\ 0 & & \lambda \end{bmatrix}, \lambda \in \mathbb{R}\}$$

Then $A\hat{A} \in S$ for all $A \in S$, $\hat{A} \in S$.

$$S = \{A \mid A = \begin{bmatrix} \lambda_1 & \lambda_2 \\ -\lambda_2 & \lambda_1 \end{bmatrix}, \lambda_1 \in \mathbb{R}, \lambda_2 \in \mathbb{R}\}$$

Then $A\hat{A} \in S$ for all $A \in S$, $\hat{A} \in S$.

Further examples of sets S:

$$S = \{A \mid A = \begin{bmatrix} 1 & & & & 0 \\ & \ddots & & & \\ & & 1 & & \\ & & & \lambda & \\ & & & & \ddots \\ 0 & & & & \lambda \end{bmatrix}\} \equiv A_1$$

$$S = \{A \mid A = \begin{bmatrix} 1 & & & & & & \\ & \ddots & & & & 0 & \\ & & 1 & & & & \\ & & & 1 & & & \\ & & & & \ddots & & \\ & & & & & 1 & \\ & & & & & & 1 \\ & & & & & & & \ddots \\ & & & & & & & & 1 \\ & & & & & & & & & 0 \\ & & & 0 & & & & & & & \ddots \\ & & & & & & & & & & & 0 \end{bmatrix}\} \equiv A_2$$

$$S = \{A \mid A = \begin{bmatrix} 1 & & & & & \\ & \ddots & & & & 0 \\ & & 1_\lambda & & & \\ & & & \ddots & & \\ & & & & \lambda_1 & \\ & & & & & \ddots \\ & & & & & & 1_\lambda \\ & 0 & & & & & & \ddots \\ & & & & & & & & \lambda \end{bmatrix}\} \equiv A_3$$

$$S = \{A \mid A = \begin{bmatrix} \lambda_1^{-1} & & & & & & & \\ & \ddots & & & & & & \\ & & \lambda_k^{-1} & & & & 0 & \\ & & & \lambda_1 & & & & \\ & & & & \ddots & & & \\ & & & & & \lambda_k & & \\ & & & & & & \lambda_1^{-1} & \\ & & & & & & & \ddots \\ & & & & & & & & \lambda_k^{-1} \\ & & 0 & & & & & & & \lambda_1 \\ & & & & & & & & & & \ddots \\ & & & & & & & & & & & \lambda_k \end{bmatrix}\} \equiv A_4$$

vi **Matrices for price indices**

TS You have the property that two matrices in the set S have to multiply to another matrix in the set S. Doesn't that mean in this case that you can't consider all these properties simultaneously?

WE Yes

TS So you can only take them one at a time?

WE Yes, that's right.

Handout **Example 3**

Statistical price indices

n = 4k

We call a function $\phi : \mathbb{R}^{4k}_{++} \to \mathbb{R}_{++}$ a (statistical) price index or purchasing power parity if and only if for all $x \in \mathbb{R}^{4k}_{++}$, $\lambda \in \mathbb{R}_{++}$, $\lambda_k \in \mathbb{R}_{++}$ $(k = 1,...,k)$

A.0 (Monotonicity)

A.1 (Linear Homogeneity)

$$\phi(x_1,...,x_k,x_{k+1},...,x_{2k},x_{2k+1},...,x_{3k},\lambda x_{3k+1},...\lambda x_{4k}) = \lambda\phi(x)$$

or $\phi(A_1x) = \lambda\phi(x)$.

A.2 (Identity)

$$\phi(x_1,...,x_k,x_{k+1},...,x_{2k},x_{2k+1},...,x_{3k},x_{k+1},...x_{2k}) = 1$$

or $\phi(A_2x) = 1$.

A.3 (Price Dimensionality)

$$\phi(x_1,...,x_k,\lambda x_{k+1},...,\lambda x_{2k},x_{2k+1},...,x_{3k},\lambda x_{k+1},...\lambda x_{2k}) = \phi(x)$$

or $\phi(A_3x) = \phi(x)$.

A4 (Commensurability)

$$\phi\left[\frac{x_1}{\lambda_1},...,\frac{x_k}{\lambda_k}, \lambda_1 x_{k+1},...,\lambda_k x_{2k}, \frac{x_{2k+1}}{\lambda_1},...,\frac{x_{3k}}{\lambda_k}, \lambda_1 x_{3k+1},..., \lambda_k x_{4k}\right] = \phi(x)$$

or $\phi(A_4x) = \phi(x)$.

Remarks on the Solution of the Equation of Measurement

Theorem (Paganoni [1987])
If ϕ is a nonconstant solution of (1), then the functions α and β have one of the following forms:

(i) $\alpha(A,p) = \mu(A)$,
where μ is a multiplicative function (ie, $\mu(A\hat{A}) = \mu(A)\mu(\hat{A})$ for all $A \in S$, $\hat{A} \in S$),

$$\beta(A,p) = \psi(A),$$

where ψ satisfies the equation

$$\psi(A\hat{A}) = \mu(A)\psi(\hat{A}) + \psi(\hat{A}) \text{ for all } \ldots$$

(ii) $\alpha(A,p) = \mu^*(A)$,

where μ^* is a multiplicative function which satisfies

$$\mu^*(\lambda I) = \lambda \text{ for all } \lambda > 0,$$

$$\beta(A,p) = b\cdot(\mu^*(A)-1) + \gamma(p),$$

where $b \in \mathbb{R}$ and $\gamma \not\equiv 0$ is a positively homogeneous additive function.

$$Y := \{0\}, \text{ ie, } p \equiv 0 \Rightarrow \text{ we can have only case (i).}$$

Hence the only functional equations of the form (1) which may have nonconstant solutions are the following ones:

(2) $\phi(Ax + p) = \mu(A)\phi(x) + \psi(A)$,

$$(3) \quad \phi(Ax + p) = \mu^*(A)\phi(x) + b\cdot(\mu^*(A)-1) + \gamma(p), \ (\gamma \cong 0),$$

where μ, μ^*, ψ, γ, b are described in the theorem.

vii On Paganoni's solution

TG There's a technical definition of a function of a matrix A which you can get in various ways by using Taylor's series expansions and so on. They are those with the same latent vectors; and with their latent roots the corresponding function fits. As a whole, they are just the matrices which commute with it. You don't have $\mu(A)$ and $\mu(\hat{A})$ being quite different things. $\mu(\hat{A})$ is just a function of A because $\hat{A}$ is already a function of A. If your class consists of this rather well defined and important class, then you wouldn't have the determinant because you can't move your $\hat{A}$ around separately from your A.

viii Solution when p = 0

AS It seems a bit strange, you've got a case where you seem to be reducing the number of constraints you're placing on it. Effectively, you're only expecting it to work for just proportional changes, not affine changes, and yet the set of solutions is being reduced. You'd think that would widen it.

WE When you have this cone and this cone has more points then you have more possibilities to play around to find solutions of this kind.

ix CJB When are both these cases satisfied, they can both hold, can't they? Or is the condition that either one should hold?

WE Yes, that's true. The consequence of this is... I answer your question by referring back to (1). The consequence is that only functional equations of the form (1) which may have nonconstant solutions are (2) and (3). These look a bit easier than the equation of measurement.

CJB Those two look as though they form a general case, put together, or is that losing information?

TG One normalises to get $\phi(0) = 0$ in the Equation of Measurement in (3) if $0 \in X$.

CB There's an either/or aspect to it. We can't have ϕ and γ both positive.

WE The result of Paganoni is: given this equation of measurement there are only solutions when μ is a multiplicative function and ψ satisfies equations (2) or (3). In all other cases there are only constant solutions.

Handout Assumptions made by Paganoni to solve equations (2) and (3):

(4) For every $A \in S$ there exists A^{-1}, and $A^{-1} \in S$.

(5) If W is the subspace of $\mathbb{R}^n$ generated by all vectors $\{Ap\}$ with $p \in Y$ and $A \in S$, then there exists a supplementary subspace of W in $\mathbb{R}^n$, W^1, which is S–invariant
(ie, $\mathbb{R}^n = W + W^1$ and $A(W^1) \subseteqq W^1$ for all $A \in S$).

x Assumptions of Paganoni

TG I take it condition (4) is required so that you can do the affine transformation the other way round. You make the old Ax to be say y and then on the right–hand side you've $A^{-1}y$ and then you can invert the affine transformation, and you now have relations between two sets of coefficients. So the reason for that is fairly clear. Now the supplementary subspace, what's the reason for it?

WE To have enough region to work in.

IJ Terence and Tony seem to think that this multiplicative thing characterises the determinant, does it?

AS Taking the set of all matrices, yes, that's one of the standard results in the last chapter of Aczel.

WE The general solution is a multiplicative function of the determinant D, multiplicative does not mean D^α because of these Hamel things for additive Cauchy equations. If you add a little bit of regularity it is D^α.

Handout IV Systems of Special Cases of the Equation of Measurement (1)

Price level measurement $(\lambda \geqq 0)$

$$\phi(\lambda x) = \lambda\phi(x) \qquad \phi(\lambda x + p) = \lambda\phi(x) + \beta(p)$$
$$\phi(x + p) = \phi(x) + \beta(p) \quad \text{[Special case of (1)]}$$

Solution: $\phi(x) = c_1x_1+...+c_nx_n(= \beta(x))$

Inequality measurement $(\lambda > 0)$

$$\phi(\lambda x) = \phi(x) \qquad \phi(\lambda x + \tau e) = \phi(x)$$
$$\phi(x + \tau e) = \phi(x) \qquad \text{[Special case of (1)]}$$

Solutions: $\phi(x) \equiv$ constant (n = 2: only solution) and

$$u(x) = \begin{cases} \text{constant if } x = \tau e \quad (e = (1,\dots,)) \\ f\left[\dfrac{x_1-\min\{x_1,\dots,x_n\}}{x_1+\dots+x_n-n\cdot\min\{x_1,..,x_n\}},\dots,\dfrac{x_n-\min\{\ \}}{x_1+,..,x_n-n\cdot\min\{\ \}}\right] \\ \text{otherwise} \end{cases}$$

f arbitrary. [All nonconstant solutions have discontinuities]

Statistical price indices

The system of their axioms, ie, the system of special cases of the Equation of Measurement which the statistical price indices fulfill, cannot be written as one equation.
[Up to now we don't know all (statistical) price indices, ie, all solutions of this system].

Some of those systems of special cases of the Equation of Measurement which can be thought of to be systems of axioms or properties of a measure in (our) theory of measurement can be solved easily.

Examples

Theorem

Linear
Homogeneity
Circularity
Identity
Commensurability
Monotonicity

$$\left.\begin{array}{l}\text{Linear}\\ \text{Homogeneity}\\ \text{Circularity}\\ \text{Identity}\\ \text{Commensurability}\\ \text{Monotonicity}\end{array}\right\} \Leftrightarrow \quad \phi(q^0,p^0,q^1,p^1) = \left[\frac{p_1^1}{p_1^0}\right]^{\alpha_1} \cdots \left[\frac{p_n^1}{p_n^0}\right]^{\alpha_n}. \quad (\alpha_k \in \mathbb{R}_{++}, \ \Sigma c_k = 1)$$

[Circularity: $\phi(q^0,p^0,q^1,p^1)\phi(q^1,p^1,q^2,p^2) = \phi(q^0,p^0,q^2,p^2)$]

Theorem

There doesn't exist any statistical price index that satisfies (in addition to the axioms A.0 – A.5) circularity and determinateness, ie, if one of the prices tends to zero, the index doesn't tend to zero or infinity.

Theorem: for all $q^0,p^0,q,p\ r^0,r$:

$$\phi(q^0,p^0,q,p+r) = \phi(q^0,p^0,q,p) + \phi(q^0,p^0,q,r)$$

$$\phi(q^0,p^0+r^0,,q,p)^{-1} = \phi(q^0,p^0,q,p)^{-1} + \phi(q^0,r^0,q,p)^{-1}$$

$\Leftrightarrow \phi(\ ) = \frac{cp}{cp^0}$ (the constant vector c depends on q^0,q)

Theorem:

Four properties of $\phi(q^0,p^0,q,p)$ $\Leftrightarrow$ $\phi(q^0,p^0,q,p) = \left[\frac{q^0p}{q^0p^0} \cdot \frac{qp}{qp^0}\right]^{\frac{1}{2}}$

Fisher's "ideal index".

Theorem:

$$\left.\begin{array}{l} \phi(A_5x) = \phi(x) \text{ (Symmetry)} \\ \text{Principle of transfers} \\ \phi(\lambda e) = 0 \\ \text{some further properties of } \phi \end{array}\right\} \Leftrightarrow \phi(x) = \text{variance}$$

Characterisations of other well–known inequality measures like the Gini index.

Why these theorems?

They show that our knowledge of the complete solution of the equation of measurement does <u>not</u> always give the idea for solving <u>systems</u> of special cases of this equation which form (part of) the system of axioms (or properties) of a measure ϕ.

Measurement and Modelling in Economics
G.D. Myles (Editor)

NONPARAMETRIC TESTS OF ADDITIVE DERIVATIVE CONSTRAINTS

by Tom Stoker*

This paper proposes nonparametric tests of additive constraints on the first and second derivatives of a model $E(y|x) = g(x)$, where the true function g is unknown. Such constraints are illustrated by the economic restrictions of homogeneity and symmetry, and the functional form restrictions of additivity and linearity. The proposed tests are based on estimates of regression coefficients, that statistically characterize the departures from the constraint exhibited by the data. The coefficients are based on weighted average derivatives, that are reformulated in terms of derivatives of the density of x. Coefficient estimators are proposed that use nonparametric kernel estimators of the density and its derivatives. These statistics are shown to be $\sqrt{N}$ consistent and asymptotically normal, and thus are comparable to estimators based on a (correctly specified) parametric model of g(x)

1 Introduction

Derivative constraints play an important role in the application of econometric methods. Derivative constraints arise from two sources; the basic modelling restrictions implied by economic theory, and standard restrictions used to simplify econometric models. For instance, standard

---

* Sloan School of Management, Massachusetts Institute of Technology, and Nuffield College, Oxford.

A revised and shortened version of this paper is forthcoming in the Review of Economic Studies. This research was supported by several National Science Foundation grants, and by the Deutsche Forschungsgemeinschaft, Sonderforschungsbereich 303. This work has benefitted from the comments of many individuals and seminar audiences. The author wishes to thank W M Gorman and J Powell for many discussions, and A R Gallant, Z Griliches, L Hansen, J Hausman, J Heckman, W Hildenbrand, D Jorgenson, D McFadden, P Robinson, J Rotemberg and A Zellner for helpful comments.

economic theory implies that costs are homogeneous in input prices and that demand functions are zero–degree homogeneous in prices and income, which are restrictions that can be written as constraints on the derivatives of cost and demand functions respectively. The symmetry restrictions inherent to optimisation provide other examples – for instance, cost minimisation implies equality constraints on the derivatives of input quantities with respect to input prices. Examples of derivative constraints not implied by basic economic theory but frequently used to simplify econometric models include constant returns–to–scale restrictions on production functions and exclusion restrictions on large demand or production systems. Such restrictions are valuable for increasing precision in estimation or facilitating applications of econometric models.

Given the importance of derivative constraints, tests used to judge their statistical validity are of great interest in assessing model specification. Rejection of a constraint representing a basic implication of economic theory suggests either a revision of model specification, or reconsideration of the applicability of the theory to the specific empirical problem. The use of restrictions to simplify empirical models is only justified when the restrictions are not in conflict with the data evidence.

The most common approach for testing derivative constraints in current practice is the parametric approach, whereby a specific functional form of behavioural equations is postulated, and the constraints on behavioural derivatives are related to restrictions on the parameters to be estimated. Tests of the derivative constraints coincide with standard hypothesis tests of the restrictions on the true parameter values. The limits of this approach arise from the initially chosen parametric form, which must be held as a maintained assumption which the restrictions are tested against. The reaction to this problem has been the development of "flexible" functional forms, as pioneered by Diewert (1971, 1973a), Christensen, Jorgenson and Lau (1971, 1973) and Sargan (1971) and developed by many others, as well as sophisticated statistical techniques for implementing them in applications. Recent proposals include Gallant (1981, 1982), Barnett and Lee (1985), and Diewert and Wales (1987).

Also related to tests of derivative constraints is the nonparametric approach to checking the restrictions of optimising behaviour of Afriat (1967, 1972a, 1972b, 1973), Diewert (1973b) and Varian (1982, 1983), among others, which is based on direct verification of the inequality constraints implied by consistency of choice. This approach involves nonlinear programming techniques to check whether any consistent behavioural model could be found in accordance with observed data. When the data is in conflict with the basic inequality constraints, statistical variants of this technique can be used to produce measures of the severity of violation of the basic inequalities, as in Varian (1984). A related approach to testing based on residual variance comparison is proposed by Epstein and Yatchew (1985), who also give a good survey of this literature.

The purpose of this paper is to propose a new nonparametric approach to testing derivative constraints. Formally, suppose that a behavioural model explaining a dependent variable y in terms of an Mvector of continuous independent variables x implies that $E(y|x) = g(x)$, where the form of g is unknown.[ii] A "derivative constraint" refers to a restriction on the derivatives of g that holds for all values of x. In particular, I consider tests of additive constraints of the form

$$G_0(x)g(x) + \sum_j G_j(x)\frac{\partial g(x)}{\partial x_j} + \sum_{j \leq k} H_{jk}(x)\,\frac{\partial^2 g(x)}{\partial x_j\,\partial x_k} = D(x) \qquad \text{(H)}$$

where $G_0(x)$, $G_j(x)$ and $H_{jk}(x)$, $j \leq k$, $j, k = 1,\ldots,M$ and $D(x)$ are known, prespecified functions of x.[iii] The constraint (H) is intrinsically linear in g(x) and its derivatives, but is otherwise unrestricted.

One obvious idea is to use a nonparametric smoothing technique to characterise g(x) and its derivatives, and study the adherence of the estimated derivatives of g to (H) over the whole data sample. But such pointwise characterisations are notoriously imprecise, converging to the true derivatives at very slow rates as sample size increases, especially when x has more than two or three components. This "curse of dimensionality" of pointwise nonparametric estimators is well studied in the statistics literature (see Stone (1980) for instance), and is discussed vis-a-vis econometric applications by McFadden (1985).

Instead, I propose a method for testing derivative constraints based on a regression analysis of the departures from (H). Suppose that the departure from (H) was observed for each observation, and one performed an ordinary least squares (OLS) regression analysis of the departures on the components of x and the squares and cross–products of the components of x. The constraint (H) could be tested on the basis of the hypothesis that all such regression coefficients vanish. This paper gives estimators of the OLS coefficients from such a regression, and proposes test statistics of the hypothesis that the coefficients vanish.

There are several attractive practical features of the proposed tests. First, the tests are nonparametric, requiring no assumptions on the functional form of g(x). Second, the tests are based on a statistical characterisation of the departures from the derivative constraint exhibited by the data. Because the characterisation is based on regression statistics of the departures, when a constraint is rejected, the source of rejection may be indicated by the procedure. Third, the test statistics are computed directly from data on y and x, and are computationally simple, requiring no iterative techniques for maximisation or other types of equation solving. Fourth, while the distribution theory for the test statistics is based on large samples, the convergence rate is $\sqrt{N}$ (where N is sample size), which is the maximal rate of convergence available for testing approaches that assume a specific parametric, nonlinear functional form of g(x). Consequently, the procedure suffers no loss of efficiency from the nonparametric treatment of the function g(x).

Section 2 presents the notation and several examples of derivative constraints of the form (H). Section 3 introduces the departure–regression approach to testing derivative constraints, illustrates its relationship to certain parametric testing approaches, and gives a general discussion of the statistical power of the approach. Section 4 presents the estimators and test statistics, which are based on the estimation of average derivatives. Section 5 contains some concluding remarks.

2 The Basic Framework and Examples

Consider an empirical problem where a dependent variable y is modelled as a function of an M–vector of predictor variables x $= (x_1,\ldots,x_M)'$. The relevant economic structure of the model is captured in the conditional expectation

$$E(y|x) \equiv g(x)\,, \tag{1}$$

so that the constraints of interest are constraints on the derivatives of g(x). The marginal density of x is denoted as f(x).

The basic assumptions are regularity properties, listed in Appendix 2. The vector x is continuously distributed, and the density f(x) vanishes on the boundary of x values. The functions g and f are assumed sufficiently differentiable, and expected values of y, x, and the derivatives of g and f exist.[i]

The data (y_i,x_i), $i = 1,\ldots,N$, represents random drawings from the underlying (joint) distribution of y and x. As a notational convention, the subscript i always denotes an observation; $i = 1,\ldots,N$; and the subscripts j and k denote components of the vector x; $j,k = 1,\ldots,M$. For instance, $\partial g/\partial x_j$ is the jth partial derivative of g, x_i is the ith observation on x, x_{ji} is the jth component of x_i.

Before proceeding to specific examples of the constraint (H), first consider the interpretation of the derivatives $\partial g/\partial x_j$ relative to the derivatives of a more primitive econometric model. In particular, suppose that y is statistically modelled using an equation of the form

$$y = \tilde{g}(x,e), \tag{2}$$

where $\tilde{g}$ is differentiable in x, and e is assumed to stochastically represent individual heterogeneity not accounted for by x, with e distributed with density $\tilde{q}(e|x)$. It is easy to see that if e is an additive disturbance (with mean 0) in (2), or if e is distributed independently of x, then $\partial g(x)/\partial x_j$ is the conditional mean of the behavioural derivatives $\partial\tilde{g}(x,e)/\partial x_j$, given the value of x. Clearly, if e represents an additive disturbance, as in $y = \tilde{g}(x) + e$, then $\tilde{g}(x) = g(x)$ and $\partial g(x)/\partial x_j = \partial\tilde{g}(x,e)/\partial x_j$ for all x. More generally, if x and e are variation free[1] and derivatives can be passed under expectations, we have that

$$\frac{\partial g(x)}{\partial x_j} = E\left[\frac{\partial\tilde{g}(x,e)}{\partial x_j}\Big|x\right] + \mathrm{Cov}\left[y,\frac{\partial \ln\,\tilde{q}(e|x)}{\partial x_j}\Big|x\right] \tag{3}$$

Equation (3) implies that $\partial g(x)/\partial x_j$ is the conditional mean of the derivative $\partial\tilde{g}(x,e)/\partial x_j$ if and only if the covariance term vanishes, which is assured if e and x are independent (since $\partial \ln\tilde{q}/\partial x_j = 0$ in this case). Moreover, under either sufficient condition it is easy to verify that $\partial^2 g/\partial x_j\partial x_k$ is the conditional mean of $\partial^2\tilde{g}/\partial x_j\partial x_k$, $j,k = 1,\dots,M$. Consequently, under such sufficient conditions, (H) is implied by the same constraint with $\tilde{g}$ replacing g, and tests of (H) coincide with tests of the same constraint on the derivatives of the primitive behavioural model $\tilde{g}$.[2]

I begin the examples with two cases of derivative constraints associated with economic properties of the function g(x), namely homogeneity (of some degree) and symmetry. For instance, demand functions derived from utility maximisation are homogeneous of degree zero in prices and income, and cost functions are homogeneous of degree one in input prices. In the analysis of production, it is often of interest to test

---

[1] x and e are variation free if the support of $q(e|x)$ has a nonempty interior. Thus x and e may be statistically dependent, but not perfectly correlated.

[2] Note that by defining $u = y - g(x)$, one has $y = g(x) + u$, so that this interpretation holds in an artificial way for all models. The important point about e is that it coincides with the specific modelling of individual differences, with the derivative $\partial\tilde{g}/\partial x$ defined holding e constant. This issue arises in the results of Zellner (1969), where y is a linear function of x, with e representing coefficient variation over observations.

whether production exhibits constant returns–to–scale, or homogeneity of degree one of output quantity with respect to input levels. Symmetry restrictions exist for virtually any model derived from optimising behaviour, and certain symmetry restrictions can be written in the form (H), such as those for models of input demand derived from cost minimisation.[iv] These examples are included in the framework as

Example 1 – Homogeneity Restrictions: For concreteness, suppose that g(x) represents the logarithm of production and x represents the vector of log–input values; input levels are $I = e^x$ and quantity produced is $\Phi(I) = e^{g(x)}$. $\Phi(I)$ is homogeneous of degree d_0 in I if $\Phi(\kappa I) = \kappa^{d_0}\Phi(I)$ for any positive scalar κ, which is valid if and only if the log–form Euler equation is valid

$$\sum_j \frac{\partial g(x)}{\partial x_j} = d_0 \tag{4}$$

Here $\partial g/\partial x_j$ is the jth output elasticity, and (4) requires the output elasticities to add to d_0. For constant returns–to–scale we have $d_0 = 1$. (4) is clearly in the form (H) where $D(x) = d_0$, $G_j(x) = 1$, $j = 1,\ldots, M$ and $G_0(x) = 0$, $H_{jk} = 0$ for all $j,k = 1,\ldots,M$. I refer to this example frequently to discuss the economic interpretation of the proposed tests.

Note that an alternative form of homogeneity constraints can be obtained from the Euler equation in level form. Specifically, suppose that g(x) represents the quantity produced and x represents the vector of variable input levels. g(x) is homogeneous of degree d_0 if and only if the following Euler equation is valid

$$\sum_j x_j \frac{\partial g(x)}{\partial x_j} = d_0 g(x) \tag{5}$$

It should be noted that (4) and (5) involve different definitions of y and x relative to the homogeneity restriction, and will imply different tests below.

Example 2 – Symmetry Restrictions of Cost Minimisation: Suppose for concreteness that $g^j(x)$, $j = 1,\ldots,M-1$, represent the demands for $M-1$ inputs, where x_j, $j = 1,\ldots, M-1$ are the prices of the inputs and x_M is the output of the firm. Then cost minimisation implies that

$$\frac{\partial g^k(x)}{\partial x_j} - \frac{\partial g^j(x)}{\partial x_k} = 0 \quad j,k = 1,\ldots,M-1 \tag{6}$$

This set of restrictions involves several behavioural equations, which are addressed in Section 4.4c.

Some other standard types of symmetry restrictions cannot be written in the additive form (H); for example, the traditional form of the Slutsky restriction on demand functions includes products of quantities and income derivatives of other quantities, which are nonlinear terms in unknown functions.

The following two examples illustrate derivative constraints associated with the specific functional form structure of g(x).

Example 3: "x_j has no effect on y": x_j does not appear as an argument of g(x) if and only if

$$\frac{\partial g(x)}{\partial x_j} = 0 \tag{7}$$

Example 4 – Additivity and Linearity: g(x) is additive in x_j, $j = 1,\ldots,M$, if $g(x) = \Sigma_j g_j(x_j)$, which is equivalent to

$$\frac{\partial^2 g(x)}{\partial x_j \partial x_k} = 0 \quad j \neq k \quad j,k = 1,\ldots,M \tag{8}$$

Moreover g(x) is linear; $g(x) = \eta_0 + \Sigma_j \eta_j x_j$; if and only if (8) is valid for all j,k = 1,...,M. Each of the equality constraints in (8) is clearly in the form (H); I indicate how to test them simultaneously in Section 4.4c.

Many other examples of the additive constraint (H) can be derived. As with (4)–(8), specific constraints will typically involve substantial simplifications of the form (H), with zero restrictions on many of the coefficient functions G_0, G_j, H_{jk} and D. Such restrictions impart analogous simplifications to the test statistics to be presented.

3 The Regression Approach to Testing Derivative Constraints

As indicated in the Introduction, the proposed tests of the constraint (H) are based on the estimation of a finite number of values, namely the coefficients of a regression analysis of departures from (H). This section introduces the departure–regression approach to testing derivative constraints. The basic approach and notation is introduced in Section 3.1. The relationship of the method to some parametric testing approaches, as well as a general discussion of its power are given in Section 3.2.

3.1 Departure–Regression Analysis

Begin by defining the departure from the constraint (H) as

$$\Delta(x) = G_0(x)g(x) + \sum_j G_j(x)\frac{\partial g(x)}{\partial x_j} + \sum_{j \leq k} H_{jk}(x)\frac{\partial^2 g(x)}{\partial x_j \partial x_k} - D(x) \tag{9}$$

The constraint (H) is then summarized by $\Delta(x) = 0$ for all x.

Suppose for the moment that one observed the departures (say up to random error) as data, namely $\Delta(x_i)$, i = 1,...,N. While one could conceive of several tests of $\Delta(x) = 0$, a natural method would be to carry out an ordinary least squares (OLS) regression analysis of $\{\Delta(x_i)\}$ to check whether $\Delta(x)$ systematically varies with x. A first step would be to compute the average of $\Delta(x_i)$, or in regression form, estimate the equation

$$\Delta(x_i) = \alpha + u_{1i} \qquad i = 1,\ldots,N \tag{10}$$

where $\alpha = E[\Delta(x)]$ is the large sample OLS constant and $E(u_1) = 0$ by definition. Next one could regress $\Delta(x_i)$ on x, by estimating the equation

$$\Delta(x_i) = \beta_c + \sum_j \beta_j x_{ji} + u_{2i} = \beta_c + x_i'\beta + u_{2i} \quad i = 1,\ldots,N \tag{11}$$

where $\beta = (\Sigma_{xx})^{-1}\Sigma_{x\Delta}$ and $\beta_c = E[\Delta(x)] - E(x)'\beta$ denote large sample OLS values, and $E(u_2) = 0$, $Cov(x,u_2) = 0$, by definition. Finally, one could add square and cross–product terms, by estimating the equation

$$\begin{aligned}\Delta(x_i) &= \gamma_c + \sum_j \gamma_{1j} x_{ji} + \sum_{j\leq k} \gamma_{jk} x_{ji} x_{ki} + u_i \\ &= \gamma_c + x_i'\gamma_1 + s_i'\gamma + u_i \qquad i = 1,\ldots,N\end{aligned} \tag{12}$$

where $s_i = (x_{1i}^2, x_{1i}x_{2i}, \ldots, x_{Mi}^2)$ denotes the $M(M+1)/2$ vector of squared and cross–product terms; γ_c, γ_1 and γ denote large sample OLS coefficient values, and $E(u) = 0$, $Cov(x,u) = 0$ and $Cov(s,u) = 0$ by definition. If estimates of any of the regression parameters of (10), (11), or (12) are significantly different from zero, then there is evidence that $\Delta(x)$ has a non-zero mean or varies statistically with x. This would be sufficient evidence to reject the constraint (H) that $\Delta(x) = 0$ for all x.[vi]

The proposal of this paper is to test the constraint (H) on the basis of the parameters γ_c, γ_1 and γ of the regression equation (12). Section 4 indicates how these parameters and their variance–covariance matrix can be estimated nonparametrically for constraints of the form (H).[v] The test statistic proposed is the Wald statistic of the joint hypothesis that $\gamma_c = 0$, $\gamma_1 = 0$ and $\gamma = 0$.[vii]

In addition to providing a test of (H), the estimated values of the regression coefficients will describe empirically the pattern of departures exhibited by the data. The procedure coincides with the fitting of equation (12) to departures, and as such can describe systematic violations of (H). Thus, the approach is advanced as not only a testing method with well-defined alternatives, but also a method for studying how the data adheres or fails to adhere to the derivative constraint.

The methods proposed can easily be used to estimate the lower order regression parameters α and β_c, β, and formulate analogous tests based on those parameters. Relative to the estimation of (12), such tests would be redundant, in view of the regression identities:

$$\begin{aligned} \alpha &= \gamma_c + E(x)'\gamma_1 + E(s)'\gamma \\ \beta_c &= \gamma_c - E(x)'B_{xs}\gamma + E(s)'\gamma \\ \beta &= \gamma_1 + B_{xs}\gamma \end{aligned} \tag{13}$$

where $B_{xs} = (\Sigma_{xx})^{-1}\Sigma_{xs}$ is the matrix of (large sample values of) the auxiliary regression coefficients of s on x. But while redundant, the use of (10), (11), or (12) will be predicated on the size of the vector x, as (12) may contain too many coefficients to be practical.[3]

3.2 Remarks and Discussion

3.2a Interpretation and Some Remarks on Parametric Approaches

It is useful at this point to motivate the approach with a more concrete example. Recall Example 1, where g(x) denotes the logproduction function (with y observed log–output values) and x the vector of log-inputs. Consider testing the constraint that g(x) exhibits constant returns–to–scale, or $\Sigma_j(\partial g/\partial x_j) = 1$, with $\partial g/\partial x_j$ the output elasticity of the jth input. The departure $\Delta(x)$ is defined as

$$\Delta(x) = \sum_j \frac{\partial g(x)}{\partial x_j} - 1 \tag{14}$$

---

[3] Contrast the case of M = 3 with that of M = 10. If M = 3, then there are 4 parameters to be estimated in (11) and 10 parameters in (12). If M = 10, there are 11 parameters to be estimated in (11) and 66 parameters to be estimated in (12).

or the "scale elasticity" (sum of the output elasticities) minus 1.

Consider the regressions (10), (11), and (12) in turn. α measures the difference between the average scale elasticity and 1. β measures whether the scale elasticities vary linearly with log–input levels. γ measures whether the scale elasticities vary in a nonlinear (quadratic) fashion with log–input levels. The hypothesis $\gamma_c = 0$, $\gamma_1 = 0$, $\gamma = 0$ says that there is no systematic variation of the scale elasticities away from 1, at least up to quadratic effects. This is the hypothesis used to test for constant returns–to–scale, or $\Delta(x) = 0$ for all x.

But suppose that this constraint was rejected. Then the estimates of equation (12) (as well as (10) and (11)) give an empirical description of how scale elasticities vary over the data, in the same sense as if one observed scale elasticities directly and estimated (12) by OLS. This is the practical advantage I alluded to above.

There is a relationship between departure regressions and some tests of constraints using parametric models, primarily because of the common use of linear and quadratic functions in econometric modelling. For instance, suppose one assumed a Cobb–Douglas production function with log–additive error:

$$y_i = \eta_0 + \sum_j \eta_j x_{ji} + e_i \quad i = 1,\ldots,N \tag{15}$$

This form constrains the jth output elasticity to be the constant parameter η_j (which constrains $\gamma_1 = 0$ and $\gamma = 0$), with the test of $\gamma_c = \alpha = 0$ coinciding with a test of $\Sigma_j \eta_j = 1$, or the standard test for constant returns-to-scale. If a translog production function with log-additive error were assumed, as in

$$y_i = \eta_0 + \sum_j \eta_j x_{ji} + (1/2) \sum_{j,k} \eta_{jk} x_{ji} x_{ki} + e_i \quad i = 1,\ldots,N \tag{16}$$

then the jth output elasticity is $\eta_j + \Sigma_k \eta_{jk} x_{ki}$ (where $\eta_{kj} \equiv \eta_{jk}$). This form constrains $\gamma = 0$, and the test of $\gamma_c = \beta_c = 0$, $\gamma_1 = \beta = 0$ coincides with the tests of $\Sigma_j \eta_j = 1$, $\Sigma_k \eta_{jk} = 0$, $j = 1,...,M$, which are the standard restrictions for testing homogeneity of a translog function. These sorts of connections are natural because of the polynomial form of these production functions.

3.2b Interpretation of Regression Coefficients via Aggregate Functions

Because of the possible generality of the constraint (H) and the departure function $\Delta(x)$, it is impossible to expect that there are global optimality properties for tests based on quadratic regressions such as (12). However, one can ask whether there is a precise sense in which departures of $\Delta(x)$ from 0 are reflected in the large sample regression parameters. After introducing some auxiliary concepts, an answer can be given, as follows.

Begin by defining the aggregate departure function from exponential family distribution changes (see Stoker (1982), (1986a) for details). Suppose that one reweighted the population generating the data, altering the density of x from f(x) to $f(x|\mu)$, where μ is an M–vector of parameters denoting the reweighted mean of x, and $f(x|\mu)$ is defined (in exponential family form) as

$$f(x|\mu) = f(x)C(\mu)\exp[\pi(\mu)'x] \tag{17}$$

Here $C(\mu) = \{\int f(x)\exp[\pi(\mu)'x]dx\}^{-1}$ is the appropriate normalizing constant, and the function $\pi(\mu)$ is uniquely determined in a neighbourhood of $\mu = E(x)$ by the equation $\mu = E(x|\mu) = \int xf(x)C(\mu)\exp[\pi(\mu)'x]dx$. When $\mu = E(x)$, there is no reweighting: $f[x|E(x)] = f(x)$; and $\pi[E(x)] = 0$.

Of interest here is the mean of the departures from (H) under population reweighting, namely

$$\phi(\mu) \equiv E[\Delta(x)|\mu] = \int \Delta(x)f(x|\mu)dx \tag{18}$$

The aggregate departure function $\phi(\mu)$ obeys $E[\Delta(x)] = \phi[E(x)]$ and is analytic in μ in an open neighbourhood of $\mu = E(x)$ (see Stoker (1982) for details).

The relation of $\phi(\mu)$ to testing for $\Delta(x) = 0$ can be seen as follows. If $\Delta(x) = 0$ for all x, then obviously $\phi(\mu) = 0$ for all μ. But if $\phi(\mu) = 0$, what does this say about $\Delta(x) = 0$? The answer is that if $\phi(\mu) = 0$ for all μ in an open neighbourhood of $\mu = E(x)$, then it must be that $\Delta(x) = 0$ a. s.; this is the completeness property of the exponential family established by Lehmann and Scheffe (1950–1955). Moreover, since $\phi(\mu)$ is analytic in μ, $\phi(\mu) = 0$ in an open neighbourhood of $\mu = E(x)$ will occur if and only if the derivatives of ϕ (of all orders) vanish at the point $\mu = E(x)$.

Thus if all derivatives of the aggregate departure $\phi(\mu)$ vanish at $\mu = E(x)$, then (H) is valid, with zero departures a.s. It is these aggregate derivatives that are connected to the regression parameters α, β, and γ of equations (10)–(12). For equation (10), α is clearly the zero–order derivative of $\phi(\mu)$, namely $\alpha = E[\Delta(x)] = \phi[E(x)]$, so that $\alpha = 0$ coincides with $\phi(\mu) = 0$ at the point $\mu = E(x)$. For equation (11), the slope coefficients β are the first derivatives (c.f. Stoker (1982), (1986)), namely $\beta = \partial\phi/\partial\mu$, so that $\beta = 0$ coincides with $\partial\phi/\partial\mu = 0$ at $\mu = E(x)$. Finally, the coefficients γ of equation (12) are uniquely connected to the second derivatives, as

Theorem 3.1: Under Assumptions A, B of Appendix 2, γ is a linear, homogeneous, invertible function of $\partial^2\phi/\partial\mu\partial\mu'$ evaluated at $\mu = E(x)$. In particular, $\gamma = 0$ if and only if $\partial^2\phi/\partial\mu\partial\mu' = 0$ for $\mu = E(x)$.

The proof of this result is given in Appendix 3.

Thus $\alpha = 0$, $\beta = 0$, $\gamma = 0$ coincides uniquely with $\phi(E(x)) = 0$, $\partial\phi/\partial\mu = 0$, $\partial^2\phi/\partial\mu\partial\mu' = 0$. In view of the correspondence (13), $\gamma_c = 0$, $\gamma_1 = 0$, $\gamma = 0$ also coincides uniquely with $\phi(E(x)) = 0$, $\partial\phi/\partial\mu = 0$, and $\partial^2\phi/\partial\mu\partial\mu' = 0$.

In sum, this development says that $\Delta(x) = 0$ a.s. occurs if and only if all derivatives of the aggregate function $\phi(\mu)$ vanish at $\mu = E(x)$. Zero values of the zero–, first– and second–order derivatives coincide uniquely with zero values of the coefficients of the quadratic departure regression (12). Consequently, the failure to reject $\gamma_c = 0$, $\gamma_1 = 0$, $\gamma = 0$ implies that departures of $\Delta(x)$ from 0 will induce at most third–order changes in the aggregate function $\phi(\mu)$ around $\mu = E(x)$. This is the precise sense in which the regression (12) can detect departures from the constraint (H) exhibited by the data.[4]

4 Estimation of Departure–Regression Coefficients

This section presents a constructive method for estimating the departure regression coefficients and forming the test that they vanish. The principle for carrying this out is straightforward – the problem easily reduces to the estimation of weighted averages of the first and second derivatives of g(x), for which nonparametric estimators can be proposed.[viii] Section 4.1 carries out the reduction. Section 4.2 describes an indirect formulation of the average derivatives, and Section 4.3 presents nonparametric estimators motivated by the reformulation. Section 4.4 discusses the approach and gives some related remarks. I have intended the exposition to communicate the main ideas of estimation and tests without becoming lost in the specific estimator formulae, which are detailed in Appendix 1.[ix]

---

[4] One should not interpret this as an argument that tests based on (12) are optimal in a global sense for this general problem. For example, if r(x) is an invertible M–vector function (transformation) of x, then the regression of $\Delta(x)$ on a constant, the components of r(x) and all squares and cross–products of components of r(x) will have analogous interpretations as measuring derivatives (up to second order) of the mean departure with respect to the mean of r(x). Consequently, the choice of regressors can be tailored to the practical questions of how the constraint is obeyed in the data, while maintaining the sense in which departure regressions detect violations of the constraint.

4.1 Reduction to Average Derivatives

Write the departure regression (12) compactly as

$$\Delta(x_i) = X_i'\Gamma + u_i \quad i = 1,\ldots,N \tag{19}$$

where $X_i' \equiv (1, x_i', s_i')'$ and $\Gamma \equiv (\gamma_c, \gamma_1', \gamma')$. Γ is written explicitly as

$$\Gamma \equiv (\Pi_{xx})^{-1}C \tag{20}$$

where $\Pi_{xx} \equiv E(XX')$ and $C \equiv E[X\Delta(x)]$. Suppose one had a consistent estimator $\hat{C}$ of C, such that $\sqrt{N}(\hat{C} - C)$ had a limiting normal distribution with mean 0 and variance–covariance matrix V_c, as well as a consistent estimator $\hat{V}_c$ of V_c. Then Γ could be estimated consistently by

$$\hat{\Gamma} \equiv (P_{xx})^{-1}\hat{C} \tag{21}$$

where $P_{xx} \equiv \Sigma_i X_i X_i'/N$, since P_{xx} is consistent for Π_{xx}. Moreover, it is a standard exercise to show that $\sqrt{N}(\hat{\Gamma} - \Gamma)$ has a limiting normal distribution, so that asymptotic inference on the value of Γ can be carried out. In particular, under the null hypothesis that $\Gamma = 0$, the Wald statistic H $= N\hat{\Gamma}'P_{xx}\hat{V}_c^{-1}P_{xx}\hat{\Gamma}$ has a limiting χ^2 distribution with $Q = 1 + M + [M(M+1)/2]$ degrees of freedom, and so can be used to test the hypothesis that $\Gamma = 0$.

Therefore, the difficulty lies in the estimation of $C = E[X\Delta(x)]$. It is easy to see that C is the sum of terms that can be estimated with sample averages and weighted average derivative terms, although some awkward notation is required to express this formally. In particular, partition C as $C = (C_0, (C_j)', (C_{jk})')'$ and write out the components as

$$C_0 = E(\Delta(x)) = c_0^0 + \Sigma_{j'} c_0^{j'} + \sum_{j' \leq k'} c_0^{j'k'}$$

$$C_j = E(x_j\Delta(x)) = c_j^0 + \Sigma_{j'} c_j^{j'} + \sum_{j' \leq k'} c_j^{j'k'} \quad j = 1,\ldots,M \tag{22}$$

$$C_{jk} = E(x_j x_k \Delta(x)) = c_{jk}^0 + \sum_{j'} c_{jk}^{j'} + \sum_{j' \leq k'} c_{jk}^{j'k'} \quad j \leq k = 1,\ldots,M$$

where the c's are individual components defined as follows. The zero order terms are:

$$c_0^0 = E[G_0(x)g(x) - D(x)] \tag{23a}$$

$$c_j^0 = E[x_j(G_0(x)g(x) - D(x))] \quad j = 1,\ldots,M \tag{23b}$$

$$c_{jk}^0 = E[x_j x_k(G_0(x)g(x) - D(x))] \quad j \leq k = 1,\ldots,M \tag{23c}$$

The first–order terms are:

$$c_0^{j'} = E\left[G_{j'}(x)\frac{\partial g(x)}{\partial x_{j'}}\right] \quad j' = 1,\ldots,M \tag{24a}$$

$$c_j^{j'} = E\left[x_j\left[G_{j'}(x)\frac{\partial g(x)}{\partial x_{j'}}\right]\right] \quad j,j' = 1,\ldots,M \tag{24b}$$

$$c_{jk}^{j'} = E\left[x_j x_k\left[G_{j'}(x)\frac{\partial g(x)}{\partial x_{j'}}\right]\right] \quad j \leq k,j' = 1,\ldots,M \tag{24c}$$

Finally the second–order terms are:

$$c_0^{j'k'} = E\left[H_{j'k'}(x)\frac{\partial^2 g(x)}{\partial x_{j'}\partial x_{k'}}\right] \quad j' \leq k' = 1,\ldots,M \tag{25a}$$

$$c_j^{j'k'} = E\left[x_j\left[H_{j'k'}(x)\frac{\partial^2 \ (x)}{\partial x_{j'}\partial x_{k'}}\right]\right] \quad j,j' \leq k' = 1,\ldots,M \tag{25b}$$

$$c_{jk}^{j'k'} = E\left[x_j x_k\left[H_{j'k'}(x)\frac{\partial^2 g(x)}{\partial x_{j'}\partial x_{k'}}\right]\right] \quad j \leq k,\ j' \leq k = 1,\ldots,M \tag{25c}$$

The zero order terms (23a–c) can each be estimated with sample averages, and the first and second order terms (25a–c, 26a–c) are weighted average derivatives, where superscripts j', k' indicate the order of the derivative and the subscripts j,k indicate multiplication by respective components of $\mathbf{x}$.

The following sections present estimators for the required average derivative terms. The estimator $\hat{C}$ of C is then constructed using (22). Finally the estimate $\hat{\Gamma}$ of Γ is constructed via (21). This is the constructive process used for the estimation of the departure–regression coefficients Γ of (18).

4.2 The Indirect Expression of Average Derivatives

Each of the terms in (24a–c) is the mean of a (first or second) derivative of g(x) multiplied by a known function of x. One could propose several direct nonparametric estimators of such terms. For instance, one could form a nonparametric estimator $\hat{g}(x)$ of g(x), and then form the sample analogue estimators of (24a–c) and (25a–c) using $\hat{g}(x)$ in place of g(x); but there is no established statistical theory for procedures of this type. In this paper I propose an approach based on nonparametric density estimation, appealing to the statistical theory given in Powell, Stock, and Stoker (1986) and Härdle and Stoker (1987). This approach uses an indirect formulation of average derivatives, derived as follows.

The indirect expression equates an average derivative term with the product–moment of y and a function of the density of x, f(x). The foundation of the approach is the application of integration by parts to average derivatives; presented as

Theorem 4.1: Given Assumptions A, B, and C of Appendix 2, suppose that G(x) is a differentiable function, then

$$E\left[G(x)\frac{\partial g}{\partial x_j}\right] = E\left[-\left[\frac{\partial G}{\partial x_j} + \frac{G(x)}{f(x)}\frac{\partial f}{\partial x_j}\right]y\right] \tag{26}$$

and if G(x) is a twice differentiable function then

$$E\left[G(x)\frac{\partial^2 g}{\partial x_j \partial x_k}\right] = E\left[\left[\frac{\partial^2 G}{\partial x_j \partial x_k} + \frac{\partial G}{\partial x_j}\frac{1}{f(x)}\frac{\partial f}{\partial x_k} + \frac{\partial G}{\partial x_k}\frac{1}{f(x)}\frac{\partial f}{\partial x_j} + \frac{G(x)}{f(x)}\frac{\partial^2 f}{\partial x_j \partial x_k}\right] y\right] \tag{27}$$

Moreover, suppose that G(x) is any function determined by f(x) and known functions $\{H_k(x)\}$. A general n^{th} order weighted average derivative can be expressed as:

$$E\left[G(x)\frac{\partial^n g(x)}{\partial x_{j_1} \cdots \partial x_{j_n}}\right] = E(\Psi(x)y) \tag{28}$$

where $\Psi(x)$ is determined by the density f(x) and the known functions $\{H_k(x)\}$. The proof is presented in Appendix 3.

Each of the first order terms of (24a–c) is of the form (26), and each of the second order terms of (25a–c) is of the form (27). Moreover, if one has general estimators of the two functionals

$$\delta_{1j} = E\left[\left[\frac{G(x)}{f(x)}\frac{\partial f}{\partial x_j}\right] y\right] \tag{29}$$

and

$$\delta_{2jk} = E\left[\left[\frac{G(x)}{f(x)}\frac{\partial^2 f}{\partial x_j \partial x_k}\right] y\right] \tag{30}$$

then each of the right–hand sides of (26) and (27) can be estimated. This is the indirect approach, based on the estimation of functionals of density derivatives.

4.3 Kernel Estimation of Average Derivatives

The functionals (29) and (30) are estimated by their sample analogues, replacing the density and its derivatives by appropriate nonparametric kernel estimators. The procedure follows the approach to estimating unweighted average derivatives of Härdle and Stoker (1987).

For the estimation of (29) and (30), the density f(x) is estimated (pointwise) using the kernel estimator

$$\hat{f}_h(x) = \frac{1}{N} \sum_{i'=1}^{N} \left[\frac{1}{h}\right]^M K\left[\frac{x-x_{i'}}{h}\right] \tag{31}$$

where the subscript i' indicates observations. The kernel estimator (31) is a local averaging estimator, where h is a bandwidth parameter controlling the size of the interval over which averaging is performed, and $K(\cdot)$ is a higher order kernel function (see Appendix 2) giving the local weights for averaging. The asymptotic theory facilitates local approximation by enforcing $h \to 0$ as $N \to \infty$.

The partial derivatives of f(x) at x are estimated by differentiating $\hat{f}_h$, with the specific formulae given in Appendix 1. Finally, note that (29) and (30) each involve dividing by f(x). To avoid inducing erratic behaviour into the estimators, all terms are dropped that exhibit very small estimated density, namely $\hat{f}_h(x_i) \leq b$, where b is a trimming bound such that $b \to 0$ as $N \to \infty$. For this define $\hat{I}_i = I(\hat{f}_h(x_i) > b)$, where $I(\cdot)$ is the indicator function.

With this setup, we define the estimators of the functionals (29) and (30) as

$$\hat{\delta}_{1j} = N^{-1} \sum_i \left[\left[\frac{G(x_i)}{\hat{f}_h(x_i)} \frac{\partial \hat{f}_h(x_i)}{\partial x_j} \hat{I}_i\right] y_i\right] \tag{32}$$

and

$$\hat{\delta}_{2jk} = N^{-1} \sum_i \left[\left[\frac{G(x_i)}{\hat{f}_h(x_i)} \frac{\partial^2 \hat{f}_h(x_i)}{\partial x_j \partial x_k} \hat{I}_i\right] y_i\right] \tag{33}$$

The statistical properties of these functional estimators are given in Theorem 4.2.

Theorem 4.2: Given Assumptions A, B, C, and D stated in Appendix 2 (and p the integer cited in D), if

(i) $N \to \infty$, $h \to 0$, $b \to 0$, $b^{-1}h \to 0$.

(ii) For some $\epsilon > 0$, $b^4 N^{1-\epsilon} h^r \to \infty$, $r = 2M + 4$.

(iii) $Nh^{2p-4} \to 0$.

then

(a) $\sqrt{N}(\hat{\delta}_{1j} - \delta_{1j})$ has a limiting normal distribution with mean 0 and variance σ_j, where σ_j is the variance of $R_j(y,x)$ given in Appendix 1. Moreover, σ_j is consistently estimated by $\hat{\sigma}_j$ of Appendix 1.

(b) $\sqrt{N}(\hat{\delta}_{2jk} - \delta_{2jk})$ has a limiting normal distribution with mean 0 and variance σ_{jk}, where σ_{jk} is the variance of $R_{jk}(y,x)$ given in Appendix 1. Moreover, σ_{jk} is consistently estimated by $\hat{\sigma}_{jk}$ of Appendix 1.

The proof (in Appendix 3) establishes the results by showing that $\hat{\delta}_{1j} - \delta_{1j}$ and $\hat{\delta}_{2jk} - \delta_{2jk}$ are $\sqrt{N}$ equivalent to the corresponding averages of $R_j(y,x)$ and $R_{jk}(y,x)$ respectively.

The procedure for estimating $C = E[X\Delta(x)]$ can now be stated precisely. First estimate the density f(x) and its derivatives at each point x_i, $i = 1,\ldots,N$. For each of the c's in equations (23a–c), form the sample analogue estimator. For each of the c's in equations (24a–c)–(25a–c), compute the indirect expression from (27)–(28), and form the sample analogue estimator using the estimated kernel densities. Sum the averages according to (22) to form $\hat{C}$. I present the formulae coinciding with these instructions in Appendix 1.

Likewise, for the variance–covariance matrix of $\hat{C}$, sum individual components to form a grand variance component $\hat{S}_i$ for each i. The variance-covariance matrix is estimated by $\hat{V}_c$, the sample covariance matrix of $\{\hat{S}_i\}$.

In summary,

Corollary 4.1: Under the conditions of Theorem 4.2, $\sqrt{N}(\hat{C} - C)$ has a limiting normal distribution with mean 0 and variance V_C. V_C is consistently estimated by $\hat{V}_C$.

Thus, the regression coefficients Γ of (19) are estimated consistently by $\hat{\Gamma}$ of (21). To test $\Gamma = 0$, the value of the Wald statistic H $= N\hat{\Gamma}' P_{xx} \hat{V}_C^{-1} P_{xx} \hat{\Gamma}$ is compared to the critical values of a χ^2 random variable with $Q = 1 + M + [M(M + 1)/2]$ degrees of freedom. This gives the regression test of the derivative constraint (H).

4.4 Remarks and Discussion

4.4a Interpretation of the Indirect Approach and Examples

This section first gives an economic interpretation to the indirect expression of average derivatives, and then gives examples taking f(x) to be the normal density.

Consider again the problem of estimating the mean departure from constant returns to scale. The departure is given in (14), and using (26) of Theorem 4.1, the mean departure is given as

$$\alpha = E[\Delta(x)] = \sum_j E\left[\frac{\partial g}{\partial x_j}\right] - 1 = \sum_j E\left[-\left[\frac{1}{f(x)}\frac{\partial f}{\partial x_j}\right]y\right] - 1 \tag{34}$$

To test for constant returns to scale, one usually considers the experiment of increasing all inputs for a firm by a factor $d\theta$, or by adding $d\theta$ to x. The log–output response for this firm is $[\Sigma_j(\partial g/\partial x_j)]d\theta$, which is statistically compared to $d\theta$. Estimation of the average departure corresponds with a different experiment, namely that of increasing all firm input levels by $d\theta$, and comparing the average log–output response, namely $E[\Sigma_j(\partial g/\partial x_j)]d\theta$, to $d\theta$.

Consider the reconfiguration of firms associated with this latter experiment. After expansion of input levels, all firms at initial log–input level x now have log–input level $x + d\theta$, or that the density of firms (after expansion) at level $x + d\theta$ is f(x). Thus, this experiment can be equivalently thought of as the adjustment of the density of firms at log-input level x by $[-\Sigma_j(\partial f/\partial x_j)]d\theta$. The average log–output response is given by $[\int[-\Sigma_j(\partial g/\partial x_j)]g(x)dx]d\theta = \Sigma_j E(-[1/f(x)][\partial f/\partial x_j]g(x))d\theta$. Thus, equation (34) arises directly from the equivalence between overall input expansion and the reconfiguration of the population of firms. Note that this equivalence would break down if there were a significant number of firms on the boundary of log–input values; this is eliminated by the condition that f(x) vanishes on the boundary of x values. In general, equations (26)–(28) of Theorem 4.1 just recast behavioural responses into the equivalent expressions based on population reconfigurations.

The following examples may be useful, each of which takes f(x) to be a normal distribution.

Example 4.1: Consider estimating the mean departure from the constant returns to scale restriction (14) where log–inputs x are multivariate normally distributed with mean E(x) and variance Σ_{xx}. In this case the vector of density derivatives is given as $\partial f/\partial x = (\partial f/\partial x_1,..,\partial f/\partial x_M)'$ $= -f(x)\Sigma_{xx}^{-1}(x - E(x))$. The vector of average output elasticities $E(\partial g/\partial x)$ is

$$E\left[\frac{\partial g}{\partial x}\right] = E\left[-\frac{1}{f(x)}\frac{\partial f}{\partial x}y\right] = \Sigma_{xx}^{-1}\mathrm{Cov}(x,y) = \Sigma_{xx}^{-1}\Sigma_{xy} \qquad (35)$$

The latter expression is consistently estimated by the OLS coefficients of y regressed on x. The test that the mean departure $\alpha = 0$ from (34) is just the test that the sum of the slope coefficients of y regressed on x equals 1. Notice that this is exactly the procedure that one would carry out if one first assumed the Cobb–Douglas form (15), and tested constant returns to scale on the basis of OLS estimates of the output elasticities η_j, j = 1,...,M. ■

Example 4.2: Consider the test of (7) of Example 3, where x is a univariate normal variable with mean 0 and variance σ_{xx}^2, and where the true function is $g(x) = x^2$. The departure is $\Delta(x) = \partial g/\partial x = 2x$, and the mean departure $\alpha = E(\Delta(x)) = E(\partial g/\partial x) = 0$. From the above example, α is estimated by the OLS coefficient of y regressed on x, which will obviously estimate 0 from the geometry of the problem: x is distributed symmetrically about 0 and $g(x) = g(-x)$ for all x. Moreover, one has that $\beta = 2$ and $\gamma = 0$. I leave as an exercise the verification of these values, by combining (20), (23), (27) and using moment formulae for the normal distribution.

4.4b On Kernel Estimation of Average Derivatives

While notationally complicated, the principles for the regression statistics are straightforward. The additive form (H) facilitates the decomposition of $E[X\Delta(x)]$ into a sum of weighted average derivative terms, which are estimated individually and then added. The implication of the asymptotic theory of Theorem 4.2 is that each nonparametric statistic is equivalent to a sample average. This latter feature facilitates the estimation of variances – to estimate the variance of the sum of components, one sums the variance terms of the components and computes the sample variance of the sum.

As mentioned above, typical examples of the constraint (H) will involve many zero restrictions, and such restrictions also simplify the estimator formulae. For instance, a constraint on first derivatives only does not require the estimation of the second derivatives of the density to form $\hat{\Gamma}$.

By themselves, the indirect formulae (26)–(27) just alter the initial problem of estimating average regression function derivatives into one of equal complexity, namely estimating average density function derivatives. Note that the use of these formulae suggests an alternative parametric approach to the estimation of the departure coefficients. In particular, if examination of the empirical density of x suggests a particular (finitely) parameterised functional form for f(x), then the density parameters could be estimated, and the derivatives formed using the estimated parameter values. It can easily be seen that this approach yields $\sqrt{N}$ consistent estimates of Γ if the assumed functional form is accurate (for instance, see

Theorem 3 of Stoker (1986b)). Parametric modelling of either g(x) or f(x) gives test statistics that have the standard convergence properties.[xi]

The main implication of Section 4.3 is that neither g(x) or f(x) needs to be parametrically modelled for $\sqrt{N}$ consistent, asymptotically normal estimators of the departure regression coefficients to be utilized. Consequently there are no efficiency costs of data usage for treating the testing problem nonparametrically. This is in contrast to the properties of pointwise kernel estimators of g(x) or f(x) (including (31)) which converge to the true pointwise values at rates that are substantially below the parametric rate $\sqrt{N}$ (for details, see Stone (1980) among others).

The fact that kernel estimators can be averaged to produce $\sqrt{N}$ rates of convergence was established by Powell, Stock and Stoker (1986) and Robinson (1988). A proper explanation of these results is beyond the scope of this paper, and so the interested reader should pursue these references together with Härdle and Stoker (1987). The main insight of this work is that convergence rates are speeded up to $\sqrt{N}$ by properly accounting for the overlap in the local averages used to estimate the density, etc, at different points.[xii]

4.4c Extensions

Three natural extensions of the procedure deserve mention. First, while the basic framework assumes that all x components are continuously distributed, discrete predictor variables can in principle be accommodated. In particular, to test constraints among the derivatives of g(x) with respect to its continuous components, the same formulae are applied, recognizing that the density derivatives (with respect to continuous components) vary over the values of the discrete variables. By partitioning the data with respect to the discrete variable values, the value of C can be estimated within each partition component, and then the estimates averaged (weighting by partition proportions) to get the overall estimate of C for use in $\hat{\Gamma}$. The main cost of recognising discrete variables is in data requirements – nonparametric density characterisation is carried out on each partition component.

The second extension is to the study of weighted departures – namely estimating the regression coefficients of (12) using $\omega(x)\Delta(x)$ as dependent variable, where $\omega(x)$ is a differentiable weighting function, assumed to vanish on the boundary of its support. When $\omega(x)$ is known, the estimation of the (weighted) departure regression coefficients is quite straightforward – all coefficient functions of (H) are just multiplied by $\omega(x)$, and since $\omega(x)$ vanishes on the boundary of its support, no boundary terms are introduced into the indirect expression of weighted average derivatives. Consequently, departures from the constraint can be studied on subsets of the data, using $\omega(x)$ as a smoothed indicator function so that it vanishes on the boundary of its support. Thus, in addition to the general interpretation of power given in Section 3.2b, the approach can be used to detect departures on specific data subsets.[5]

The final extension concerns constraints involving several equations or the simultaneous test of several constraints. Since the constraints are additive, terms arising from different equations can again be estimated individually, with tests based on the sum of individual estimates. The appropriate variances are again computed by the sample variance of the sum of individual variance components. For a simple example, consider the constraint (6), where y_i^j, y_i^k denote observations on the jth and k^{th} input quantities, respectively. The mean departure from the constraint (6) is $\alpha = E[\Delta(x)] = E[\partial g^k/\partial x_j - \partial g^j/\partial x_k]$, which is expressed in indirect form as

$$\alpha = E\left[-\frac{1}{f(x)}\frac{\partial f}{\partial x_j}y^k\right] - E\left[-\frac{1}{f(x)}\frac{\partial f}{\partial x_k}y^j\right] \tag{36}$$

5 It should be noted that to estimate the mean departure from a linear first derivative constraint, weighted by the density $\omega(x) = f(x)$, the computationally simpler (density–weighted) average derivative estimator of Powell, Stock, and Stoker (1986) can be utilised. If the constraint to be tested is of the form $\Sigma G_j[\partial g/\partial x_j] = D$, where G_j, $j = 1,\ldots,M$ and D are constants, then the density weighted mean departure can be written as $E[f(x)\Delta(x)]/E[f(x)] = \Sigma G_j E[f(x)\partial g/\partial x_j]/E[f(x)] - D$. Powell, Stock, and Stoker (1986) give a $\sqrt{N}$ consistent, asymptotically normal estimator of $E[f(x)\partial g/\partial x]/E[f(x)]$, which can be used to estimate the mean departure.

Each of the terms can be individually estimated, and the required variance statistics formed through the adding process. The same remarks apply to estimating several constraints simultaneously – the regression statistics applicable to each constraint can be computed individually and will be joint normally distributed, and the joint variance-covariance matrix of the regression statistics is estimated by the sample covariance matrix of the corresponding estimated variance component.

5 Conclusion

This paper has proposed a new nonparametric technique for testing additive derivative constraints. The technique is based on a regression characterisation of the departures from the constraint exhibited by the data. Rejection of the constraint corresponds with rejection of zero values for all coefficients from the departure regressions. Consequently, rejection of a constraint occurs when systematic departures are detected by a standard regression analysis. Moreover, the coefficient estimates themselves provide useful information – the coefficient estimates coincide in a large sample with the coefficients of the least squares equation fit to the departures.

The method proposed for estimating the regression coefficients is based on estimating weighted average derivatives, which are reformulated in terms of derivatives of the density of the predictor variables. Explicit statistics based on nonparametric characterisation of the density and its derivatives are proposed. The regression statistics exhibit the same efficiency properties (rates of convergence) as available in parametric modelling approaches. Moreover, the statistics are computed directly from the data. I have not dealt with the practical issue of how to set (small sample) values for the bandwidth and trimming bound parameters, however this will be studied via simulation analysis as part of future research.

The main limitation of the approach is that it is not directly applicable to nonlinear derivative constraints. I have not established tests involving products of derivatives of unknown functions or other nonlinear combinations. This eliminates some important economic hypotheses of interest, such as the traditional form of the Slutsky equations in demand analysis. Further research is warranted to see whether analogous testing technique can be applied to such nonlinear constraints.

Appendix 1: Estimator Formulae

The marginal density f(x) of x is estimated at $x = x_i$ using the kernel estimator

$$\hat{f}_h(x) = \frac{1}{N}\sum_{i'=1}^{N}\left[\frac{1}{h}\right]^{M} K\left[\frac{x-x_{i'}}{h}\right]$$

where h is a bandwidth parameter and K(u) is a higher order kernel function (see Appendix 2). Let $K^{(j)} = \partial K/\partial u_j$ denote the jth first partial derivative of K, and $K^{(jk)} = \partial^2 K/\partial u_j \partial u_k$ the j,k^{th} second partial derivative of K. The jth partial derivative of f is estimated by

$$\frac{\partial \hat{f}_h(x)}{\partial x_j} = \frac{1}{N}\sum_{i'=1}^{N}\left[\frac{1}{h}\right]^{M+1} K^{(j)}\left[\frac{x-x_{i'}}{h}\right]$$

and the j,k^{th} second partial derivative of f at $x = x_i$ is estimated by

$$\frac{\partial^2 \hat{f}_h(x)}{\partial x_j \partial x_k} = \frac{1}{N}\sum_{i'=1}^{N}\left[\frac{1}{h}\right]^{M+2} K^{(jk)}\left[\frac{x-x_{i'}}{h}\right].$$

Finally, recall that $\hat{I}_i = I[\hat{f}_h(x_i) > b]$.

Now the estimator $\hat{\delta}_{1j}$ of (30) is written compactly as

$$\hat{\delta}_{1j} = N^{-1}\sum_i T_j(G, y_i, x_i)$$

where T_j is the estimator component defined as

$$T_j(G, y_i, x_i) = \left[\left[\frac{G(x_i)}{\hat{f}_h(x_i)}\ \frac{\partial \hat{f}_h(x_i)}{\partial x_j}\ \hat{I}_i\right] y_i\right].$$

Also define the "true variance component" for $\hat{\delta}_{1j}$ as

$$R_j(G,y,x) = \frac{G(x)}{f(x)} \frac{\partial f}{\partial x_j} (y - g(x)) \frac{\partial [Gg]}{\partial x_j}$$

which is taken (by Theorem 4.2) to mean that $(\hat{\delta}_{1j} - \delta_{1j})$ is first–order equivalent to $N^{-1}\Sigma_i\{R_j(G,y_i,x_i) - E[R_j]\}$, so that $\sqrt{N}(\hat{\delta}_{1j} - \delta_{1j})$ has a limiting normal distribution with mean 0 and variance $\sigma_j = Var[R_j(G,y,x)]$. Finally define the "estimated variance component" for $\hat{\delta}_{1j}$ as

$$\hat{R}_j(G,y_i, x_i) = \frac{G(x_i)}{\hat{f}_h(x_i)} \frac{\partial \hat{f}_h(x_i)}{\partial x_j} \hat{I}_i y_i$$

$$+ \frac{1}{N} \sum_{i'=1}^{N} \left[h^{-M-1} K^{(j)} \left[\frac{x_i - x_{i'}}{h}\right] \right.$$

$$\left. + h^{-M} K\left[\frac{x_i - x_{i'}}{h}\right] \left[\frac{\frac{\partial \hat{f}_h(x_{i'})}{\partial x_j}}{\hat{f}_h(x_{i'})}\right] \right] \frac{G(x_{i'}) y_{i'} \hat{I}_{i'}}{\hat{f}_h(x_{i'})}$$

which means that the limiting variance σ_j of $\sqrt{N}(\hat{\delta}_{1j} - \delta_{1j})$ is estimated consistently by $\hat{\sigma}_j = [N^{-1}\Sigma_i[\hat{R}_j(G,y_i,x_i) - \overline{R}_j]^2 I_i]$.

With these conventions, the estimator $\hat{\delta}_{2jk}$ of (33) is written compactly as

$$\hat{\delta}_{2jk} = N^{-1}\sum_i T_{jk}(G,y_i,x_i)$$

where T_{jk} is the estimator component

$$T_{jk}(G,y_i,x_i) = \left[\left[\frac{G(x_i)}{\hat{f}_h(x_i)} \frac{\partial^2 \hat{f}_h(x_i)}{\partial x_j \partial x_k} \hat{I}_i\right] y_i\right]$$

The true variance component for $\hat{\delta}_{2jk}$ is

$$R_{jk}(G,y,x) = \frac{G(x)}{f(x)}\frac{\partial f}{\partial x_j \partial x_k}(y - g(x)) + \frac{\partial [Gg]}{\partial x_j \partial x_k}$$

or that $\sqrt{N}(\hat{\delta}_{2jk} - \delta_{2jk})$ has a limiting normal distribution with mean 0 and variance $\sigma_{jk} = \text{Var}[R_{jk}(G,y,x)]$. The estimated variance component for $\hat{\delta}_{2jk}$ is

$$\hat{R}_{jk}(G,y_i,x_i) = \frac{G(x_i)}{\hat{f}_h(x_i)}\frac{\partial^2 \hat{f}_h(x_i)}{\partial x_j \partial x_k}\hat{I}_i y_i$$

$$+ \frac{1}{N}\sum_{i'=1}^{N}\left[h^{-M-2}K^{(jk)}\left[\frac{x_i - x_{i'}}{h}\right]\right.$$

$$\left. + h^{-M}K\left[\frac{x_i - x_{i'}}{h}\right]\left[\frac{\frac{\partial^2 \hat{f}_h(x_{i'})}{\partial x_j \partial x_k}}{\hat{f}_h(x_{i'})}\right]\right]\frac{G(x_{i'})y_{i'}\hat{I}_{i'}}{\hat{f}_h(x_{i'})}$$

which again means that the limiting variance σ_{jk} is estimated consistently by $\hat{\sigma}_{jk} = [N^{-1}\Sigma_i[\hat{R}_{jk}(G,y_i,x_i) - \overline{R}_{jk}]^2\hat{I}_i]$. The components T, R and $\hat{R}$ above are used to define all of the remaining estimators and their variances.

We can now define the estimators of the c's of (23)–(25), as well as their true and estimated variance components, which are denoted by s and $\hat{s}$ with the same super– and sub–scripts. First define the estimators of the zero order terms (23a–c) as

$$\hat{c}_0^0 = N^{-1}\Sigma[G_0(x_i)y_i - D(x_i)]$$

$$\text{with } s_0^0 = G_0(x)y - D(x);\ s_{0,i}^0 = s_0^0(y_i,x_i)$$

$$\hat{c}_j^0 = N^{-1}\Sigma x_{ji}[G_0(x_i)y_i - D(x_i)]$$

$$\text{with } s_j^0 = x_j[G_0(x)y - D(x)];\ s_{j,i}^0 = s_j^0(y_i,x_i)$$

$$\hat{c}^{0}_{jk} = N^{-1}\Sigma x_{ki}x_{ji}[G_0(x_i)y_i - D(x_i)]$$

$$\text{with } s^{0}_{jk} = x_k x_j[G_0(x)y - D(x)];\ s^{0}_{jk,i} = s^{0}_{jk}(y_i,x_i)$$

For the first–order terms of (24) define

$$\hat{c}^{j'}_{0} = N^{-1}\Sigma\left[\frac{\partial G_{j'}(x_i)}{\partial x_{j'}} y_i - T_{j'}(G_{j'},y_i,x_i)\right]$$

with true variance component

$$s^{j'}_{0} = \left[-\frac{\partial G_{j'}}{\partial x_{j'}} y - R_{j'}(G_{j'},y,x)\right]$$

which simplifies to a compact form as

$$s^{j'}_{0} = \left[G_{j'}(x)\frac{\partial g(x)}{\partial x_{j'}} - \frac{1}{f(x)}\frac{\partial[G_{j'}f]}{\partial x_{j'}}(y - g(x))\right]$$

The estimated variance component is

$$\hat{s}^{j'}_{0,i} = \left[-\frac{\partial G_{j'}(x_i)}{\partial x_{j'}} y_i - \hat{R}_{j'}(G_{j'},y_i,x_i)\right]$$

The other first order terms are estimated as follows: for (24b)

$$\hat{c}^{j'}_{j} = N^{-1}\Sigma\left[-\frac{\partial[x_j G_{j'}(x_i)]}{\partial x_{j'}} y_i - T_{j'}(x_j G_{j'},y_i,x_i)\right]$$

with

$$s^{j'}_{j} = \left[x_j G_{j'}(x)\frac{\partial g}{\partial x_{j'}} - \frac{1}{f(x)}\frac{\partial[x_j G_{j'}f]}{\partial x_{j'}}(y - g(x))\right]$$

and

$$\hat{s}^{j'}_{j,i} = \left[-\frac{\partial[x_j G_{j'}(x_i)]}{\partial x_{j'}} y_i - \hat{R}_{j'}(x_j G_{j'}, y_i, x_i) \right]$$

---

and for (24c)

$$\hat{c}^{j'}_{jk} = N^{-1}\Sigma \left[-\frac{\partial[x_j x_k G_{j'}(x_i)]}{\partial x_{j'}} y_i - T_{j'}(x_j x_k G_{j'}, y_i, x_i) \right]$$

with

$$s^{j'}_{jk} = \left[x_j x_k G_{j'}(x) \frac{\partial g}{\partial x_{j'}} - \frac{1}{f(x)} \frac{\partial[x_j x_k G_{j'} f]}{\partial x_{j'}} (y - g(x)) \right]$$

and

$$\hat{s}^{j'}_{jk,i} = \left[-\frac{\partial[x_j x_k G_{j'}(x_i)]}{\partial x_{j'}} y_i - \hat{R}_{j'}(x_j x_k G_{j'}, y_i, x_i) \right]$$

For the second order terms of (25a) we have

$$\hat{c}^{j'k'}_0 = N^{-1}\Sigma \left[\frac{\partial^2[H_{j'k'}(x_i)]}{\partial x_{j'} \partial x_{k'}} y_i + T_{j'}(\partial[H_{j'k'}]/\partial x_{j'}, y_i, x_i) \right.$$

$$\left. + T_{k'}(\partial[H_{j'k'}]/\partial x_{k'}, y_i, x_i) + T_{j'k'}(H_{j'k'}, y_i, x_i) \right]$$

with

$$s^{j'k'}_0 = \left[H_{j'k'}(x) \frac{\partial^2 g}{\partial x_{j'} \partial x_{k'}} - \frac{1}{f(x)} \frac{\partial^2[H_{j'k'} f]}{\partial x_{j'} \partial x_{k'}} (y - g(x)) \right]$$

and

$$\hat{s}^{j'k'}_{0,i} = \left[\frac{\partial^2[H_{j'k'}(x_i)]}{\partial x_{j'}\partial x_{k'}} y_i + \hat{R}_{j'}(\partial[H_{j'k'}]/\partial x_{j'}, y_i, x_i)\right.$$

$$\left. + \hat{R}_{k'}(\partial[H_{j'k'}]/\partial x_{k'}, y_i, x_i) + \hat{R}_{j'k'}(H_{j'k'}, y_i, x_i)\right]$$

For the terms of (25b)

$$\hat{c}^{j'k'}_{j} = N^{-1}\Sigma\left[\frac{\partial^2[x_{j'}H_{j'k'}(x_i)]}{\partial x_{j'}\partial x_{k'}} y_i + \hat{T}_{j'}(\partial[x_{j'}H_{j'k'}]/\partial x_{j'}, y_i, x_i)\right.$$

$$\left. + T_{k'}(\partial[x_j H_{j'k'}]/\partial x_{k'}, y_i, x_i) + T_{j'k'}(x_j H_{j'k'}, y_i, x_i)\right]$$

with

$$s^{j'k'}_{j} = \left[x_j H_{j'k'}(x) \frac{\partial^2 g}{\partial x_{j'}\partial x_{k'}} - \frac{1}{f(x)} \frac{\partial^2[H_{j'k'}f]}{\partial x_{j'}\partial x_{k'}}(y - g(x))\right]$$

and

$$\hat{s}^{j'k'}_{j,i} = \left[\frac{\partial^2[x_j H_{j'k'}(x_i)]}{\partial x_{j'}\partial x_{k'}} y_i + \hat{R}_{j'}(\partial[x_j H_{j'k'}]/\partial x_{j'}, y_i, x_i)\right.$$

$$\left. + \hat{R}_{k'}(\partial[x_j H_{j'k'}]/\partial x_{k'}, y_i, x_i) + \hat{R}_{j'k'}(x_j H_{j'k'}, y_i, x_i)\right]$$

---

and finally for (25c)

$$\hat{c}_{jk}^{j'k'} = N^{-1}\Sigma\left[\frac{\partial^2[x_j x_k H_{j'k'}(x_i)]}{\partial x_{j'}\,\partial x_{k'}} y_i + T_{j'}(\partial[x_j x_k H_{j'k'}]/\partial x_{j'}, y_i, x_i)\right.$$

$$\left. + T_{k'}(\partial[x_j x_k H_{j'k'}]/\partial x_{k'}, y_i, x_i) + T_{j'k'}(x_j x_k H_{j'k'}, y_i, x_i)\right]$$

with

$$s_{jk}^{j'k'} = \left[x_j x_k H_{j'k'}(x)\frac{\partial^2 g}{\partial x_{j'}\partial x_{k'}} - \frac{1}{f(x)}\frac{\partial^2[x_{jk} H_{j'} x_{k'} f]}{\partial x_{j'}\partial x_k}(y - g(x))\right]$$

and

$$\hat{s}_{j,i}^{j'k'} = \left[\frac{\partial^2[x_j x_k H_{j'k'}(x_i)]}{\partial x_{j'}\partial x_{k'}} y_i + \hat{R}_{j'}(\partial[x_j x_k H_{j'k'}]/\partial x_{j'}, y_i, x_i)\right.$$

$$\left. + \hat{R}_{k'}(\partial[x_j x_k H_{j'k'}]/\partial x_{k'}, y_i, x_i) + \hat{R}_{j'k'}(x_j x_k H_{j'k'}, y_i, x_i)\right]$$

With these component estimators, the vector $E[X\Delta(x)] = C = (C_0,(C_j)',(C_{jk})')'$ is estimated by $\hat{C} = (\hat{C}_0,(\hat{C}_j)',(\hat{C}_{jk})')'$ where

$$\hat{C}_0 = \hat{c}_0^0 + \sum_{j'} \hat{c}_0^{j'} + \sum_{j' \le k'} \hat{c}_0^{j'k'}$$

$$\hat{C}_j = \hat{c}_j^0 + \sum_{j'} \hat{c}_j^{j'} + \sum_{j' \le k'} \hat{c}_j^{j'k'} \qquad j = 1,\ldots,M$$

$$\hat{C}_{jk} = \hat{c}_{jk}^0 + \sum_{j'} \hat{c}_{jk}^{j'} + \sum_{j' \le k'} \hat{c}_{jk}^{j'k'} \qquad j \le k = 1,\ldots,M$$

and clearly $\sqrt{N}(\hat{C} - C)$ has a limiting normal distribution with mean 0 and variance-covariance matrix $V_C = \text{Var}(S)$, where S is formed by adding the true variance components in the same fashion as above. Finally, the estimated variance component $\hat{S}_i = (\hat{S}_{0,i},(\hat{S}_{j,i})',(\hat{S}_{jk,i})')'$ where

$$\hat{S}_{0,i} = \hat{s}^{0}_{0,i} + \sum_{j} \hat{s}^{j'}_{0,i} + \sum_{j' \leq k'} \hat{s}^{j'k'}_{0,i}$$

$$\hat{S}_{j,i} = \hat{s}^{0}_{j,i} + \sum_{j'} \hat{s}^{j'}_{j,i} + \sum_{j' \leq k'} \hat{s}^{j'k'}_{j,i} \quad j = 1,\ldots,M$$

$$\hat{S}_{jk,i} = \hat{s}^{0}_{jk,i} + \sum_{j'} \hat{s}^{j'}_{jk,i} + \sum_{j' \leq k'} \hat{s}^{j'k'}_{jk,i} \quad j \leq k = 1,\ldots,M$$

is such that $\hat{V}_C = N^{-1}\Sigma_i(\hat{S}_i - \overline{S})(\hat{S}_i - \overline{S})'\hat{I}_i$ is a consistent estimator of V_C.

Appendix 2: Assumptions

The assumptions are blocked into four groups, corresponding to their use in the text. Some of the regularity conditions assume exactly what is needed (A4, B1, C2–4, and D3–5), because a primitive restriction assuring these conditions is not used further.

A Basic Smoothness Conditions:

A1 The random variable (y,x) is distributed with distribution Υ that is absolutely continuous with respect to a σ–finite measure ν, with density $F(y,x) = q(y|x)f(x)$. The third moments of y and and the components of x exist, and the fourth moments of components of x exist. The support Ω of f(x) is a convex subset of $\mathbb{R}^M$ with nonempty interior. The measure ν can be written as $\nu = \nu_y x \nu_x$, where ν_x is Lebesgue measure.

A2 The functions $g(x) = E(y|x)$ and f(x) are twice continuously differentiable in the components of x.

A3 $G_j(x)$ is continuously differentiable, and $H_{jk}(x)$ is twice continuously differentiable for all $x \in \Omega$, j,k = 1,..,M.

A4 The expectations in (23a–c), (24a–c), and (25a–c) exist.

B **Aggregate Departure Function:** For the development of Section 3.2b, assume

B1 The expectation $E[\Delta(x)|\mu] = \phi(\mu)$ of (18) exists for all $\mu \in \Pi$, where Π contains an open neighbourhood of $\mu = E(x)$.

C **Indirect Formulation:** For Theorem 4.1, assume

C1 $f(x) = 0$ for all $x \in d\Omega$, where $d\Omega$ is the boundary of the support of $f(x)$. $G_j(x)f(x)$ and $H_{jk}(x)f(x)$ vanish on the boundary of their respective supports, for all $j,k = 1,..,M$.

C2 The right hand side expectation of (26) exists, when applied to (24a–c).

C3 The right hand side expectation of (27) exists, when applied to (25a–c).

C4 For (28), $G(x)f(x)$ vanishes on the boundary of its support. For the n–1 to n step of the proof, the expectations comprising (***) exist.

D **Kernel Estimation:** For Theorem 4.2, let p be an integer, $p > M + 3$, and assume

D1 All derivatives of $f(x)$ of order p exist.

D2 The kernel function $K(u)$ has bounded support $S = \{u| \ |u| \leq 1\}$, $j = 1,...,M$. $K(u)$ is of degree p:

$$\int K(u)du = 1$$

$$\int u_1^{\ell_1} \ldots u_M^{\ell_\rho} K(u)du = 0 \qquad \ell_1 + \ldots + \ell_\rho < p$$

$$\int u_1^{\ell_1} \ldots u_M^{\ell_\rho} K(u)du \neq 0 \qquad \ell_1 + \ldots + \ell_\rho = p$$

Assumptions D3 through D5 are regularity conditions required for the estimation of δ_{1j} of (29) and δ_{2jk} of (30) for given values of j and k. Let (') denote differentiation with respect to the jth component and (") denote differentiation with respect to the jth and k^{th} components, so for instance, we have that $f' = \partial f/\partial x_j$ and $f'' = \partial^2 f/\partial x_j \partial x_k$.

D3 (Gg), (Gg)', (Gg)", f, f ', f " are Lipschitz continuous.

D4 R_j and R_{jk} (of Appendix 1) are bounded in probability, and their variances exist.

D5 Tail Conditions: Let $A_N = \{x | f(x) > b)$ and $B_N = \{x | f(x) \leq b\}$. As $N \to \infty$,

$$\int_{B_N} Ggf' \, dx = o(N^{-1/2})$$

$$\int_{B_N} Ggf'' \, dx = o(N^{-1/2})$$

If $f_\iota^{(p)}$ denotes any p^{th} order partial derivative of f, then $f_\iota^{(p)}$ is Hölder continuous: there exists $\eta > 0$ such that $|f_\iota^{(p)}(u) - f_\iota^{(p)}(v)| \leq c|u - c|^\eta$. The $p + \gamma$ moments of $K(\cdot)$ exist. moreover as $N \to \infty$, the following integrals are bounded;

$$\int_{A_N} G \, g \, f_\iota^{(p)} \, dx \leq C_1 < \infty$$

$$h^\gamma \int_{A_N} G \, g \, dx \leq C_2 < \infty$$

$$h^2 \int_{A_N} G \, g \frac{f'}{f} f_\iota \, dx \leq C_3 < \infty$$

$$h^{\gamma+2}\int_{A_N} G\, g\, \frac{f'}{f}\, dx \leq C_4 < \infty$$

$$h^2\int_{A_N} G\, g\, \frac{f''}{f}\, f_\iota dx \leq C_5 < \infty$$

$$h^{\gamma+2}\int_{A_N} G\, g\, \frac{f''}{f}\, dx \leq C_6 < \infty$$

These conditions rule out explosive behaviour in the tails of the distribution – see Härdle and Stoker (1987, Appendix 1) for a discussion.

Assumptions D3 through D5 are taken to hold for all functionals of the form (29) or (30) that are estimated using estimators of the form (4.14) and (4.15) respectively. In this connection, the variances and covariances exist of all variance component terms (the s terms) of Appendix 1.

Appendix 3: Proofs

Proof of Theorem 3.1: Define the large sample residual functions

$$\xi = \Delta(x) - \beta_c - x'\beta$$

$$\zeta = s - B_c - B_{xs}'x$$

where $B_c = E(s) - B_{xs}'E(x)$, so that ξ,ζ are the residuals of y,s regressed on x and a constant. Now, by the definitions of OLS regression coefficients, we have that the quadratic coefficients γ are the (large sample) slope coefficients obtained by regressing ξ on ζ, or

$$\gamma = \Sigma_{\zeta\zeta}{}^{-1}\Sigma_{\zeta\xi}.$$

But, $\Sigma_{\zeta\xi} = \Sigma_{s\xi}$, since $\Sigma_{x\xi} = 0$ by construction. Consequently, we have that

(*) $\Sigma_{s\xi} = \Sigma_{\zeta\zeta}\gamma.$

Now, from Theorem 7 of Stoker (1982), the matrix of second derivatives of the aggregate function $E[\Delta(x)|\mu] = \phi(\mu)$ can be written

$$\frac{\partial^2\phi}{\partial\mu\partial\mu'} = \Sigma_{xx}^{-1}\Sigma_{xx\Delta}\Sigma_{xx}^{-1} - \Sigma_{xx}^{-1}\Sigma_{xxx}[\Sigma_{xx}^{-1}\Sigma_{x\Delta}\otimes\Sigma_{xx}^{-1}]$$

where $\Sigma_{xx\Delta}$ is the M×M matrix with j,k element $E[(x_j - E(x_j))\,(x_k - E(x_k))\,(\Delta(x) - E(\Delta))]$, and Σ_{xxx} is the M × M^2 matrix $\Sigma_{xxx} = [\Sigma_{1xx};..;\Sigma_{Mxx}]$, with $\Sigma_{\ell xx}$ the M×M matrix with j,k element $E[(x_\ell - E(x_\ell))(x_j - E(x_j))(x_k - E(x_k))]$. Now, if I_M is the M × M identity matrix, the second derivative matrix can be rewritten as

$$\frac{\partial^2\phi}{\partial\mu\partial\mu'} = \Sigma_{xx}^{-1}\Sigma_{xx\Delta}\Sigma_{xx}^{-1} - \Sigma_{xx}^{-1}[\Sigma_{xxx}(\beta\otimes I_M)]\Sigma_{xx}^{-1}$$

$$= \Sigma_{xx}^{-1}[\Sigma_{\Delta xx} - \Sigma_{xxx}(\beta\otimes I_M)]\Sigma_{xx}^{-1}$$

$$= \Sigma_{xx}^{-1}[\Sigma_{xx\xi}]\Sigma_{xx}^{-1}$$

where $\Sigma_{\xi xx}$ is the M×M matrix with j,k element $E[x_j - E(x_j))(x_k - E(x_k))\xi] = \mathrm{Cov}(x_j x_k, \xi)$, the latter equality using $E(x_j\xi) = E(x_k\xi) = E(\xi) = 0$. Consequently

(**) $\Sigma_{xx}\frac{\partial^2\phi}{\partial\mu\partial\mu'}\Sigma_{xx} = \Sigma_{xx\xi}$

The result follows by noting the correspondence between (*) and (**). First, γ a linear, homogeneous, nonsingular transformation of $\Sigma_{s\xi}$ by (*). Second, $\Sigma_{xx\xi}$ is the M × M symmetric matrix uniquely constructed from the elements of the $[M(M+1)/2] \times 1$ matrix $\Sigma_{s\xi}$. Finally, $\partial^2\phi/\partial\mu\partial\mu'$ is a linear, homogeneous, nonsingular transformation of $\Sigma_{xx\xi}$ by (**). Obviously $\partial^2\phi/\partial\mu\partial\mu' = 0$ if and only if $\gamma = 0$. QED

Proof of Theorem 4.1: The proof follows from application of integration by parts (c.f. Billingsley (1979)) to the average derivatives. Equation (26) follows as

$$E\left[G(x)\frac{\partial g}{\partial x_j}\right] = \int G(x)\,\frac{\partial g}{\partial x_j}\,f(x)dx = -\int\left[\frac{\partial G}{\partial x_j}\,f(x) + G(x)\,\frac{\partial f}{\partial x_j}\right]g(x)dx$$

$$= E\left[-\left[\frac{\partial G}{\partial x_j} + \frac{G(x)}{f(x)}\frac{\partial f}{\partial x_j}\right]y\right]$$

where the boundary terms vanish by Assumption C1 using for unbounded sets an argument analogous to that of Theorem 1 of Stoker (1986b), and the final equality follows by iterated expectation. Equation (27) results from applying integration by parts twice, as in

$$E\left[G(x)\frac{\partial^2 g}{\partial x_j \partial x_k}\right] = \int G(x)\frac{\partial^2 g}{\partial x_j \partial x_k}\,f(x)dx$$

$$= -\int\left[\frac{\partial G}{\partial x_j}\,f(x) + G(x)\frac{\partial f}{\partial x_j}\right]\frac{\partial g}{\partial x_k}\,dx$$

$$= \int\left[\frac{\partial^2 g}{\partial x_j \partial x_k}\,f(x) + \frac{\partial G}{\partial x_j}\,\frac{\partial f}{\partial x_k} + \frac{\partial G}{\partial x_k}\,\frac{\partial f}{\partial x_j} + g(x)\,\frac{\partial^2 g}{\partial x_j \partial x_k}\right]g(x)\,dx$$

$$= E\left[\left[\frac{\partial^2 g}{\partial x_j \partial x_k} + \frac{\partial G}{\partial x_j}\,\frac{1}{f(x)}\,\frac{\partial f}{\partial x_k} + \frac{\partial G}{\partial x_k}\,\frac{1}{f(x)}\,\frac{\partial f}{\partial x_j} + \frac{G(x)}{f(x)}\,\frac{\partial^2 f}{\partial x_j \partial x_k}\right]y\right]$$

where again all boundary terms vanish by Assumption C1. For the remainder, note that equation (28) is established for $n = 1,2$ by the above formulae. The validity of equation (28) is then established by induction, where the result is assumed for $n - 1$ and shown to be valid for n. In particular, we have

$$(***) \qquad \int G(x) \frac{\partial^n g(x)}{\partial x_{j_1} \cdots \partial x_{j_n}} f(x)\, dx$$

$$= \int \frac{\partial^{n-1} g(x)}{\partial x_{j_1} \cdots \partial x_{j_{n-1}}} \left[-\frac{\partial G}{\partial x_{j_n}} - G(x) \frac{\partial f}{\partial x_{j_n}} \frac{1}{f(x)} \right] f(x)\, dx$$

where the boundary terms vanish as before. The latter integral represents an expectation in the form (28) for derivatives of order n − 1, for which the result is assumed. Consequently, by induction, the result is true for all positive n. QED.

Proof Sketch of Theorem 4.2: The proof follows closely the proof of Theorem 3.1 of Härdle and Stoker (1987), which gives the properties of $\hat{\delta}_{1j}$ when $G(x) = 1$. Replacing y by $G(x)y$ in that proof gives properties of $\hat{\delta}_{1j}$ stated here (recognising the derivative of $E(G(x)y|x) = G(x)g(x)$ in the variance expression). The stated properties of $\hat{\delta}_{2jk}$ follow from a modification of the proof of the above theorem, applying it to kernel estimation of the second derivatives of f(x). I sketch the main steps below, as the details are easy to fill in from a reading of the above reference.

I focus on the estimator $\hat{\delta}_{2jk}$ for given j,k. Use a double prime (″) to indicate differentiation with respect to the jth and k^{th} argument; as in $f'' = \partial^2 f/\partial x_j \partial x_k$, $\hat{f}_h'' = \partial^2 \hat{f}_h/\partial x_j \partial x_k$ and for $K = K(u)$, $K'' = \partial^2 K/\partial u_j \partial u_k$. The estimator $\hat{\delta}_{2jk}$ can then be written as

$$\hat{\delta}_{2jk} = N^{-1} \sum_{i=1}^{N} \frac{\hat{f}_h''(x_i)}{\hat{f}_h(x_i)} G(x_i) y_i \hat{I}_i$$

The proof involves comparing $\hat{\delta}_{2jk}$ to two auxiliary "estimators", first the estimator based on trimming with the true density value

$$\bar{\delta} = N^{-1} \sum_{i=1}^{N} \frac{\hat{f}_h''(x_i)}{\hat{f}_h(x_i)} G(x_i) y_i I_i$$

where $I_i \equiv I[f(x_i) > b]$, and second, to remove division by $\hat{f}_h(x_i)$, the "linearised" estimator

$$(****) \qquad \tilde{\delta} = N^{-1} \sum_{i=1}^{N} \hat{\lambda}_h(x_i) G(x_i) y_i I_i$$

where

$$\hat{\lambda}_h = \frac{f''}{f} + \frac{\hat{f}_h''}{f} - \frac{\hat{f}_h f''}{f^2} .$$

The proof of $\sqrt{N}$ consistency and asymptotic normality now consists of four steps, sketched below:

Step 1: **Linearisation:** $\sqrt{N}(\tilde{\delta} - \bar{\delta}) = o_p(1)$.

Step 2: **Asymptotic Normality:** $[\tilde{\delta} - E(\tilde{\delta})]$ is $\sqrt{N}$ equivalent to $N^{-1}\Sigma[R_{jk}(y_i, x_i) - E(R_{jk})]$, so that $\sqrt{N}[\tilde{\delta} - E(\tilde{\delta})]$ has a limiting normal distribution with mean 0 and variance $\sigma_{jk} = \mathrm{Var}(R_{jk})$.

Step 3: **Bias:** $[E(\tilde{\delta}) - \delta_{2jk}] = o(N^{1/2})$.

Step 4: **Trimming:** $\sqrt{N}(\hat{\delta}_{2jk} - \delta_{2jk})$ has the same limiting distribution as $\sqrt{N}(\bar{\delta} - \delta_{2jk})$.

Step 1 Linearisation: Following the remarks in Härdle and Stoker (1987), condition (iii) implies that the pointwise mean–squared–errors of $\hat{f}_h$ and $\hat{f}_h''$ are dominated by their variances. Since the set $\{x \mid f(x) \geq b\}$ is compact and $b^{-1}h \to 0$, the arguments of Collomb and Härdle (1986) or Silverman (1979) imply

$$\sup |\hat{f}_h(x) - f(x)| I[f(x) > b] = 0_p[(N^{1-(\epsilon/2)} h^M)^{-1/2}]$$

$$\sup |\hat{f}_h''(x) - f''(x)| I[f(x) > b] = 0_p[(N^{1-(\epsilon/2)} h^{M+4})^{-1/2}]$$

for any $\epsilon > 0$. Now, the Taylor expansion of $\bar{\delta}$ with respect to $\hat{f}_h''$ and $\hat{f}_h$ implies that with high probability

$$\sqrt{N}\,|\tilde{\delta} - \bar{\delta}| = 0_p(\sqrt{N} b^{-2} \sup[|f - \hat{f}_h|I] \ \sup[|\hat{f}_h'' - f''|I])$$

$$= 0_p(b^{-2} N^{-(1/2)+(\varepsilon/2)} h^{-(2M+4/2)}) = o_p(1)$$

the latter equality by condition (ii).

Step 2: Asymptotic Normality: From (****), we have that

$$\tilde{\delta} = \tilde{\delta}_0 + \tilde{\delta}_1 + \tilde{\delta}_2$$

where

$$\hat{\delta}_0 = N^{-1}\sum_i \frac{f''(x_i)}{f(x_i)} G(x_i) y_i I_i$$

$$\hat{\delta}_1 = N^{-1}\sum_i \frac{\hat{f}_h''(x_i)}{f(x_i)} G(x_i) y_i I_i$$

$$\hat{\delta}_2 = -N^{-1}\sum_i \frac{\hat{f}_h''(x_i)}{f^2(x_i)} f''(x_i) G(x_i) y_i I_i.$$

Note that $\tilde{\delta}_0$ is the average $\tilde{\delta}_0 = N^{-1}\sum r_{0_N}(z_i)$, where $z_i = (y_i, x_i)$ and $r_{0_N}(z_i) = f''(x_i)G(x_i)y_iI_i/f(x_i)$, $i = 1,\ldots,N$.

To analyse $\tilde{\delta}_1$ and $\tilde{\delta}_2$, we note that they each can be approximated by U–statistics. As in Härdle and Stoker (1987), we have $\sqrt{N}(\tilde{\delta}_1 - \hat{U}_1) = o_p(I)$ where

$$\hat{U}_1 = \binom{N}{2}^{-1} \sum_{i=1}^{N-1} \sum_{i'=i+1}^{N} p_{1_N}(z_i, z_{i'})$$

where $z_i = (y_i, x_i)$ and

$$p_{1_N}(z_i, z_{i'}) = \frac{1}{2}\left[\frac{1}{h}\right]^{M+2} K''\left[\frac{x_i - x_{i'}}{h}\right]\left[\frac{G(x_i)y_i I_i}{f(x_i)} + \frac{G(x_{i'})y_{i'} I_{i'}}{f(x_{i'})}\right]$$

where $K'' \equiv \partial^2 K/\partial u_j \partial u_k$. Similarly, $\sqrt{N}(\hat{\delta}_2 - \hat{U}_2) = o_p(I)$ where

$$U_2 = \begin{bmatrix} N \\ 2 \end{bmatrix}^{-1} \sum_{i=1}^{N-1} \sum_{i'=i+1}^{N} p_{2_N}(z_i, z_{i'})$$

where

$$p_{2_N}(z_i, z_j) = -\frac{1}{2}\left[\frac{1}{h}\right]^{M} K\left[\frac{x_i - x_{i'}}{h}\right]\left[\frac{f''(x_i)G(x_i)y_i I_i}{f^2(x_i)} + \frac{f''(x_{i'})G(x_{i'})y_{i'} I_{i'}}{f^2(x_{i'})}\right].$$

To prove asymptotic normality of $\hat{\delta}_1$ and $\hat{\delta}_2$, a modification of the standard Central Limit Theorem for U–statistics, due to Powell, Stock and Stoker (1986), is used (see Serfling (1980) for references to U-statistics). In particular, for $\hat{\delta}_1$, define $r_{1_N}(z_i) = E[p_{1_N}(z_i, z_{i'}) | z_i]$, so that $E(r_{1_N}) = E(p_{1_N})$, and define

$$\overline{U}_1 = \frac{2}{N} \sum_{i=1}^{N} [r_{1_N}(z_i) - E(r_{1_N})].$$

By Lemma 3.1 of Powell, Stock and Stocker (1986), if $E[|p_{1_N}(z_i, z_{i'})|^2] = o(N)$, then $\sqrt{N}[\hat{U}_1 - E(\hat{U}_1) - \overline{U}] = o_p(1)$. Since $\overline{U}_1$ is a sample average, under the assumptions here a standard Central Limit Theorem will apply. Thus if $R_{1_N} = 4\ \mathrm{Var}(r_{1_N})$ is nonzero and $\lim R_{1N} = R_1$, then $\sqrt{N}[\hat{\delta}_1 - E(\hat{\delta}_1)]$ has a limiting normal distribution with mean 0 and covariance matrix R_1.

It is easy to verify that $E[|p_{1_N}(z_i, z_{i'})|^2] = 0(N)$ if $b^2 N h^{M+4} \to \infty$, which is implied by condition (ii). By analogous methods applied to $\hat{\delta}_2$, it is easy to show that $[\hat{\delta}_2 - E(\hat{\delta}_2)]$ is $\sqrt{N}$ equivalent to $2N^{-1}\Sigma_i[r_{2_N}(z_i) - E(r_{2_N})]$, where $r_{2_N}(z_i) = E[p_{2_N}(z_i, z_{i'}) | z_i]$.

Thus, if

$$r_N(z_i) \equiv r_{0N}(z_i) + 2r_{1N}(z_i) + 2r_{2N}(z_i)$$

then the above arguments have shown that $[\hat{\delta} - E(\hat{\delta})]$ is first–order equivalent to $N^{-1}\Sigma_i[r_N(z_i) - E(r_N)]$, and consequently that $\sqrt{N}[\hat{\delta} - E(\hat{\delta})]$ has a limiting normal distribution, with mean 0 and variance $R = \lim R_N$, where $R_N = Var(r_N)$. Moreover, by methods analogous to Appendix 2 of Härdle and Stoker (1987), we have that

$$r_{0N}(z_i) \to \frac{f''(x_i)}{f(x_i)} G(x_i)y_i$$

$$r_{1N}(z_i) \to (1/2)\left[\frac{f''(x_i)}{f(x_i)}G(x_i)y_i + G''(x_i)g(x_i) + \frac{\partial G}{\partial x_j}\frac{\partial g}{\partial x_k} + \frac{\partial G}{\partial x_k}\frac{\partial g}{\partial x_j} + G(x_i)g''(x_i)\right]$$

$$r_{2N}(z_i) \to -(1/2)\left[\frac{f''(x_i)}{f(x_i)}G(x_i)y_i + \frac{f''(x_i)}{f(x_i)}G(x_i)g(x_i)\right]$$

so that $r_N(z_i) \to R_{jk}(G,y_i x_i)$ as listed in Appendix 1.

Step 3: Bias: The bias of $\hat{\delta}$ is examined term by term as

$$E(\hat{\delta}) - \delta = \tau_{1N} + \tau_{2N} - \tau_{3N}$$

where

$$\tau_{1N} = E\left[N^{-1}\sum_{i=1}^{N}\frac{f''(x_i)}{f(x_i)}G(x_i)y_i I_i\right] - \delta$$

$$\tau_{2N} = E\left[N^{-1}\sum_{i=1}^{N}[\hat{f}_h''(x_i) - f''(x_i)]\frac{G(x_i)y_i I_i}{f(x_i)}\right]$$

$$\tau_{3N} = E\left[N^{-1}\sum_{i=1}^{N}[\hat{f}_h(x_i) - f(x_i)]\,\frac{f''(x_i)G(x_i)y_i I_i}{f^2(x_i)}\right].$$

Each of these terms of $o(N^{-1/2})$. $\tau_{1N} = o(N^{-1/2})$ is assumed by the first equation of Assumption D5. τ_{2N} and τ_{2N} are examined by writing out the expectations in integral form, and applying the fact that K is of order p. From the remaining conditions of Assumption D5, τ_{2_N} and τ_{3_N} are each shown to be $0(h^{p-2})$, so that $\rho_{2_N} = 0(N^{-1/2})$ and $\tau_{3_N} = 0(N^{-1/2})$ follow from condition (iii).

Step 4: Trimming: The trimming argument is identical to that used in Härdle and Stoker (1987).

The estimated variance components represent direct estimation of the U-statistic components used in the proof of asymptotic normality. There components clearly converge uniformly over the set $f(x) > b$, and the triangle inequality is applicable to show that $\hat{\sigma}_{jk}$ is a consistent estimator of σ_{jk}.

End of Proof Sketch, Theorem 4.2

References

Afriat, S (1967): "The Construction of a Utility Function from Expenditure Data", *International Economic Review*, 8, pp 67–77.

Afriat, S (1972a): "The Theory of International Comparison of Real Income and Prices", *International Comparisons of Prices and Output*, D J Daly, ed, New York, National Bureau of Economic Research.

Afriat, S (1972b): "Efficiency Estimates of Production Functions", *International Economic Review*, 13, pp 568–598.

Afriat, S (1973), "On a System of Inequalities in Demand Analysis: An Extension of the Classical Method", *International Economic Review*, 14, pp 460–472.

Barnett, W A & Y W Lee (1985): "The Global Properties of the Minflex Laurent, Generalised Leontief, and Translog Flexible Functional Forms", *Econometrica*, 53, pp 1421–1437.

Billingsley, P(1979): *Probability and Measure*, New York, Wiley.

Christensen, L R, D W Jorgenson & L J Lau (1971): "Conjugate Duality and the Transcendental Logarithmic Production Function", *Econometrica*, 39, pp 355–266.

——— (1973): "Transcendental Logarithmic Production Frontiers", *Review of Economics and Statistics*, pp 28–45.

Collomb, G and W Härdle (1986): "Strong Uniform Convergence Rates in Robust Nonparametric Time Series Analysis and Prediction: Kernel Regression Estimation from Dependent Observations", *Stochastic Processes and Their Applications*; 23, pp 77–89.

Diewert, W E (1971): "An Application of the Shephard Duality Theorem: A Generalised Leontief Production Function", *Journal of Political Economy*, 79, pp 481–507.

——— (1973a): "Functional Forms for Profit and Transformation Functions", *Journal of Economic Theory*, 6, pp 284–316.

——— (1973b): "Afriat and Revealed Preference Theory", *Review of Economic Studies*, 40, pp 419–426.

——— & T J Wales (1987): "Flexible Functional Forms and Global Curvature Conditions", *Econometrica*, 55, pp 43–68.

Epstein, L G & A Yatchew (1985): "Nonparametric Hypothesis Testing Procedures and Applications to Demand Analysis", *Journal of Econometrics*, 30, pp 149–169.

Gallant, A R (1981): "On the Bias in Flexible Functional Forms and an Essentially Unbiased Form: The Fourier Flexible Form", *Journal of Econometrics*, 30, pp 149–167.

——— (1982): "Unbiased Determination of Production Technologies", *Journal of Econometrics*, 20, pp 285–323.

Härdle, W & T M Stoker (1987): "Investigating Smooth Multiple Regression by the Method of Average Derivatives", forthcoming, *Journal of the American Statistical Association.*

Lehmann, E L & H Scheffe (1950): "Completeness, Similar Regions and Unbiased estimation, Part I", *Sankya*, 10, pp 305–340.

——— & ——— (1955): "Completeness, Similar Regions and Unbiased Estimation, Part II", *Sankya*, 15, pp 219–236.

McFadden, D (1985): "Specification of Econometric Models", Presidential Address to the Fifth World Congress of the Econometric Society, Cambridge, Massachusetts, August.

Powell, J L, J H Stock & T M Stoker(1986): "Semiparametric Estimation of Index Coefficients", forthcoming *Econometrica.*

Robinson, P M (1987): "Hypothesis Testing in Semiparametric and Nonparametric Models for Economic Time Series", Discussion Paper, London School of Economics.

——— (1988): "Root–N Consistent Semiparametric Regression", *Econometrica*, 56, pp 931–954.

Sargan, J D (1971): "Production Functions", in *Qualified Manpower and Economic Performance*, P R G Layard, J D Sargan, M E Ager & D J Jones, Eds, pp 145–204, London, Penguin Press.

Serfling, F J (1980): *Approximation Theorems of Mathematical Statistics*, John Wiley and Sons.

Silverman, B W (1979): "Weak and Strong Uniform Consistency of the Kernel Estimate of a Density Function and its Derivatives", *Annals of Statistics*, 6, pp 177–184. (Addendum 1980, *Annals of Statistics*, 8, pp 1175–1176.)

Stoker, T M (1982): "The Use of Cross–Section Data to Characterise Macro Functions", *Journal of the American Statistical Association*, 77, pp 369–380.

——— (1986a): "Aggregation, Efficiency and Cross–Section Regression", *Econometrica*, 54, pp 171–188.

——— (1986b): "Consistent Estimation of Scaled Coefficients", *Econometrica*, 54, pp 1461–1481.

——— (1988): "Equivalence of Direct, Indirect and Slope Estimators of Average Derivatives", MIT School of Management Working Paper No 1961–87, revised May 1988.

Stone, C J (1980): "Optimal Rates of Convergence for Nonparametric Estimators", *Annals of Statistics*, 8, pp 1348–1360.

Varian, H (1982): "The Nonparametric Approach to Demand Analysis", *Econometrica*, 50, pp 945–973.

——— (1983): "Nonparametric Tests of Consumer Behaviour", *Review of Economic Studies*, 50, pp 99–110.

——— (1984): "The Nonparametric Approach to Production Analysis", *Econometrica*, 52, pp 579–598.

Zellner, A (1969): "On the Aggregation Problem, A New Approach to a Troublesome Problem", in *Economic Models, Estimation and Risk Programming*, K Fox, J K Sengupta & G V L Narasimham, New York, Springer–Verlag.

Conference Discussion

Handout Setup:

(y_i,x_i), $i=1,...,N$ – i.i.d. sample of observations, x_i is an M–vector.
$g(x) \equiv E(y|x)$ – conditional expectation of y given x.
$f(x)$ – marginal density of x.

• Major Assumptions:

i) $f(x) = 0$ on boundary of x values.

ii) $g(x)$, $f(x)$ appropriately differentiable but otherwise unknown functions.

i **Dummy variables**

SP So there are no dummy variables?

TS No dummy variables at this stage of the setup. First of all, it's not clear how you'd define derivatives with respect to dummy variables, and secondly, dummy variables can be included as predictor variables. Let me address that later.

TG I don't see why not. Why shouldn't your corresponding G_j in equation (H) below just be zero for a dummy variable?

SP That's right. If you had a mixture of dummies and then, say, an income variable.

TS The precise generalisation is that you can test derivatives with respect to the continuous variables and I can tell you, at least formally, how the dummy variables enter into the formula. But there is a cost – whatever data requirements are required for a case without dummy variables, to be fully non-parametric you have to do the approximation for each value of the dummy variables, so there is a data cost. In the basic set up everything is continuous.

SP And there's no censoring either? You can't have continuous variables that have a probability mass at zero?

TS Yes and no. No in the sense that I'm not going to make explicit reference to censoring. The point is everything rests on the conditional expectation of y given x; for instance, start with a linear model, then, with censoring, g(x) may not be a linear model any more, and so you might reject linearity. So in that sense the censoring is not taken care of. But given g(x), whether the data is censored or not, the test of the constraint on g(x) is still correct.

Handout **Object**: To test the constraint

$$\text{(H)} \quad G_0(x)g(x) + \sum_j G_j(x)\frac{\partial g(x)}{\partial x_j} + \sum_{j \leq k} H_{jk}(x)\frac{\partial^2 g(x)}{\partial x_j \partial x_k} = D(x)$$

without specifying g(x), where the coefficients $G_0(x)$, $G_j(x)$ and $H_{jk}(x)$, j, k = 1,...,M and D(x) are known, prespecified functions of x.

ii TG You're assuming at the beginning that the expectation includes all the economic variables?

TS Yes.

TG Which really means that the normalisation you're using somewhere is critical. I mean, if you happen to use the square of those variables, the expectation of those would be the square of the expectation plus the variance?

TS Well, in terms of the y's, yes. There's no scaling issue here but nonlinear transformations are a problem.

iii **On Second Derivatives**

SP So are we talking about a system of restrictions here or just a single one?

TS I'm talking about a single restriction, but these extensions are all trivial once you see what I'm trying to do.

- Examples of such constraints:
  a) Constant returns to scale
  b) Symmetry
  c) Independence

iv During examples

JM So you have in mind a cross–section context?

TS Yes, pretty much so. The properties of the statistics I talk about will not necessarily be valid when y and x arise from an autocorrelated process. I'm not sure whether they can be extended.

JM The problems of simultaneity, incomplete specification of the inputs, and measurement errors in the inputs and outputs, are all abstracted from?

TS Yes, they're all abstracted from by me putting everything in the form of a conditional expectation of y given x. I have a little bit more to say about this but I think the point you're making is exactly right.

TG One thing is not exactly right: that there can't be measurement errors in the output.

TS There certainly can be measurement errors in the output, suitably random and uncorrelated with the level of inputs and so forth, but measurement errors in the inputs, as with any errors in variables case, will change the regression function which is, of course, what you end up estimating. Those problems are serious problems. In the time series context I believe everything I say here just requires some qualified changes.

SP Is stratified sampling a problem, because most cross–section work is not i.i.d?

TS Which definition of stratified sampling do you mean?

SP Well, where you have independent but not necessarily identically distributed observations because you're drawing from sub-populations.

TS It will be a problem for features of this approach and I'll say what they are when we get there. Basically, you're talking about varying the density from which the x's are drawn because you have a mixture of populations. In the aproach, and other more direct approaches I can conceive of, you still end up having to estimate the density of the x's, and for that reason I am troubled by this point.

SP So you'd need to assume that there's a finite number of sub-populations?

TS Oh, you'd have to know something about where they are, which sub-populations there are. For instance, if you knew that there were six different ones and your data points came in six different colours then you could assign dummy variables to take care of that and so that throws it into the arena of your last question. But that one can be taken care of, certainly; theoretically.

v JM In the context of data again, there isn't very much data on prices, perhaps 20 observations if you have 20 years of data, and the variation of relative prices is usually very small, so the whole nonparametric approach becomes rather limited in that case. I think it's rather more sensible to assume a flexible functional form, in prices at least, parameterize it, and use conventional econometric techniques.

TS With only the data you suggest, assuming a functional form is right, but not assuming a flexible functional form. I think you'd have trouble saying anything specific about how flexible that form is. This approach is a fully flexible one, and I guess will suffer from a small number of data points, but I'm not convinced anything but a highly restricted functional form will give you enough information to learn anything from those 20 data points. The weak data means you have to get information from something and assuming a functional form is the way to do that. Even if you were good at picking functional forms you'll be losing a lot of the description of the data by a restricted functional form.

CJB Isn't there a parallel between the two cases? If we succeed don't we essentially reject the null hypothesis, with a parametric approach that means we're testing the line that a particular function works and its parameters satisfy any side restrictions? You're testing that a particular functional form works and that the side derivative constraints work.

TS No, what I'm doing is different. Let's go to my particular example again. Consider when you model (H) with a Cobb-Douglas function so g(x) is a constant plus the sum of the input elasticities times the input level plus some error. That would be your model and what would you be doing for constant returns? You'd find the G_j, add them up, and see if they sum to one. What you are holding as your maintained hypothesis is that this functional form is right. To the extent that the functional form is not right you may be getting a departure of the sum of these things from 1, even though the true output elasticities sum to one everywhere, because it is not adequately accommodating the non-linearities. What I'm talking about are techniques to check this directly.

Handout • Connection to Primitive Model Derivatives

The Departure–Regression Approach to Testing Derivative Constraints: Define departures from (H) as

$$\Delta(x) = G_0(x)g(x) + \sum_j G_j(x)\frac{\partial g(x)}{\partial x_j} + \sum_{j\leq k} H_{jk}(x)\frac{\partial^2 g(x)}{\partial x_j \partial x_k} - D(x)$$

so that (H) corresponds with $\Delta(x) = 0$ for all x. The approach is to estimate the (large sample) OLS regression coefficients of

$$(12) \quad \Delta(x_i) = \gamma_c + \sum_j \gamma_{1j}x_{ji} + \sum_{j\leq k} \gamma_{jk}x_{ji}x_{ki} + u_i$$

$$= \gamma_c + x'\gamma_1 + s_i'\gamma + u_i \quad i=1,\ldots,N$$

where $s_i = (x_{1i}^{\;2}, x_{1i}x_{2i}, \ldots, x_{Mi}^{\;2})$, and test (H) by testing $\gamma_c = 0$, $\gamma_1 = 0$ and $\gamma = 0$ of (12).

vi On (12)

SP You would have to do something like this anyway because your equation (H) here is a very stringent restriction. If you're saying it must hold for all x, you're testing an increasing number of hypotheses as you observe more and more points. What you're doing is kind of boiling it down to one restriction.

TS Yes, that's right. In fact, that raises something I hadn't really thought about. You see, the objections to using non-parametric statistics directly is that point–wise they have poor convergence properties. What these results suggest is that if you combine them, but in a specific way, that you improve their properties. I haven't actually shown that yet.

SP Just as a practical problem, you claim to establish that x never plays any role in determining y. All you can say is that there's no evidence in this sample as you're always looking at local properties.

TS Yes, these kinds of parameters are defined as what you'd get in a large sample and so it's clear that is evidence against. If they're all zero it's still conceivable that the constraint is violated. There's a method in the paper of making that more precise.

TG If x was a scalar then $\Delta(x)$ might be orthogonal to 1, x and x^2.

TS Exactly. There is a story in the paper about making that statement a little more precise but that's what's missing out of this approach.

TG But isn't it a little bit like choosing an appropriate flexible functional form? You'd try to choose your x's in such a way that a second–order approximation would pick up what you think are possible $\Delta(x)$'s if, in fact, the thing is wrong?

TS That's how I think about this justification; that you're looking at the departures directly as opposed to starting with a functional form and relating the derivatives to those parameters.

TG No, no. I'm saying that everyone knows about a Taylor expansion and the problem with a flexible functional form is choosing the right normalisation for the variables in it. It seems to me that you have precisely the same thing there. You have to choose your x's, and maybe your G's, but let's say your x's in any case, in such a way that you hope the Δ's will be picked up in the quadratic expansion because that's now a flexible functional form of these variables. So it seems to me that there is a close parallelism between the two.

TS Well, I think there is a difference in the kinds of properties from this. The estimators I'll talk about are estimators of these objects, and the choice of the x's determines what these objects are, but they're not restricted by requiring G to be of any form.

vii SP Suppose you got rid of x and the squared term from (12) and just looked at the γ's. Is that in any sense the same thing as saying that we're going to test the original restriction at a point x that corresponds to the sample mean?

TS No, no. That's not it. Unless there's linearity that is an existence of an aggregate function type of problem. There are two other regressions involved which follow below, so I'll use that notation. We can kick out β and suppose we just estimate the parameter α. Here α is the mean departure and that is typically not what you would consider the departure at the mean of x. Depending on what you want to say about how you let the x's vary, this would only be true under linearity if the x's were freely varying without distribution restrictions.

SP Oh yes, I realise that.

TS That's the point, though.

SP I suppose I'm trying to anticipate what you're about to go on to – whether the statistical methods you end up with are essentially different to those that would be implied if you were testing that derivative constraint at a single point which you'd nominate.

TS The examples I've been choosing obscure that issue. The Cobb-Douglas function makes the derivatives depend only on the parameters, so there you're testing it at all points in essence: but that's just by choosing the derivatives to be solely determined by the parameters. Typically, if you fitted a functional form and computed the derivatives at the mean of the data, that would be different fundamentally from what I am doing. That would not be the same procedure.

SP Suppose you were being non–parametric both ways and wanted to test your restriction at a particular point x which you nominate, the statistical methods you would use for testing that hypothesis are completely unrelated to the methods you're proposing even if you suppress the x and squared term?

TS Not completely unrelated. The idea of this regression is completely unrelated to the idea of testing it at a point. The techniques you would use to test it at a point are used in the construction of these estimators.

Handout • Equivalence to $\alpha = 0$, $\beta = 0$, $\gamma = 0$, where $\alpha = E[\Delta(x)]$, $\beta = [\mathrm{Var}(x)]^{-1}\mathrm{Cov}[x,\Delta(x)]$ are (large sample) OLS coefficients of the lower order equations:

$$\Delta(x_i) = \alpha + u_{1i}$$

$$\Delta(x_i) = \beta_c + \sum_j \beta_j x_{ji} + u_{2i} = \beta_c + x_i'\beta + u_{1i}$$

• Interpretation and Relation to Parametric Approaches.

viii JM Do the actual estimates of the $\partial g/\partial x$'s depend on the fineness of the grid that you employ?

TS This is more motivation of the $\partial g/\partial x$'s. What I'm going to do is to take an indirect approach to computing these statistics where you don't actually use this approach.

That doesn't mean it's not possible, I just don't know the results.

Handout • Interpretation of Coefficients via Aggregate Functions.

ix TG Can I push this a little bit further at what appears to be a natural break? By the way, I think it's a lovely paper, so if I say things they are meant in no sense other than that. You may think with a flexible functional form that you have some notion whether it's likely to be a quadratic thing, maybe in logs, or it might be a better fit in the x's or the square roots, say. You're used to thinking in those terms, and you might have graphs of the functions people have estimated before, so in choosing the particular form of flexible functional form you have some past evidence. Here you do not have such evidence. That must make it more difficult so to normalise your variables that the quadratic terms pick up most of the variation.

TS Well, I think you're right; obviously at this level of generality you can certainly construct examples which would make your point exactly. But on the other hand I think there is some appeal to attacking the derivatives directly. There's a sense at which going at the level of this constraint has some practical value. Basically the idea of testing subject to a finite number of values is always subject to your criticism. One could generally conjecture that because you can estimate the finite number of values at the same convergence rate as parametric rates, you could generally estimate any functional form for the departures at parametric rates.

Handout Estimation of Departure Regression Coefficients:

1 Reduction to Average Derivatives – (*) is written compactly as

$$(*) \qquad \Delta(x_i) = X' \, \Gamma + u_i$$

where $X_i' \equiv (1, x_i', s_i')'$ and $\Gamma \equiv (\gamma_c, \gamma_1', \gamma')'$, with

$$\Gamma \equiv (\Pi_{xx})^{-1} C \qquad \text{where } \Pi_{xx} = E(XX') \text{ and } C = E[X\Delta(x)]$$

Π_{xx} is estimated by $P_{xx} \equiv \Sigma_i X_i X_i'/N$, so the problem is in estimating C. $C = E[X\Delta(x)]$ consists of sums of two types of terms: expectations of known functions of y and x, and weighted first– and second– average derivatives: terms of the following form, with G(x) known

$$E\left[G(x)\frac{\partial g}{\partial x_j}\right] \text{ and } E\left[G(x)\frac{\partial^2 g}{\partial x_j \, \partial x_k}\right]$$

2 The Indirect Expression of Average Derivatives – by applying integration by parts, average derivatives are expressed in terms of density derivatives as:

$$E\left[G(x)\frac{\partial g}{\partial x_j}\right] = E\left[-\left[\frac{\partial G}{\partial x_j} + \frac{G(x)}{f(x)}\frac{\partial f}{\partial x_j}\right]y\right]$$

and

$$E\left[G(x)\frac{\partial^2 g}{\partial x_j \partial x_k}\right] = E\left[\left[\frac{\partial^2 g}{\partial x_j \partial x_k} + \frac{\partial G}{\partial x_j}\frac{1}{f(x)}\frac{\partial f}{\partial x_k} + \frac{\partial G}{\partial x_k}\frac{1}{f(x)}\frac{\partial f}{\partial x_j} + \frac{G(x)}{f(x)}\frac{\partial^2 g}{\partial x_j \partial x_k}\right]y\right]$$

x **In discussion**

JM The constraint that you use as an example is the constant returns to scale one, the sum of the derivatives equal to 1, but in principal you could have worked to see if the effect of the first input on output was zero.

TS Certainly, but the ingredients that you'd use for testing that are the same as you'd use for estimation of average returns to scale.

JM I wanted to be reassured that by not considering all inputs at the same time you're not using some averaging as opposed to looking at only the first input and ignoring the effect of the other inputs.

TS I'm not quite sure what your question is but two things occur to me. One is: that would be subject to bias, one would think.

JM I'm sorry that's not the question.

TS That's not the question?

JM No. Take a general specification of g and test only whether the elasticity with respect to the first input is zero.

TS That is entirely possible. That would be associated in this testing terminology with $\partial g/\partial x_1$ as zero. So in this framework the departure is just literally $\partial g/\partial x_1$, and what I'm talking about estimating is the regression coefficient that you would get in a large sample from regressing $\partial g/\partial x_1$ on a constant, x_1 and x_1^2. The point is you'd be measuring these coefficients by the procedure and the test is that all three vanish.

Handout More broadly, we have

$$E\left[G(x)\frac{\partial^n g(x)}{\partial x_j \ldots \partial x_i}\right] = E(\Psi(x)y)$$

where $\Psi(x)$ is determined by G(x) and the density f(x).
Consequently, an estimator of C can be based on estimators of the functionals:

$$\delta_{1j} = E\left[\left[\frac{G(x)}{f(x)}\frac{\partial f}{\partial x_j}\right]y\right] \text{ and } \delta_{2jk} = E\left[\left[\frac{G(x)}{f(x)}\frac{\partial^2 f}{\partial x_j \partial x_k}\right]y\right]$$

- Interpretation of the Indirect Approach and Normal Examples.

3 Kernel Estimation of Average Derivatives: Estimate the density f(x) at x_i via the kernel estimator

$$\hat{f}_h(x) = \frac{1}{N}\sum_{i'=1}^{N}\left[\frac{1}{h}\right]^M K\left[\frac{x-x_i}{h}\right]$$

where K(.) is a kernel density (of order p) and h is a bandwidth parameter. Estimate δ_{1j} and δ_{2jk} by average kernel estimators:

$$\hat{\delta}_{1j} = N^{-1}\sum_i\left[\left[\frac{G(x_i)}{\hat{f}_h(x_i)}\frac{\partial\hat{f}_h(x_i)}{\partial x_j}\hat{I}_i\right]y_i\right]$$

and

$$\hat{\delta}_{2jk} = N^{-1}\sum_i\left[\left[\frac{G(x_i)}{\hat{f}_h(x_i)}\frac{\partial^2\hat{f}_h(x_i)}{\partial x_j \partial x_k}\hat{I}_i\right]y_i\right]$$

where $\hat{I}_i = I[\hat{f}_i(x_i)>b]$ excludes terms with tiny estimated density. Under the asymptotic conditions

(i) $N \to \infty$, $h \to 0$, $b \to 0$, $b^{-1}h \to 0$.

(ii) For some $\epsilon > 0$, $b^4N^{1-\epsilon}h^{2M+2} \to \infty$.

(iii) $Nh^{2p-4} \to 0$.

one can show that $\sqrt{N}(\hat{\delta}_{1j} - \delta_{1j}) \to n(0,\sigma_j)$ and $\sqrt{N}(\hat{\delta}_{2jk} - \delta_{2jk}) \to n(0,\sigma_{jk})$

By writing out $C = E[X\Delta(x)]$ as above, and inserting average kernel estimators, one gets $\hat{C}$ such that $\sqrt{N}(\hat{C}-C) \to \mathcal{N}(0,V_C)$, and a consistent estimator $\hat{V}_C$ of V_C.

From $\hat{C}$ form $\hat{\Gamma} = (P_{xx})^{-1}\hat{C}$. To test $\Gamma = 0$, the value of the Wald statistic $H = N\hat{\Gamma}' P_{xx} \hat{V}_C^{-1} P_{xx} \hat{\Gamma}$ is compared to the critical values of a χ^2 random variable with $Q = 1 + M + [M(M+1)/2]$ degrees of freedom.

xi On convergence and Monte Carlos

SP Have you compared this technique with a semi–parametric approach where you specify g( ) as some arbitrary function of $\beta'x$ and estimate the β's. That's a sort of intermediate approach isn't it?

TS I worked on this project and another simultaneously and it was the estimation of β's in those models that was the other problem. Let me just indicate the connection. If you recall, everything in this story depends on some form of weighted average derivative or second derivative. The other problem that you're referring to is to suppose we have a semi-parametric approach where g(x) can be written $F(\beta'x)$. This is semi–parametric due to the effects of x being straight–jacketed by using this linear form, but the function F is general. For any of these problems, this would be an overly–restricted formulation but it is useful in many empirical problems. The nature of this restriction in terms of the kind of contexts I've discussed is: what do the derivatives look like? Well, they look like the derivative of f with respect to $\beta'x$, this is just a scalar value, times β. In particular, each of these derivatives is proportional to the β. In fact this was one of the first motivations for studying average derivatives that I stumbled on to. If you look at this, it's the average of the scalar functions times β or a constant times β. My answer is that this particular form is something you exploit by just using the first derivatives.

Handout • Further Remarks and Extensions

xii SP I'm just wondering if you have any practical experience with this perfectly general approach. You would think that you need a vast amount of data.

TS In models of this form we set up a discrete choice model and numbers like four x's and 200 data points is where the statistics and approximation work well. Even going up to ten x's and 200 data points give you Monte Carlo means that are close to the right values, but are fairly variable still at that stage. So something like 200 data points is too small for a 10 dimensional problem.

SP But what degree of population fit? I am thinking of things like cross–section Engel curve where you have a population R^2 of about .2 at best, in which parametric modelling is difficult. You might have 5000 observations and still not be estimating things terribly precisely.

TS There's a sense in which I sidestep some of these issues by just focusing attention on an average derivative. Well, one way of approaching this is to use this object to estimate these coefficients and then fit the f function.

SP How fast is the convergence, and how is it related to the dispergence of y about the regression?

TS That's a statement about the variance. I have the formulas for the variance. They're definitely related to those things. The formulas for the variance for this object, well the variance of the estimator of this, the thing I call Δ, turns out to be the sample average of a component that would be the right component if you just observed these derivatives directly plus y–g, the residual times the density term. This is the form of the true variance of this parameter, so, basically, you pick up a term here but it's nullified by the derivative of log–density.

TG We did something like this in the old days when we had lots of data and only one independent variable, dividing the data into thin slices, averaging the dependent variable in each, measuring how well the resulting 'curve' fitted by the correlation ratio ρ, of the appropriately weighted standard deviation of the sample means to that of the raw data. It seemed an intuitively acceptable idea. You got quite a lot of observations in each slice, so the average was likely to be tolerably well behaved, and if you'd no idea what the shape was, the shape of the curve traced out by averages appropriately smoothed, was liable to be a good approximation to any 'true' relative. So I don't think the R^2 is awfully important here.

TS It may not be, but it is the case that these things are likely to differ somewhat. Another way of thinking about this is to consider $\tilde{g}$ here. The derivative of $\tilde{g}$ with respect to x is something that depends on the error, and if that error is bouncing about all over the place you may have the right expectational properties but it's not clear that you've got a very adequate description of the derivatives themselves by this regression.

Measurement and Modelling in Economics
G.D. Myles (Editor)

POVERTY INDICES AND DECOMPOSABILITY

by

James E Foster* and Anthony F Shorrocks#

A large number of alternative forms of poverty indices have been suggested in the decade since Sen's (1976) influential paper.[1] Among the most popular in recent years is the family of indices proposed by Foster, Greer and Thoerbecke (1984) – henceforth FGT – which may be written

$$P_\alpha(x;z) := \frac{1}{n(x)} \sum_{i=1}^{q(x;z)} \left[1 - \frac{\hat{x}_i}{z}\right]^\alpha \quad \alpha \geq 0, \tag{1}$$

where x is a vector of incomes; n(x) is the dimension of x; z is the poverty line; q(x;z) is the number of persons with incomes at or below the poverty line; and $\hat{x}_i$ is the i–th lowest income in the distribution x. For convenience, we will assume that incomes are positive, so $D^n := \{(x_1,..,x_n) | x_i \in D := (0,\infty)\}$ denotes the set of n–person income distributions and $\mathscr{D} := \{x | x \in D^n \text{ for some } n \geq 1\}$ denotes the set of all feasible income distributions. The class of indices represented by (1) includes two well known poverty indices as special cases. When $\alpha = 0$ we obtain the <u>headcount ratio</u>

$$H(x;z) := \frac{q(x;z)}{n(x)}, \tag{2}$$

which indicates the proportion of the population lying at or below the poverty line, while setting $\alpha = 1$ yields the <u>income gap ratio</u>

* Purdue University, West Lafayette, Indiana 47907.
\# University of Essex.

1 The text of this paper follows the presentation made at the Nuffield conference in 1987. Foster and Shorrocks (1988) contains a more detailed discussion of the assumptions, together with proofs of the Propositions and a number of additional results.

$$I(x;z) := \frac{1}{n(x)} \sum_{i=1}^{q(x;z)} \left[\frac{z-\hat{x}_i}{z}\right], \tag{3}$$

which represents the minimum aggregate income needed to eliminate poverty, expressed as a proportion of $zn(x)$. Both $H(\cdot)$ and $I(\cdot)$ fall foul of the criticisms raised by Sen (1976). But "distribution sensitive" poverty indices of the type favoured by Sen may be obtained by choosing any $\alpha \geq 2$.

The objective of this paper is to identify the desirable features of the FGT measures, so we may then investigate whether other types of indices also share these properties. We begin by noting that all indices of the form given in (1) satisfy five restrictions commonly imposed on poverty indices: symmetry; replication invariance; the focus axiom; monotonicity; and continuity in the incomes of the poor. For any given poverty line $z \in D$, an index $P(x;z)$ is said to be symmetric if

$$P(y;z) = P(x;z) \qquad \text{whenever y is a permutation of x,} \tag{A1}$$

and replication invariant if

$$P(y;z) = P(x;z) \qquad \text{whenever y is a replication of x.} \tag{A2}$$

The symmetry property (A1) allows us to focus attention on distributions which can be partitioned in the form $x = (x_p,x_r)$, where x_p is a vector containing the incomes of those who are poor, and x_r is the vector of incomes above z. For any given poverty line $z \in D$, we will denote the set of poor income distributions by $\mathscr{D}_p(z) := \{x \in \mathscr{D} \mid x_i \leq z \text{ for all } i\}$, and the set of non-poor income distributions by $\mathscr{D}_r(z) := \{x \in \mathscr{D} \mid x_i > z \text{ for all } i\}$. The focus axiom, which requires a poverty index to be invariant to changes in incomes above the poverty line, may then be stated as

$$P(x_p,x_r;z) = P(x_p,x_r';z) \quad \text{whenever } x_r,x_r' \in \mathscr{D}_r(z) \text{ and } n(x_r) = n(x_r'), \tag{A3}$$

while P is monotonic, and hence does not decrease when the income of any poor person is raised, if

$$P(x_p',x_r;z) \leq P(x_p,x_r;z) \quad \text{whenever } x_p,x_p' \in \mathscr{D}_p(z) \text{ and } x_p' > x_p. \tag{A4}$$

Finally, P is continuous in poor incomes if

$$P(x_p,x_r;z) \text{ is continuous in } x_p \text{ on } \mathscr{D}_p(z). \tag{A5}$$

We will use the term poverty index to refer to any function $P: \mathscr{D} \times D \rightarrow \mathbb{R}$ which satisfies (A1)–(A5). It should be noted that there are several other common properties of poverty indices which we choose not to include at this stage. For example, we do not require that P is continuous in all incomes and hence satisfies

$$P(x;z) \text{ is continuous in } x \text{ on } \mathscr{D}, \tag{A6}$$

since this property would be violated by the headcount ratio, or any other index which experiences an abrupt change as one person's income is raised through the poverty line. Nor do we impose the so–called transfer axiom, formulated by Sen (1976) as[i]

$$P(x_p,x_r;z) < P(x_p',x_r;z) \quad \text{whenever } x_p \in \mathscr{D}_p(z) \text{ and } x_p' \text{ is obtained from } x_p \text{ by a regressive transfer.} \tag{A7}$$

This property, designed to ensure that a poverty index is sensitive to the distribution among the poor, is satisfied by the FGT measures (1) when $\alpha \geq 2$, but is violated by both the headcount ratio (2) and the income gap ratio (3).

Assumptions (A1)–(A5) are satisfied by the majority of poverty indices that have been proposed.[ii] So these properties alone cannot explain the special appeal of the FGT measures. What really distinguishes the FGT indices from most of the other suggestions is the fact that they are decomposable in the sense that if the population is divided into K subgroup income distributions x^k (k = 1,...,K), then

$$P(x^1,\dots,x^K;z) = \sum_{k=1}^{K} \frac{n(x^k)}{n(x)} P(x^k;z). \tag{4}$$

So overall poverty is a weighted average of subgroup poverty, with weights equal to the subgroup population shares. Decomposability is an attractive feature since it ensures that the overall poverty value will always move in the same direction as poverty levels within subgroups. In other words, if poverty increases within one subgroup, and remains the same elsewhere, then overall poverty must increase. We refer to this property as subgroup consistency, defined as follows:

Definition P is subgroup consistent if, for every $z \in D$ and any x^1, x^2, y^1, $y^2 \in \mathscr{D}$ such that $n(x^1) = n(x^2)$ and $n(y^1) = n(y^2)$, we have

$$P(x^1,y^1;z) > P(x^2,y^2;z) \tag{5}$$

whenever

$$P(x^1;z) > P(x^2;z) \text{ and } P(y^1;z) = P(y^2;z). \tag{6}$$

Subgroup consistency is clearly a weaker condition than decomposability, and appears to be a highly desirable feature.[iii] Yet it is violated by almost all of the poverty indices proposéd in recent years, the main exceptions being the class of decomposable indices due to Chakravarty (1983) and the second family of indices suggested by Clark, Hemming and Ulph (1981).

We now consider the implications of the subgroup consistency condition in the context of a poverty index satisfying the basic properties (A1)–(A5). To simplify the notation we regard the poverty line z as fixed and omit this argument of the poverty index and the other functions and sets defined earlier. A subgroup consistent index induces an ordering on each D^n that is strictly separable (in the sense of Gorman (1968)) in each partition of incomes. This allows us to appeal to standard results on separability (Gorman (1968); Blackorby et al (1978)) to establish

Proposition 1 If P is a subgroup consistent poverty index, there exist functions ϕ and F such that

$$P(x_p,x_r) = F[P\phi(x_p),n(x_p),x_r] \text{ for all } x_p \in \mathscr{D}_p \text{ and } x_r \in \mathscr{D}_r, \tag{7}$$

where $F[P\phi,n,x_p]$ is continuous and increasing in $P\phi$;

$$P\phi(x) := \frac{1}{n(x)} \sum_{i=1}^{n(x)} \phi(x_i) \quad \text{for } x \in \mathscr{D}, \tag{8}$$

ϕ is continuous and non–increasing; and $\phi(t) = 0$ for all $t \geq z$.

Notice that the expression $P\phi$ given in (8) is itself a poverty index, which is both continuous and decomposable.[iv] Furthermore, if P is continuous in all incomes, as described in (A6), it may be shown that (7) simplifies to

$$P(x) = F[P\phi(x)] \quad \text{for all } x \in \mathscr{D}. \tag{9}$$

Hence we may deduce

Corollary 1 P is a continuous, subgroup consistent poverty index if and only if P is a continuous, increasing transformation of a continuous, decomposable poverty index.

Thus subgroup consistency provides a means of justifying the use of decomposable poverty measures. For, corresponding to each continuous subgroup consistent index, there is a continuous decomposable index which ranks distributions in precisely the same way.

Applying the focus axiom (A3) to (7) yields

$$P(x_p,x_r) = F[P\phi(x_p),n(x_p),x_r] = \Pi_1[P\phi(x_p),n(x_p),n(x_r)] \tag{10}$$

for all $x_p \in \mathscr{D}_p$ and $x_r \in \mathscr{D}_r$. Replication invariance then allows us to write

$$P(x_p,x_r) = \Pi_2[P\phi(x_p),n(x_p)/n(x_r)] = \Pi_3[P\phi(x_p),H(x_p,x_r)] \tag{11}$$

and since $P\phi(x_p,x_r) = H(x_p,x_r)P\phi(x_p)$, we obtain

Proposition 2 If P is a subgroup consistent poverty index, there exist functions ϕ and π such that

$$P(x) = \pi\left[\frac{1}{n(x)}\sum_{i=1}^{n(x)} \phi(x_i),\ H(x)\right] \quad \text{for all } x \in \mathscr{D}, \tag{12}$$

where ϕ is continuous and non–increasing; $\phi(t) = 0$ for all $t \geq z$; and $\pi[P\phi,H]$ is continuous and increasing in $P\phi$, and non–decreasing in H.[v,vi]

The analysis so far has assumed a fixed poverty line. When we allow for variations in the poverty line, the results take on a slightly different form. For example, equation (12) in Proposition 2 becomes

$$P(x;z) = \pi\left[\frac{1}{n(x)}\sum_{i=1}^{n(x)} \phi(x_i;z),\ H(x;z);z\right] \quad \text{for all } x \in \mathscr{D} \text{ and } z \in D, \tag{13}$$

where, at each specific z, the functions $\pi[\cdot;z]$ and $\phi(\cdot;z)$ have the properties listed in Proposition 2. To ensure some resemblance between the indices associated with different poverty lines, two invariance properties are typically used. A function $P(x;z)$ is said to be <u>scale invariant</u> if

$$P(\lambda x;\lambda z) = P(x;z) \text{ for all } x\epsilon \mathscr{D}, z\epsilon D \text{ and } \lambda > 0, \tag{A8}$$

and translation invariant if

$$P(x + \lambda \mathbf{1}; z + \lambda) = P(x;z) \text{ for all } x\epsilon \mathscr{D}, z\epsilon D \text{ and } \lambda \geq 0, \tag{A9}$$

where **1** is an appropriate sized vector of 1's. Employing the usual terminology, we will say that P(x;z) is a relative poverty index if it is scale invariant,[viii] and an absolute poverty index if it is translation invariant. All members of the FGT class (1) are scale invariant, together with almost all the other poverty indices that have been proposed. In contrast, no member of the FGT family apart from the headcount ratio satisfies translation invariance (A9). However, it may be noted that P_α is "almost" translation invariant, in the sense that $z^\alpha P_\alpha(x;z)$ is an absolute poverty index for every $\alpha \geq 0$.

Combining the scale invariance condition (A8) with the results of Proposition 2 allows us to derive

Proposition 3 If P is a subgroup consistent, relative poverty index, there exist functions ϕ and π such that

$$P(x;z) = \pi\left[\frac{1}{n(x)} \sum_{i=1} \phi(x_i/z), H(x;z)\right] \quad \text{for all } x\epsilon \mathscr{D} \text{ and } z\epsilon D, \tag{14}$$

where ϕ is continuous and non–increasing; $\phi(t) = 0$ for all $t \geq 1$; and $\pi[P\phi,H]$ is continuous and increasing in $P\phi$, and non–decreasing in H.[vii]

Similarly, combining translation invariance (A9) with the results of Proposition 2 yields

Proposition 4 If P is a subgroup consistent, absolute poverty index, there exist functions ϕ and π such that

$$P(x;z) = \pi\left[\frac{1}{n(x)}\sum_{i=1}\psi(z-\hat{x}_i),H(x;z)\right] \quad \text{for all } x\in\mathscr{D} \text{ and } z\in D, \qquad (15)$$

where ψ is continuous and non–increasing; $\psi(t) = 0$ for all $t \leq 0$; and $\pi(P^{\psi},H)$ is continuous and increasing in P^{ψ}, and non–decreasing in H.

An obvious question to ask at this stage is whether a subgroup consistent index can be both a relative and an absolute index of poverty. It is relatively easy to establish:

Proposition 5 P is a subgroup consistent poverty index that is both relative and absolute if and only if P is either an increasing transformation of the headcount ratio or else a constant function.[ix]

Since the requirement that an index be both relative and absolute is particularly restrictive, we might instead ask whether a pair of relative and absolute poverty indices can be compatible, in the sense that any change in the income distribution which raises the level of poverty for the relative index also increases the absolute poverty value.

Definition The poverty indices P and P′ are compatible if for all $z \in D$ and $x, y\in\mathscr{D}$,

$$P(x;z) \leq P(y;z) \text{ if and only if } P'(x;z) \leq P'(y;z). \qquad (16)$$

Our final result indicates that any subgroup consistent, relative poverty index which is compatible with some subgroup consistent, absolute poverty index must be a combination of the headcount ration and some member of the FGT family (1).

Proposition 6 The poverty indices P_R and P_A are subgroup consistent, and form a pair of compatible relative and absolute poverty indices if and only if there is some $\alpha \geq 0$ such that

$$P_R(x;z) = \pi_R[P_\alpha(x;z), H(x;z)] \quad (17)$$

and

$$P_A(x;z) = \pi_A[z^\alpha P_\alpha(x;z), H(x;z)]. \quad (18)$$

REFERENCES

Blackorby, C and D Donaldson (1980): "Ethical Indices for the Measurement of Poverty", *Econometrica*, 48, pp 1053–1060.

Blackorby, C, D Primont and R Russell (1978): *Duality, Separability, and Functional Structure: Theory and Economic Applications*, New York and Amsterdam, North–Holland.

Chakravarty, S R (1983): "A New Index of Poverty", *Mathematical Social Sciences*, 6, pp 307–313.

Clark, S, R Hemming and D Ulph (1981): "On Indices for the Measurement of Poverty", *Economic Journal*, 91, pp 515–526.

Donaldson, D and J A Weymark (1986): "Properties of Fixed-Population Poverty Indices", *International Economic Review*, 27, pp 667–668.

Foster, J E (1984): "On Economic Poverty: A Survey of Aggregate Measures", in R L Basmann & G F Rhodes (eds), *Advances in Econometrics, Volume 3*, Connecticut: JAI Press.

Foster, J E, J Greer & E Thorbecke (1984): "A Class of Decomposable Poverty Measures", *Econometrica*, 52, pp 761–766.

Foster, J E & A F Shorrocks (1988): "Subgroup Consistent Poverty Indices", mimeo.

Gorman, W M (1968): "The Structure of Utility Functions", *Review of Economic Studies*, 35, pp 367–390.

Sen, A K (1976): "Poverty: An Ordinal Approach To Measurement", *Econometrica*, 44, pp 219–231.

Conference Discussion

Handout First consider an existing index:

Poverty Index Foster, Greer, Thorbecke, *Econometrica* 1984

$$P_\alpha(x;z) = \frac{1}{n(x)} \sum_{i=1}^{q(x,z)} \left[1 - \frac{\hat{x}_i}{z}\right]^\alpha \quad \alpha \geq 0,$$

where z = poverty line; $\hat{x}_i$ = incomes below z; x = vector of incomes.

This class includes

$$P_0(x;z) = \frac{q(x;z)}{n(x)} \quad \text{headcount ratio } H(x;z)$$

$$P_1(x;z) = \frac{1}{n(x)} \sum_{i=1}^{q(x,z)} (z - \hat{x}_i) \quad \text{income gap ratio}$$

$P_\alpha(x;z)$ satisfies properties (A)–(H).

Properties

(A) Symmetry

$$P_\alpha(x;z) = P_\alpha(\hat{x};z) = P_\alpha(x_p,x_r;z)$$

x_p = vector of incomes below poverty line.
x_r = vector of incomes above poverty line.

(B) Focus Axiom

$$P_\alpha(x_p,x_r;z) = P_\alpha(x_p,x_r';z) \quad \text{if } n(x_r) = n(x_r')$$

(C) Monotonic

$$P_\alpha(x_p,x_r;z) \geqq P_\alpha(x_p',x_r;z) \quad \text{if } x_p' > x_p \in \mathscr{D}_p(z)$$

$\mathscr{D}_p(z)$ = space of possible vectors x_p.

Strictly Monotonic

$$P_\alpha(x_p,x_r;z) > P_\alpha(x_p',x_r;z) \qquad (\text{only } \alpha > 0)$$

(D) Continuous

$P_\alpha(x_p,x_r;z)$ is continuous in $x_p \in \mathscr{D}_p(z)$

(E) Replication Invariant

$P_\alpha(x;z) = P_\alpha(y;z)$ if y is a replication of x

(F) Transfer Axiom

$$P_\alpha(x;z) \geqq (>) \; P_\alpha(x';z)$$

if x, $x' \in \mathscr{D}_p(z)$ and x′ is obtained from x by a progressive transfer.

i **Transfer axiom**

CJB I don't see how it could be a criticism of the transfer axiom that it's violated by the headcount ratio rather than it being an objection to the headcount ratio that it violates the transfer axiom.

AS I was using that as a justification. As it happens, Amartya wrote this axiom down, derived his poverty index, and then found out his measure didn't satisfy this property. He then had to weaken it.

Handout (G) Scale Invariant

$$P_\alpha(\lambda x;\lambda z) = P_\alpha(x;z)\ \forall\ \lambda > 0$$

(H) (Almost) Translation Invariant

$$P_\alpha(x + \lambda 1;\ z + \lambda) = \frac{z^{\alpha}}{(z+\lambda)^{\alpha}}\ P(x;z), \forall\ \lambda > 0$$

ii **On the Axioms**

WE All these properties are properties of Foster's functional form. Are there other functional forms that satisfy these properties?

AS That is our paper.

WE Are these properties independent in the logical sense?

AS Yes, I think so.

CB They can't be.

AS They're not all independent, no. But I'm not sure which ones imply others.

WE The point to my question is that a much smaller class of properties may be enough to get all the others.

AS In fact, as I just said to you the transfer axiom is not one we're going to use at all.

CB He (Foster) thinks they are independent.

WE We always show it by giving examples satisfying n–1 of them but not the n–th.

Handout The important property under discussion here is:

(I) Subgroup Consistency

P is <u>subgroup consistent</u> if, for every $z \in D$ and any $x^1, x^2, y^1, y^2 \in \mathscr{D}$ such that $n(x^1) = n(x^2)$ and $n(y^1) = n(y^2)$, we have

$$P(x^1,y^1;z) > P(x^2,y^2;z)$$

whenever

$$P(x^1;z) > P(x^2;z) \text{ and } P(y^1;z) = P(y^2;z).$$

iii **On subgroup consistency**

TB It rules out the possibilities of relativities mattering.

GM Relative poverty is also ruled out by the focus axiom, isn't it?

AS In a sense, yes. But this is independent of the focus axiom and we'd still want it to be true if we relaxed the focus axiom.

Handout

Proposition 1 If P is a subgroup consistent poverty index, there exist functions ϕ and F such that

$$P(x_p,x_r) = F[P\phi(x_p),n(x_p),x_r] \text{ for all } x_p \in \mathscr{D}_p \text{ and } x_r \in \mathscr{D}_r,$$

where $F[P\phi,n,x_p]$ is continuous and increasing in $P\phi$;

$$P\phi(x) := \frac{1}{n(x)} \sum_{i=1}^{n(x)} \phi(x_i) \quad \text{for } x \in \mathscr{D},$$

ϕ is continuous and non–increasing; and $\phi(t) = 0$ for all $t \geq z$.

iv More on Subgroup Consistency

TG Although I absolutely agree with your result and think it's a very nice result, I think one should face the fact that subgroup consistency turns out to have stronger implications than maybe you had in mind when you assumed it. I'm sure you had in mind additivity, but it also gets rid of a possible class of poverty measure in which the weight given to a person depends on his ranking, so you might give the highest weight to the poorest person.

AS But that was deliberate, in the sense that once you do that you get very bad violations of the consistency axiom. You not only get violations of the consistency axiom stated here, but you can get situations such that the distribution changes, poverty goes down in every single subgroup of the population, and yet overall it goes up.

TG Yes, certainly.

CB You could have proceeded by doing this in rank–order subsets, a weaker place to work.

AS You could do. I'm quite happy for you to do so.

TG I'm just pointing out that rank independence came out of the subgroup consistency axiom rather than the symmetry axiom.

AS Yes, that's quite clear.

WE Is the converse also true, that every function of this kind satisfies the five properties (A1)–(A5) plus subgroup consistency?

AS Well, it's clearly not the case at the moment because we haven't used all the properties yet. We could certainly state line be fixed? It should be dependent on mean income or some other possibility.

AS That's one of the basic presumptions here, that there's a poverty line given exogenously. If you then say the poverty line ought to be a function of incomes, that's interesting, and we're working on it.

Proposition 2 If P is a subgroup consistent poverty index, there exist functions ϕ and π such that

$$P(x) = \pi\left[\frac{1}{n(x)}\sum_{i=1}^{n(x)} \phi(x_i), H(x)\right] \quad \text{for all } x \in \mathscr{D},$$

where ϕ is continuous and non–increasing; $\phi(t) = 0$ for all $t \geq z$; and $\pi[P\phi, H]$ is continuous and increasing in $P\phi$, and non–decreasing in H.

v On Proposition 2

TS I think the fixed poverty line is what differentiates this kind of index from an inequality index. If you're lowering inequality in groups without a fixed poverty line, I think this point about increasing inequality ibetween groups is of concern in that kind of index.

AS That's the view I take. You have something here – the poverty line – which acts as a reference point, whereas in inequality measurement you don't have a reference point.

GM Why should the poverty line be fixed? It should be dependent on mean income or some other possibility.

AS That's one of the basic presumptions here, that there's a poverty line given exogenously. If you then say the poverty line ought to be a function of incomes, that's interesting, and we're working on it.

vi **On Proposition 2**

TG Here it doesn't say that it must be increasing in the head-count ratio, is that right?

AS There doesn't seem to be any reason why it necessarily would be increasing in the head–count ratio.

TG Suppose you had someone just below z, then you move him into the rich group. Should there not be something that decreases the index? I thought that poverty was decreased if you increased the income of someone in the lower group.

AS As long as you don't move them over the poverty line. We haven't said anything about what happens if you move people between the two groups.

TG So if you strengthen that to include moving out I believe you would get strict monotonicity in H(x).

Handout Proposition 3 If P is a subgroup consistent, relative poverty index, there exist functions ϕ and π such that

$$P(x;z) = \pi\left[\frac{1}{n(x)} \sum_{i=1} \phi(x_i/z), H(x;z)\right] \quad \text{for all } x \in \mathscr{D} \text{ and } z \in D,$$

where ϕ is continuous and non–increasing; $\phi(t) = 0$ for all $t \geq 1$; and $\pi[P\phi,H]$ is continuous and increasing in $P\phi$, and non–decreasing in H.

vii **On Proposition 3**

TB I'm not quite sure of the sense in which we're meant to understand this as a measure of relative poverty. You tend to think of scale invariance as just meaning that if you change the units of measurement nothing else changes.

AS I never accept that argument, because it seems to me that if you were just saying we want a measure such that if you just change the units then the value doesn't change, that's not really very helpful. Because you can just insist that you're measuring incomes and the poverty line in some fixed units.

CJB P is defined for a particular set of units.

GM There's an obvious set of units here because you can only measure people in one set of units, so there's at least one normalisation involved and that's the number of people.

AS I would prefer to forget about the units argument completely and assume everything is measured in 1987 pounds.

TG In fact the point I understand Tim as making is that this has a deeper meaning?

TB It's just that I can't see what the deeper meaning is.

AS I think we're saying that poverty is a relative phenomenon and if it's the case that everyone's income changes, say, doubles, and that the poverty standard which we assign to that society in these changed circumstances has also doubled, then we would want to say that the relative poverty in that society hasn't changed.

CB There's a kind of illusion going on here, though. We recently had a bunch of Americans go into East Berlin and say how terribly poor they looked. Someone else said they looked about as poor as my parents did in America in 1955. Do you think they were really poor? You're saying yes, but I think there's a sense in which it's not true.

CJB But that account hasn't brought in a poverty standard. How's the z functioning in that example?

CB Well, you're saying you're going to move this up, the number of apples you need to eat.

AS I'm not particularly arguing that one would want to approve this condition. I'm merely saying if it had this property, I'm going to label it a measure of relative poverty.

CB That's certainly an allowable statement.

viii **Invariance to relative changes**

TG British writers tend to use current social security benefits in defining poverty. Their measures do satisfy this property in a general way.

AS All the standard poverty indices satisfy this property.

Handout Proposition 4 If P is a subgroup consistent, absolute poverty index, there exist functions ϕ and π such that

$$P(x;z) = \pi\left[\frac{1}{n(x)} \sum_{i=1} \psi(z - \hat{x}_i), H(x;z)\right] \quad \text{for all } x \in \mathscr{D} \text{ and } z \in D,$$

where ψ is continuous and non–increasing; $\psi(t) = 0$ for all $t \leq 0$; and $\pi(P^{\psi},H)$ is continuous and increasing in P^{ψ}, and non–decreasing in H.

Proposition 5 P is a subgroup consistent poverty index that is both relative and absolute if and only if P is either an increasing transformation of the headcount ratio or else a constant function.

ix CB There is a sense in which the headcount ratio is a very good number. If you think about a policy maker taking a first pass at what he's got to do with his budget, the headcount ratio is a good thing to have. It tells him something about the amount of funds he's got to think about allocating.

AS Yes.

GM No, it doesn't tell you that at all if you think about two situations, one where everyone in poverty is 5p below the poverty line, and one where everyone is almost on zero.

CB True.

JF It's the aggregate gap measure that would give the sum that is needed to lift everyone up to the poverty line.

Handout Proposition 6 The poverty indices P_R and P_A are subgroup consistent, and form a pair of compatible relative and absolute poverty indices if and only if there is some $\alpha \geq 0$ such that

$$P_R(x;z) = \pi_R[P_\alpha(x;z), H(x;z)]$$

and

$$P_A(x;z) = \pi_A[z^\alpha P_\alpha(x;z), H(x;z)].$$

x TS I want to ask what you know about the properties of these indices when you move people from the poor class to the rich class. Have you thought about a property where you take people from the poor class, raise their income and move them across the boundary? I wondered if that always reduced the poverty measure.

TG I don't see why your poverty movement axiom which says making the poor a bit richer, but still keeping them poor reduces poverty, but making the poor richer, and taking them into the rich group, doesn't necessarily reduce poverty. I can't actually see any sense in that.

AS We just didn't go that far.

TG I don't think there's anything that would lead you to the first conclusion, that wouldn't lead you to the second conclusion and I think the second conclusion would very likely allow you to get rid of the headcount ratio in (14) and (15).

AS But we're okay here.

TG If you have continuity as well.

TS In (12), suppose you take somebody that's just slightly below the poverty line and move them up. What you have done is to just take a little bit out of the sum, which would seem to suggest that if the index itself had to go down you could force this to be independent of H.

AS I think you're probably right here. We're still working on this.

TG You then have to have continuity in the whole of x, not just x_p.

AS You obviously lose the headcount ratio.

CB Another way to think about this which I think makes the problem you've solved more important than you stress, is to think about it as the following: begin with two arbitrary income vectors of different dimensions and ask yourself what you're trying to do. This is something economists very rarely think about, and that's when we have to order things of different sizes, one in n space and one in q. Well, what you've done is a special case of that, you've adopted a set of axioms that actually allow you to link valuations of vectors in different spaces. P(x), as you've written it here, is in fact a special case of that.

AS That's what replication replication invariance does. As soon as you get down to additive type functions it is sufficient to collapse the thing.

CB If you went back and thought about the problem of ordering a vector in n and n+1, it would be natural to not just use your poverty axiom as you've used it, but to use it as Terence and Tom have been suggesting for kicking somebody out, it gives a way to link n and n+1 in a natural way.

TG I'm now going to argue on Tony's side. H is inherently a discontinuous variable so anything that depends on H can't really be continuous at this critical point. It now seems to me, however, that if you set things up in terms of infinite populations and integrals, that argument is going to disappear and your position becomes perfectly agreeable. Then the headcount ratio becomes a proportion which is smoothly variable.

TS Then you have to worry about where additivity comes from, in other words you start with expectations.

TG No, it's very easy to do these additively separable things over a continuum.

GM It doesn't make sense to assume a continuum variable in any way in this problem

CJB But as an approximation to a large population why not?

GM Because the population isn't large in this problem and the size of the population is of great importance. If you were asked a simple question: 'how many people are below the poverty line?' you couldn't claim 'we have a continuum so there's obviously an infinite number'.

TG But you can take the proportion under the poverty line.

GM Yes, but you can't answer just the question – how many?

AS Yes, I think that is just sweeping the problem under the carpet, it's not really tackling it, just concealing it.

TG I do think this discontinuity is a disagreeable thing. One of the things is that we're dealing with measures of income which are really rather inaccurate, and the classification of whether they're just above or just below this poverty line is quite a difficult thing to do. If your thing is jumping all around the place as you make small errors of measurement, you're in a real mess.

AS Well, that's true.

Measurement and Modelling in Economics
G.D. Myles (Editor)

OPTIMAL UNIFORM TAXATION AND THE STRUCTURE OF CONSUMER PREFERENCES

by

Tim Besley* and Ian Jewitt#

Abstract

The well–known conditions for proportional commodity taxation to be optimal: that commodities be implicitly homogeneously separable from the single leisure good are sufficient conditions. It is not so clear from the published literature whether they are necessary or not. Mirrlees in an unpublished paper gives a form of direct utility function which is shown to be sufficient for proportional taxation to be optimal but which is not implicitly homogeneously separable. This paper generalizes the conditions and relates them to each other.

1 Introduction

This paper is concerned with the problem of deriving a class of preferences which imply the optimality of uniform taxes. The model we consider is the simplest single consumer one. We shall derive conditions under which it is optimal to tax all commodities (not including labour) uniformly as a proportion of producer prices. More generally, we provide conditions under which it is optimal for a subset of commodities, say food, to bear the same tax rates. The procedure is to derive the first-order conditions, impose the uniformity condition and then apply the separability techniques of Leontief (1947) to solve for a preference structure. This procedure, however, may lead to a variety of different preference structures. A number of authors have already derived conditions: Simmons (1974), and Deaton (1979, 1981) show that it suffices that consumption

* All Souls College, Oxford.
# University of Bristol.

consumption goods be separable in the consumer's cost function. Deaton calls this implicit separability, although Gorman (1970) in which it was introduced calls it quasi–separability. Homogeneous implicit separability would seem to be more descriptive of the actual situation. In an unpublished note, Mirrlees presents a different condition expressed in terms of the consumer's direct utility function. We are interested here in deriving conditions on preferences which make proportional taxation optimal regardless of the parameters of the problem; that is, independently of the producer prices.

In the next two sections we outline the model and derive the "Deaton" and "Mirrlees" conditions, in section four we derive a condition which encompasses both Deaton's and Mirrlees' as special cases, and section five concludes.

2 The "Deaton" Condition

We choose to represent preferences by the consumer's cost function

$$c(p,w,u) = \min\{px - wy \,|\, u(x,y) = u\}$$

where $u(x,y)$ is the utility of consuming a vector of commodities x and supplying an amount of labour y. The planner's problem is to

maximise	u
subject to	$c(p,w,u) = 0$
and	$\Sigma q_i c_i(p,w,u) + \omega c_w(p,w,u) = R$

where (p,w) and (q,ω) are consumer and producer prices respectively. The first of these constraints is that the consumer has no unearned income and that there are no lump sum taxes. If there were, it would be possible to attain first best. The model is of course somewhat artificial in this respect. Although obstacles to lump sum taxation arise in many consumer economies, it is difficult to imagine what they might be in a single consumer one. The second constraint is that the consumer chooses a feasible consumption plan. We assume constant returns to scale for simplicity; none of the results depend on it.

The first–order conditions for this problem include

$$\Sigma q_i c_{ik} + \omega c_{wk} = \lambda c_k \text{ for each k,} \tag{1}$$

where a k subscript denotes differentiation with respect to the price of the k–th commodity, and a w subscript denotes differentiation with respect to the wage w. It follows from the homogeneity of the cost function that if at the optimum consumer and producer prices are collinear, then there is a scalar μ such that

$$c_{wk} = \mu c_k \text{ for all k} \tag{2}$$

hence,

$$\frac{\partial c_i(p,w,u)/c_j(p,w,u)}{\partial w} = 0 \quad \begin{array}{l}\text{at the optimal (p,w,u) and}\\ \text{for all consumption goods i,j}\end{array} \tag{3}$$

Condition (3) says simply that <u>at the optimum</u> the ratio of Hicksian demands for any pair of commodities is independent of the wage.[i] Condition (3) is necessary, is it also sufficient? It will be under the usual assumption of Slutsky negativity for then the minor of the Slutsky matrix appearing in (2) is nonsingular (see e.g. Afriat (1980)). Given the existence of the inverse, the proof follows easily: suppose the optimum is reached at p, w, λ, then there is a unique q satisfying the equations (1). Since with separability some q collinear with p will satisfy (1), this must be the only one and is therefore the solution. Condition (3) is recognizable as a Leontief (1947)/Sono (1961) separability condition and it is tempting to conclude from it that a sufficient and also <u>necessary</u> condition for uniform taxation to be optimal is that the consumer's cost function be separable in the form:[ii]

$$c(p,w,u) = f(g(p,u),w,u). \tag{4}$$

Condition (4) is well known from Deaton (1979, 1981). Deaton formulates the problem in terms of leisure rather than labour and conducts his analysis in terms of the "distance function".[1] Stern (1986) points out some drawbacks of this approach.[2] It is of some interest therefore to try and get the benefits of the Deaton approach without the associated costs. We can do this by replacing the distance function with the "partial distance function" which is defined as

$$D(x,y,u) = \{\lambda \,|\, u(x/\lambda,y) = u\}. \tag{5}$$

The partial distance function is dual to the "restricted cost function"

$$C(p,y,u) = \min\{px \,|\, u(x,y) = u\} = \min\{px \,|\, D(x,y,u) = 1\}. \tag{6}$$

By differentiating the identity

$$D(x,y,u(x,y)) \equiv 1$$

---

1 Introduced in Gorman (1970), (1976).

2 "A disadvantage of using the distance function and the Antonelli matrix is that it requires positive total expenditure x. ...the derivative property of the distance function...obviously breaks down at x = 0. ...The way out is to define all consumptions to be non–negative, by replacing labour supply q_0 by leisure $(T - q_0)$ where T is large enough to ensure non-negativity of leisure. This is quite often the procedure adopted in labour supply analysis and is that followed by Deaton. It has a number of unsatisfactory aspects. First, the total time available T is hard to define let alone measure. Secondly, elasticities of leisure demand and elasticities of leisure supply are difficult subjects for our intuition if we don't know how to define and measure leisure. Thirdly, the great convenience of factors as negative demands in general equilibrium theory is lost. Fourthly, the absence of lump-sum income is a central feature of the commodity tax problem and it is partly obscured by making expenditure strictly positive."

we obtain

$$\partial D/\partial x_i(\Sigma \partial u/\partial x_j) = \partial u/\partial x_i \tag{7}$$

$$\partial D/\partial y_i(\Sigma \partial u/\partial x_j) = \partial u/\partial y_i. \tag{8}$$

Hence, at D = 1 the derivatives of the partial distance function are simply prices deflated by expenditure on commodities. The optimal tax problem can now be re–written as

maximise	u
subject to	$D(x,y,u) = 1$
	$\Sigma x_i \partial D/\partial x_i + y\partial D/\partial y = 0$
and	$\Sigma x_i q_i - y\omega = R.$

Since D is homogeneous of degree one in x, we can simplify somewhat to get:

maximize	u
subject to	$D(x,y,u) = 1$
	$1 + y\partial D/\partial y = 0$
and	$\Sigma x_i p_i - yw = R.$

Upon forming the Lagrangean and differentiating we obtain the following first-order conditions

$$\partial D/\partial x_k = \lambda y \partial^2 D/\partial y \partial x_k + \mu p_k \text{ for each } x_k. \tag{9}$$

This can be written as[v]

$$\frac{t_i}{p_i} - \frac{t_j}{p_j} = \lambda \frac{\partial \log(D_i/D_j)}{\partial \log y} \tag{10}$$

which is essentially as in Deaton (1979, 1981) except that we have used the partial distance function to better accommodate the no lump sum income problem raised by Stern.[iii, iv] The simplification of the optimal tax formulae arises from the homogeneity properties of the functions used to represent preferences.[vi, vii] We now turn to an alternative approach.

3 The Mirrlees Condition

James Mirrlees[3] has shown that another condition is also sufficient: this is that the utility function be of the form[viii]

$$u(x,y) = \nu(\phi(x/y),y). \tag{11}$$

Formulating the optimal tax problem in terms of the direct utility function, the planner must:

maximize	$u(x,y)$
subject to	$\Sigma u_i(x,y)x_i + u_y(x,y)y = 0$
and	$\Sigma q_i x_i - \omega y = R.$

At the optimum there are scalars μ and λ such that

$$u_k = \mu(\Sigma u_{ik}x_i + u_{yk}y) + \lambda q_k \text{ for each k.} \tag{12}$$

Hence, if marginal utilities are proportional to producer prices for consumption goods at the optimum, then

$$(\Sigma u_{ik}x_i + u_{yk}) \,/\, u_k = \eta \text{ for each k.} \tag{13}$$

Condition (13) is equivalent to

---

3 In an unpublished one page remark.

$$\frac{\partial \log(u_k/u_j)}{\partial \log y} + \Sigma \frac{\partial \log(u_k/u_j)}{\partial \log x_i} = 0. \tag{14}$$

Equation (14) is a separability condition of sorts, and is equivalent, when <u>it holds throughout a neighbourhood</u>, to the form of the utility function in (11). This is established in the following lemma.

Lemma 1: Consider the function u(x,y). If

$$\partial u/\partial x_i \big/ \partial u/\partial x_j \text{ is homogeneous of degree zero in } (x,y) \tag{15}$$

throughout some neighbourhood with $y >> 0$, then and only then, u can be written in the form

$$u(x,y,u) = f(\phi(x,y),y) \tag{16}$$

with ϕ homogeneous of degree zero in (x,y).

Proof: Define $u(\lambda x,\lambda y) = g(x,y,\lambda)$. Hence,

$$\partial u(\lambda x,\lambda y)/\partial x_i \big/ \partial u(\lambda x,\lambda y)/\partial x_j = \partial g(x,y,\lambda)/\partial x_i \big/ \partial g(x,y,\lambda)/\partial x_j \tag{17}$$

and the homogeneity condition on u becomes a separability condition on g. It follows that g can be written in the form

$$g(x,y,\lambda) = h(\gamma(x,y,),y,\lambda) \tag{18}$$

hence, for any λ, μ

$$u(\lambda\mu x,\lambda\mu y) = h(\gamma(\mu x,\mu y),\mu y,\lambda), \tag{19}$$

so setting $\lambda = \mu^{-1} = y_1$, yields

$$u(x,y) = h(\gamma(xy_1^{-1}, yy_1^{-1}), yy_1^{-1}, y_1) = f(\phi(x,y),y) \tag{20}$$

with homogeneity in (x,y) as required. Sufficiency follows immediately upon differentiation. □

Remark: Note that the proof does not rest on y being a scalar. The vector case applies if we examine the uniform taxation of a subset of commodities. This we do in section 5.

The "Mirrlees" and "Deaton" preference structures can be compared, perhaps most easily, by considering their restricted cost functions,

$$C(p,y,u) = \min_x \{px \,|\, u(x,y) = u\}. \tag{21}$$

The "Mirrlees" restricted cost function is

$$C(p,y,u) = \min_x \{px \,|\, \nu(\phi(x/y),y) = u\} \tag{22}$$

multiplying and dividing by y this becomes

$$C(p,y,u) = y.\min_{x/y} \{px/y \,|\, \nu(\phi(x/y),y) = u\}$$

making the change of variable $x/y = x'$, and dropping the " $'$ " we have

$$C(p,y,u) = y.\min_x \{px \,|\, \nu(\phi(x),y) = u\}$$

$$= y.\min_x \{px \,|\, \phi(x) = \Phi,\ \nu(\Phi,y) = u\}.$$

Assuming non–satiation in the x goods, it must be possible to solve $\nu(\Phi,y) = u$ for Φ in terms of y, and u, giving $\Phi = \theta(y,u)$, say, this yields

$$C(p,y,u) = y.\min_{x}\{px \mid \phi(x) = \theta(y,u)\} \tag{23}$$

$$= y\psi(p,\theta(y,u)).$$

The "Deaton" restricted cost function is

$$C(p,y,u) = \max_{w}\{c(p,w,u) + wy\} \tag{24}$$

$$= \max_{w}\{f(g(p,u),w,u) + wy\}$$

$$= h(g(p,u),y,u), \text{ say,} \tag{25}$$

setting (23) and (25) equal to each other:

$$\psi(p,\theta(y,u)) = h(g(p,u),y,u)/y. \tag{26}$$

Suppose that $\partial\theta(y,u)/\partial y \neq 0$, given continuity this inequality holds throughout a neighbourhood of (p,y,u) since ψ has p separable from y, and as y and u both enter ψ only through θ it must also have p separable from u. Hence, in this neighbourhood the restricted cost function is of the form

$$C(p,y,u) = h(f(p),y,u). \tag{27}$$

It follows from this that the full cost function takes the form

$$c(p,y,u) = \Psi(f(p),w,u) \tag{28}$$

or equivalently, the direct utility function is homogeneously separable in x:

$$u = \nu(h(x),y) \text{ with h homogeneous of degree one.} \tag{29}$$

This is reminiscent of Sandmo (1974). We cannot conclude from these "separability conditions" that preferences must be homothetically separable for proportional taxation to be optimal. Indeed we already know weaker sufficient conditions. What this does imply is that neither the Mirrlees form nor the Deaton form of preferences is necessary for proportional taxation to be optimal.

From the perspective of the necessary and sufficient condition given in (3), it is not immediately apparent why the Mirrlees condition works. The reason can be seen by using the following argument. The Mirrlees form of utility implies that the full cost function can be written in terms of the restricted cost function as follows

$$c(p,w,u) = \min_{y}\{C(p,y,u) - wy\} = \min_{y} y[\psi(p,\Phi(u,y)) - w].$$

At the cost minimising choice of y,

$$[\psi(p,\Phi(u,y)) - w] + y\psi_{\Phi}(p,\Phi(u,y))\Phi_{y} = 0.$$

Hence, if there is no lump sum income $c = 0 = \psi - w$, then

$$\psi_{\Phi}(p,\Phi(u,y))\Phi_{y} = 0.$$

Since, $c_{\Phi} \neq 0$ (otherwise cost is independent of utility), we have

$$c(p,w,u) = 0 \Leftrightarrow \Phi_{y} = 0. \tag{30}$$

From the form of the cost function

$$C_i/C_j = c_i(p,\Phi(u,y))/c_j(p,\Phi(u,y)).$$

Hence,

$$c = 0 \Rightarrow \partial(c_i/c_j)/\partial w = 0, \tag{31}$$

that is, with zero lump sum income condition (3) holds. This reconciles the Mirrlees and Deaton results.[ix]

4 A New Condition

A class of preferences which encompasses both the "Deaton" and "Mirrlees" ones, and for which uniform taxation is optimal is given in the following theorem.

Theorem 1: A sufficient condition for uniform commodity taxation to be optimal is that the direct utility can be defined implicitly by a function of the form

$$F(x,y,u) = f(\phi(x,y,u),y,u) = 1 \tag{32}$$

with $\phi(x,y,u)$ homogeneous of degree zero in (x,y).[x]

Proof: Suppose that preferences are represented implicitly by $F(x,y,u) = 1$, the optimal tax problem can be written as

maximize	u
subject to	$F(x,y,u) = 1$
	$\Sigma x_i \partial F/\partial x_i + y\partial F/\partial y = 0$
and	$\Sigma x_i q_i - y\omega = R.$

This yields the following first-order conditions

$$F_k = \mu(\Sigma F_{ik} x_i + F_{yk} y) + \lambda q_k \text{ for all k.} \tag{33}$$

If the condition of the theorem is satisfied then for some scalar ρ

$$\Sigma F_{ik}x_i + F_{yk}y = \rho F_k \text{ for all k,} \tag{34}$$

and this gives

$$F_k = kq_k \text{ as desired.} \quad \square \tag{35}$$

The form of preferences in theorem 1 includes both the Deaton and Mirrlees conditions as special cases. The Mirrlees case occurs when ϕ is independent of u, the Deaton case occurs when F is homogeneous of degree one in x. Another simple case arises when u enters F only through ϕ. This gives

$$F(\phi(x/y,u),y) = 1.$$

Solving $F(\Phi,y) = 1$ for Φ defines $\Phi = K(y)$, say. K satisfies

$$F_{\Phi}K'(y) + F_y = 0$$

and since $F_{\Phi} \neq 0$, we have

$$K'(y) = 0 \Leftrightarrow F_y = 0.$$

With no lump sum income

$$\Sigma u_i x_i + u_y y = 0 \Leftrightarrow \Sigma F_{\Phi}\phi_i x_i/y + (F_y - \Sigma F_{\Phi}\phi_i x_i/y^2)y = F_y y = 0.$$

Hence, $K'(y) = 0$ which defines the chosen level of work independently of prices. The utility function is of the form

$$u = u(x/y, K(y)) \tag{36}$$

with u strictly increasing in K.

An equivalent representation for the preferences in the theorem is that the restricted cost function should take the form

$$C(p,y,u) = y\psi(p,\Phi(y,u),u) \tag{37}$$

with $\psi(p,\Phi,u)$ increasing in Φ. This is obtained as follows:

$$\begin{aligned} C(p,y,u) &= \min\{px \mid f(\phi(x,y,u),y,u) = 1\} \\ &= y.\min\{px/y \mid f(y\phi(x/y,1,u),y,u) = 1\} \\ &= y.\min\{px \mid f(y\phi(x,1,u),y,u) = 1\} \\ &= y.\min\{px \mid \phi(x,1,u) = \Phi,\ f(y\Phi,y,u) = 1\} \\ &= y.\min\{px \mid \phi(x,1,u) = \Phi(y,u)\} \\ &= y\psi(p,\Phi(y,u),u). \end{aligned}$$

It is not immediately apparent that (37) imposes any restriction on preferences at all. In any neighbourhood where $\Phi_y \neq 0$, the mapping u,y $\to \Phi$, y is one–to–one and can therefore be inverted by suitable choice of the ψ function. Given the structure (37), the cost minimising choice of y at zero unearned income is determined as the solution to

$$\Phi_y(y,u) = 0, \tag{38}$$

which is independent of prices. Hence, if the structure (37) applies, then the wage compensated supply of labour function is independent of commodity prices. The reason that this yields proportional commodity taxes is plain to see; the (x,y) that the planner can reach on any particular indifference surface all have the same value of y. The utility level at the optimum therefore fixes y, and conditional on this u and y the problem is essentially a first–best one. That is, if

$$\Sigma x_i \partial F/\partial x_i + y\partial F/\partial y = 0 \;\Leftrightarrow\; y = f(u),$$

then the program appearing in the proof of theorem 1 becomes

maximise u
subject to $F(x,f(u),u) = 1$
and $\Sigma x_i q_i - f(u)\omega = R.$

Evidently, at the solution to this program consumer and producer prices must be collinear. The condition is moreover necessary: consider an indifference surface, the one reached at the optimum, and let the optimal labour supply be y^*. For the sake of geometrical intuition consider the two commodity case. We can trace out two lines on this surface. The first is the intersection of the surface with the $y = y^*$ plane, ie, the x_1, x_2 indifference curve. The second is the set of tangencies with planes passing through the origin, ie, the decentralisable consumption bundles. These two curves have y^* in common. If they have different tangents at y^* then the two tangent lines determine a plane which is the tangent plane to the indifference surface. This passes through the origin by construction and implies that the first–best is attainable.

Theorem 2: A necessary and sufficient condition for proportional taxation to be optimal is that the wage compensated supply of labour is independent of commodity prices.

Proof: See above text.

Theorem 2 can be stated in purely geometric terms. For the optimal level of u, consider the indifference surface in x_1, x_2, y space. Arranging co–ordinates so that the y axis is vertical, this indifference surface will have the appearance of a hill when viewed from the origin. It is necessary and sufficient for proportional taxation to be optimal for the horizon as viewed from the origin to be a contour of this "hill". The sufficient condition of Theorem 1 obviously implies that of Theorem 2 but it is not necessary. The condition of Theorem 1 implies a global restriction on preferences rather than simply at those points which can be reached as demands with zero lump–sum income. In view of the statement following (37) the global restriction is very weak indeed; in any neighbourhood with $\Phi_y \neq 0$, there is <u>no</u> restriction.

5 Extensions

Theorem 1 in this paper gives a condition for uniform commodity taxation to be optimal when there is a single "other good" – work. Consider a set of aggregates, like food, clothing, and so on, and denote them by the vectors $x^1, x^2, \ldots, x^T$, and let the remaining goods, including work, geiger counters, etc, be denoted by the vector y. It is optimal to tax uniformly <u>within</u> the aggregates if preferences can be represented implicitly by a function of the form

$$F(\phi^1(x^1,y,u),\ldots,\phi^T(x^T,y,u),y,u) = 1, \tag{39}$$

where each of the ϕ^t is homogeneous of degree zero in (x^t,y). For each t let z^t be the vector of <u>all</u> goods not in subgroup t. We can define a conditional cost function for each t, $g^t(p^t,z^t,u)$. Its derivatives are compensated conditional demands. When the preferences are as in (39), then,

$$\Sigma z_k^t \frac{\partial (g_i^t/g_j^t)}{\partial z_k^t} = 0 \text{ for each t.} \tag{40}$$

Hence, the ratio of compensated conditional demands are independent of small proportional changes in the quantity of all goods outside group t. Condition (40) is necessary and sufficient in the same way that (3) is in the scalar case.

6 Concluding Remarks

The Leontief separability conditions are sufficient for separable structures only if they hold throughout a neighbourhood. In the optimal tax problem, since there is no lump–sum income, feasible allocations only occupy a surface of lower dimension than the commodity space. Hence, any Leontief conditions derived from the first–order optimizing conditions do not necessarily hold throughout a neighbourhood making integration to an equivalent condition on preferences impossible.[xi] However, it is possible to integrate to get sufficient conditions; at least two of which have appeared in the literature. We obtain a closed form representation of preferences which encompasses the existing ones and derive a necessary and sufficient condition to which it can be compared.

References

Afriat, S (1980): *Demand Functions and the Slutsky Matrix*, Princeton University Press.

Deaton, A S (1979): "The Distance Function and Consumer Behaviour with Applications to Index Numbers and Optimal Taxation", *Review of Economic Studies*, 46, pp 391–405.

——— (1981): "Optimal Taxes and the Structure of Preferences", *Econometrica*, 49, pp 1245–1260.

Gorman, W M (1970a): "The Structure of Utility Functions", *Review of Economic Studies*, 37, pp 391–405.

——— (1970b): "Quasi Separable Preferences, Costs and Technologies", unpublished.

——— (1976): "Tricks with Utility Functions", in Artis, M and R Nobay (eds), *Essays in Economic Analysis*, Cambridge University Press.

Leontief, W W (1947): 'Introduction to a Theory of the Internal Structure of Functional Relationships', *Econometrica*, 15, pp 361-373.

Mirrlees, J (1984): "Taxing Work (Correctly)?"; unpublished, Nuffield College.

Sandmo, A (1974): "A Note on the Structure of Optimal Taxation", *American Economic Review*, 64, pp 701–706.

Simmons, P (1974): "A Note on Conditions for the Optimality of Proportional Taxation", unpublished, University of York.

Sono, M (1961): 'The Effect of Price Changes on the Demand and Supply of Separable Goods'; *International Economic Review*, 2, pp 239–271.

Stern, N (1986): "A Note on Commodity Taxation: The Choice of Variable and the Slutsky, Hessian and Antonelli Matrics (SHAM)", *Review of Economic Studies*, 53, pp 293–299.

Conference Discussion

i **Procedure**

TG I'm a bit worried about this (equation 3) because these things don't have interiors.

IJ I'm saying there's a procedure here which we're following rather blindly.

ii TG I don't think you can make statements like this without an interior, but maybe I'm wrong.

IJ Well, I'll shut you up for now.

iii **Formulation of problem with distance function**

TG Has the level of a lump–sum income being constant any meaning? It means you've had to define a numeraire somewhere

IJ Here it would create difficulties. The sort of issues involved would have to be dealt with in a different way.

TB But you do need to use up the normalisation earlier on. If you're going to have lump–sum income in, we need to fix it, but here we don't need to make a normalisation yet, whereas with lump–sum income you do.

iv TG Lump–sum income is treated in exactly the same way if it's now taken as an endowment of y, you just define the new y to be the deviation from that and the lump–sum income you're keeping constant.

IJ I can believe that.

v **First–order condition with distance function**

TG Again, in my belief, this doesn't go well with your Leontief-Sono conditions. If you had separability and it hadn't happened to be conical separability in a certain, very specific, sense, then it would not have occurred as separability in the distance function and would have remained hidden from you. So that seems to be an argument against this formulation.

IJ Can I come back to that?

vi **On taxing complements to leisure**

CJB Why can it have nothing to do with that? Supposing something was strictly consumed in fixed proportions with leisure, you've got to have a bicycle to enjoy any leisure, and we could tax leisure by taxing bicycles.

IJ In what way can't you tax leisure?

CJB By taxing everything?

IJ You can alter the price of it in the same way you can alter the price of everything else so, in a way, the problem's symmetric between leisure and other things. You've got an endowment of leisure and you tend not to have an endowment of other things.

vii **Equation (10)**

AS Perhaps I misunderstand you but the standard argument is you want to impose a lump–sum tax on someone: you want to tax the whole of their endowment of time. But in fact you can only tax the bit they work so anything that you can do to tax the other bit will compensate for not taxing leisure directly. Isn't that the argument?

IJ That's an argument I've used.

TB At least one drawback is the fact that you're relegating leisure to a certain status because you have an endowment. But if you have an endowment of another good...

AS You'd want exactly the same, if it's the case you couldn't observe.

IJ Having got endowments in here though, we have got some things are goods and some things that are bads, I guess, but there's no actual endowment of leisure here; you're selling the bit you work. Preferences are represented over trades and it's symmetrical in that, but we're getting a special effect over here which is an artifice.

viii CJB Where is the Mirrlees' paper? It's not cited.

TB It's an unpublished one page note.

AnB It's also in two Oxford M Phil exam papers, but the model there is much more general. He has m consumers, n producers and not necessarily a leisure endowment, the consumers get profit shares. The conditions for uniform taxation, with a Cobb-Douglas utility function, turns out to be that the covariance of the expenditure shares with income has to be constant irrespective of the good. Even when you have a leisure endowment and labour, the condition becomes the covariance of the shares with leisure.

IJ It's the covariance because...

AB It's the covariance because you have n consumers and can use the Diamond covariance result.

ix TG Again, I think Angus did this with the distance function because he'd just come across it and was thinking in terms of it. I believe it to be quite a considerable mistake. It has several disadvantages, one it shares with the cost function: if you have separability anywhere you won't get it unless it happens to be conical separability.

CB Can I add a footnote? It is possible that you can write down necessary and sufficient conditions on the cost function for non–conical separability. They're quite clean.

TG So you can certainly carry it across ... the point is simple conditions ...

CB They're not so complex.

TG We'll come back to that, I have to keep quiet in any case.

x Theorem 1

CB I just don't understand this. On the top line of the board $F(\phi(\lambda x,1,\lambda u),y,u)$ is homogeneous of degree zero in x and y, so if you multiply by λ the thing remains unchanged?

IJ Yes

CB So you let λ equal 1/y and you get F. This just says x is implicitly separable from y given u, which was your original result more or less?

IJ Say that again.

CB Well, you can more or less eliminate y from inside ϕ and so you've said that for an arbitrary implicit representation x is pseudo–separable from y.

AS Why can't you now replace x/y with something else?

IJ Well, you can't, these things have got meaning.

xi Terence's solution in 'Comment'

TG You've got a set of measure zero. All you can know about is what happens on that set and, to whatever degree is required for maximisation, close to it. So you can make the function, subject to concavity conditions, whatever you like anywhere else. That seems to me to make it peculiar to look for conditions on the whole function that just give you conditions in this very small set.

IJ I agree with you. That's the point of the paper really. It's a comment on a procedure rather than doing the thing correctly. It is a procedure that it's tempting to follow; you see something that looks like Leontief–type conditions and you get to the end of the paper by writing down those conditions. However the preferences we get depend on the procedure. We obtain three types, two are contained in the other, but none is necessary.

TB The point of looking at global conditions that preferences should satisfy is that these conditions are very hard to get a handle on if you just see them in terms of very local things. To understand economic implications it's better to have global properties and their implications for demand systems.

TG I'd obviously claim that if I'm right, which is a very large assumption, that this is a global sufficient condition as well as a global necessary condition.

IJ Which one is?

TG The thing (equation 15, 'Comment') that it is at a certain point the order of zero. Any function that has that property will solve this, and if I'm not wrong, only those functions.

IJ Do you take my point that what you're saying is that here's the class of all preferences we're interested in ...

TG It's measure zero so it's just a curve not an area.

IJ You're saying the class we're interested in is up to second-order.

TG Oh, yes, I see, I'm talking nonsense.

IJ You can add a bit on to this class and that's got to be second-order, but you're also saying we can do the same for that class. The answer you gave is basically the Deaton form but you can add a bit on. Any of the other conditions you derive might be of that form but there's more work to be done.

TG One can be certain if you extend it beyond this local condition you'll lose anything like necessity.

A Comment on Tim and Ian's paper on Optimal Proportional Taxation

by

Terence Gorman*

Tim and Ian find

$$c_{jw} = \mu c_j,\ c = c(p,w,u) = 0;\ w \text{ scalar}, \tag{1}$$

as a necessary condition for uniform taxation of the goods (X_j), with prices $(p_j) = p$ to be optimum. This implies

$$dc_w - c_{ww}\,dw - c_{wu}\,du = \mu(dc - c_w\,dw - c_u\,du), \text{ say} \tag{2}$$

$$dc_w = \mu dc + \nu dw + \rho du, \text{ say}, \tag{3}$$

so that given arc–connectivity which does not seem likely to be a problem here,

$$c_w = \varphi(c,w,u) = \varphi(0,w,u) = -\sigma(u), \text{ say}, \tag{4}$$

since c_w is homogeneous of degree zero in p,w, and $c = 0$ throughout this analysis.

Hence

$$y = -c_w(p,w,u) = \sigma(u), \text{ when } c = p\cdot x - y = 0. \tag{5}$$

Let us now set this problem up in the primal and in a slightly different way.

---

* Nuffield College, Oxford.

$$y = f(x,u): \underline{\text{tastes}} \tag{6}$$

$$\Sigma f_k x_k = p \cdot x = y = f(x,u); \underline{\text{budget constraint}}\ x \in X(u) \tag{7}$$

$$z = g(x): \underline{\text{technology}} \tag{8}$$

$$\text{Max: } R = y - z = f(x,u) - g(x), \text{ given } u; \underline{\text{government target}}, \tag{9}$$

where we will allow $g(\cdot)$ to be any well behaved input requirement function to get

$$f_j - g_j = \lambda\ \Sigma_k f_{jk} x_j, \tag{10}$$

$$\begin{aligned} \theta_j &:= (f_j - g_j)/f_j = \lambda\ \Sigma_k\ \partial \log f_j / \partial \log x_k, \\ &= \lambda\ \partial \log f_j / \partial \log \rho, \end{aligned} \tag{11}$$

where ρ is an equiproportional increase in each x_k. Here

$$\theta_j = t_j/(1 + t_j), \tag{12}$$

in the normal notation. Hence taxes are best uniform iff

$$\partial \log (f_j/f_k)/\partial \log \rho = 0 \quad \text{each } j,k, \tag{13}$$

that is to say iff the slopes f_1:,.. f_j:.. of the $f(x,u)$ = constant map remain constant when one changes the x_j in the same small proportion. That is to say

$$f(\cdot,u) \text{ is homothetic}, \tag{14}$$

to the second order, I believe, for small changes in x leading out of $X(u)$, <u>as defined for the budget constraint (7)</u>. That is to say, I suppose,

$$\Sigma_k f_k x_k = \psi(f(x,u),u) + o(|x-X(u)|^2), \tag{15}$$

where $|x-X(u)|$ is the Euclidean distance from x to X(u). When $x \in X(u)$, therefore,

$$f(x,u) = \psi(f(x,u),u), \tag{16}$$

yielding

$$f(x,u) = \sigma(u), \text{ when } \Sigma f_k x_k = f(x,u), \tag{17}$$

again, assuming solubility which, I think, should be easily proven under normal assumptions.

Equally well

$$f(x,u) = \sigma(u) \text{ when } \Sigma f_k x_k = f, \tag{18}$$

would seem to imply the homotheticity condition (15). Since homogeneity of degree zero is ruled out on normal assumptions, (15) implies

$$f(x,u) = F(\psi(x,u),u) + o(|x - X(u)|^2) \tag{19}$$

where

$$\psi(\cdot,u) \text{ is conical.} \tag{20}$$

Hence

$$\Sigma f_k x_k = F'(\psi(x,u),u)\, \psi(x,u), \tag{21}$$

$$= F(\psi(x,u),u)\,,$$

both on X(u), yielding, one hopes,

$$\psi(x,u) = \alpha(u), \text{ say; } f(x,u) = F(\alpha(u),u) = \sigma(u), \tag{22}$$

on X(u), again.

Second order homotheticity about the surface X(u) in (x,u) space, which looked to me like the appropriate condition, can therefore be replaced by the simpler version (5).

Note that neither X(u), nor

$$Z = \{(x,y)y = f(x,u) = \Sigma_k f_k x_k, \text{ some } u\} = \cup\,[X(u) \times \{\phi(u)\}]$$

is open in the appropriate space, given strictly concave preferences, so that the best we can hope for is local approximations, such as (15), on the tastes.

To be slightly more precise, I think uniform taxes are uniformly best iff something like the following is true for 'well behaved' tastes:

There are functions $\psi(\cdot)$, $F(\cdot)$, such that $\psi(\cdot,u)$ is conical, and (19) holds, where X(u) is now defined by,

$$X(u) = \{x \mid F'(\psi(x,u),u)\ \psi(x,u) = F(\psi(x,u),u)\}. \tag{23}$$

Note, by the way, that the government first taxes profits at 100%. For our purposes, it must insist on at least that revenue.

Measurement and Modelling in Economics
G.D. Myles (Editor)

AGGREGATE PRODUCTION FUNCTIONS AND PRODUCTIVITY MEASUREMENT: A NEW LOOK

by

John Muellbauer*

Many economists currently take a somewhat jaundiced view of the estimation of aggregate production functions. Three problems seem particularly troublesome: the unobservables problem, especially with regard to utilization, the aggregation problem and the simultaneous equation problems. This paper presents theoretical arguments and empirical evidence from British manufacturing for the view that the first of these is the most serious with important dimensions in the measurement of capital and output as well as that of utilization.

New light is shed on two classic questions. One was first raised by Feldstein (1967) who observed in a cross-section context that the elasticity of output with respect to average observed hours of work significantly exceeded the elasticity with respect to employment. Craine (1973) observed a similar result for time–series data. The other question is one with which most researchers on productivity have struggled: how to correct productivity for cyclical variations in the utilization of inputs. A novel answer based on the use of overtime hours data is found to give excellent empirical results.

1 Introduction

One of the main applications of estimated aggregate production functions is to the measurement of productivity. This paper takes a new look at the underlying methodology. Though there have been waves of enthusiasm in the past for the estimation of production functions on aggregate time series data, many economists currently take a somewhat jaundiced view of such activity. Three potential problems could be particularly

* Nuffield College, Oxford.

troublesome. These are the unobservables problem, especially with regard to utilization, the aggregation problem and the simultaneous equations problem. The view taken below is that the first of these is by far the most important though aggregation plays an important role in coping with it. The theoretical arguments which are presented are supported by a substantive piece of empirical work on quarterly British manufacturing data for 1956–83.

New light is shed on two classic questions. One was first raised by Feldstein (1967) who observed in a cross–section context that the elasticity of output with respect to average hours of work significantly exceeded the elasticity with respect to employment. Feldstein suggested fixed costs as an explanation. He argued plausibly that part of the daily or weekly time input is taken up with starting up or winding down production. A similar effect arises from an increase in hours increasing the intensity of utilization of the capital stock without adding to fixed interest costs. Craine (1973) and others have supported Feldstein's cross–section results with aggregate time series evidence.

The other question is one with which most researchers on productivity have struggled at some time. This is how to correct productivity for the pronounced cyclical variations, or to put it another way, the variations in utilization of inputs, to which it is subject. Among the solutions which have been proposed are to use general distributed lags in estimating employment or production functions to pick up short term fluctuations in utilization, to use the unemployment rate or the rate of profit as a cyclical indicator, to survey firms on whether they are working at full capacity and use the survey mean as an indicator and to measure electricity consumption as a percentage of the installed wattage to indicate the level of utilization. There are difficulties with all these proposals and I believe that the arguments against the solution proposed below are less severe.

The paper is structured as follows. Section 2 deals with aggregation problems abstracting from the utilization issue. Section 3 is devoted to measurement problems for labour input with the main emphasis on the measurement of labour utilization. I argue that there are no good direct measures of 'under–utilization' but that overtime hours of operatives are a good indicator of 'over–utilization'. In aggregate, under–utilization should be low when over–utilization is high and vice versa. By a statistical aggregation argument I show how to construct a measure of average utilization from a nonlinear function of average overtime hours per operative scaled by normal hours. Appendix 1 derives an alternative utilization measure.

Section 4 discusses measurement problems for capital. There is a widely held position that the capital stock should not be adjusted for utilization. One potential source of information on capital utilization is a survey of firms' opinions on capacity utilization and this suggests a test of the hypothesis that such data contain no information additional to that in labour utilization. Appendix 2 shows how a capacity utilization measure can be derived from such data. The main focus is on the fact that in most countries gross capital stock data are constructed from gross investment data under fixed service lives assumptions. These assumptions fly in the face of the responses economic theory would predict for retirement decisions when prices, wages, taxes and demand conditions vary. Without data on retirement decisions, there seems little prospect in applying theories of scrapping to derive proxies for unobserved scrapping. The proposal to deal with this problem through the use of shifts in time trends is thus inevitably crude.

Section 5 discusses measurement problems in output. Methods of measuring output differ somewhat in different countries but British practice is probably quite representative. Particular emphasis is given to four problems. The first arises from approximating changes in real value added by applying fixed value added weights to changes in gross output volumes. This results potentially in what I call the 'gross output bias'. The other three biases arise because of problems with the price deflators which are used to deflate the current price data which are the major source for the output index. Thus there is potentially a 'domestic price bias' because, in the absence of reliable export price indices, the CSO uses domestic wholesale prices to deflate the exported component of output.

There is potentially a 'list–price bias' because the price data which are collected may not fully reflect transactions prices. Finally, there is potentially a 'price control bias' because of the incentives faced by firms in periods of price controls to distort the prices or specifications of goods in order to bypass these controls. Darby (1984) has argued that such an effect was important for the U.S. Observable proxies are proposed for each of these biases.

Section 6 describes the empirical application to quarterly data for 1955–83 for British manufacturing. The basic equation which best embodies these ideas is subjected to a battery of econometric tests and comparisons with alternatives. These include tests of parameter stability and of whether the residuals are well behaved, tests of the exogeneity of employment and overtime hours and a test of the hypothesis that the utilization of capital as well as that of labour is important. In addition, an explanation is given of how the Feldstein–Craine result arises for such data.

Section 7 summarises the conclusions for the methodology of aggregate production function estimation. The substantive implications of this approach for measuring and understanding British manufacturing productivity are discussed in Mendis and Muellbauer (1984).

2 Aggregation in the Absence of Variations in Utilization

This section pursues what is essentially a Divisia index approach (see Divisia (1952) and, for example, Jorgenson and Griliches (1967)) to measuring the relation between changes in outputs and inputs. The analysis of aggregation is fairly standard, following the pioneering work of Theil (1954). For firm i, I assume there exists a constant returns production function linking value added output, labour and capital. I abstract from aggregation problems over types of labour and types of capital within the firm, though the techniques used can in principle easily be extended to deal with them. The production function is time dependent reflecting the state of technology, established practices governing the allocation of labour within firms and is meant to hold at a normal rate of utilization:

$$q_{it} = F_i(\bar{h}_{it}\ell_{it}, K_{it}, t) \tag{1}$$

where q_i = real value added, ℓ_i = employment, K_i = gross capital stock, and $\bar{h}_i$ = normal hours of work. Thus, in rates of change, suppressing time subscripts and assuming $\bar{h}_i$ constant,

$$d\ln q_i = \alpha_i d\ln \ell_i + (1 - \alpha_i) d\ln K_i + \theta_i\, dt \tag{2}$$

where θ_i reflects changes in technology and work practices and α_i is the elasticity of output with respect to employment. For cost minimizing firms, α_i is the share of labour in factor payments. The fact that the weights on $d\ln \ell_i$ and $d\ln K_i$ add to unity reflects the constant returns to scale assumption. Aggregating across firms with the same value of α, one can define

$$d\ln \bar{q} \equiv \Sigma_i w_i^q d\ln q_i \tag{3}$$

where w_i^q is the share of the ith output in total output for the aggregate in question. Then, suppressing time and sector subscripts,

$$d\ln \bar{q} = \alpha d\ln \bar{\ell} + (1 - \alpha) d\ln \bar{K} + (\Sigma w_i^q \theta_i) dt$$

$$+ \alpha \Sigma (w_i^q - w_i^\ell) d\ln \ell_i + (1 - \alpha) \Sigma (w_i^q - w_i^k) d\ln K_i \tag{4}$$

where w_i^ℓ is the share of the ith employment level in total employment and w_i^k is the analogous capital share. Identifying these aggregates with observable sectoral aggregates within manufacturing, it is clear that the last two terms in (4) are aggregation biases about which one can do nothing without access to individual firm data. For sector I, let us therefore write (4) as

$$\mathrm{dln}\bar{q}_I = \alpha_I \ln\bar{\ell}_I + (1-\alpha_I)\mathrm{dln}\bar{K}_I + \theta_I \mathrm{dt} \tag{5}$$

where the last term in (5) is the sum of the last three terms in (4).

Aggregating across sectors with different α_I:

$$\mathrm{dln}\bar{q} = (\Sigma w_I^q \alpha_I)\mathrm{dln}\bar{\ell} + (1 - \Sigma w_I^q \alpha_I)\mathrm{dln}\bar{K} + \Sigma w_I^q \theta_I \,\mathrm{dt}$$

$$+ \Sigma w_I^q \alpha_I(\mathrm{dln}\bar{\ell}_I - d\ell n\bar{\ell}) + \Sigma w_I^q(1-\alpha_I)(\mathrm{dln}\bar{K}_I - \mathrm{dln}\bar{K}) \tag{6}$$

The last two terms are aggregation biases which are measurable. The first is a value added weighted covariance between labour shares in value added α_I and the sectoral growth rates of employment and the second term an analogous expression for capital. Therefore, these aggregation biases are zero when the sectoral rates of change are identical or, more generally, when deviations in sectoral rates of change are distributed independently from sectoral factor shares.

With constant utilization rates and cost minimizing firms, (6) is the basis for accounting for growth by chaining together period to period changes. This is essentially the kind of technique recommended and used by Jorgenson and Griliches (1967). The residual $\Sigma w_I^q \theta_I$ is a weighted average of the firm θ_i's and of the inescapable aggregation biases represented by the last two terms in (4). A major reason for deriving (6) is to understand how far one would be likely to go wrong if utilization rates were constant in fitting a Cobb–Douglas production function in the form[ii]

$$\ln\bar{q} = \mathrm{const.} + \alpha\ln\bar{\ell} + (1-\alpha)\ln\bar{K} + \theta t \tag{7}$$

Empirically, it would appear that apart from cyclical fluctuations associated with variations in utilization rates, the α_I's for British manufacturing are remarkably stable over time. Since, by the construction of the aggregate output index, the w_I^q are constant weights, this would suggest that the constancy of $\alpha \approx \Sigma w_I^q \alpha_I$ would be a good approximation.

The evidence for British manufacturing suggests that the aggregation biases over sectors are small.[1] Thus there are good reasons for believing that the aggregation problems in fitting (7) to aggregate data are unlikely to be resolved by using industry group data.

Of course, if utilization rates were constant there would be no difficulty about measuring labour productivity by chaining together $d\ln\bar{q} - d\ln\bar{\ell}$, perhaps taking the aggregation bias over sectors, $\Sigma w_I^q \alpha_I (d\ln\bar{\ell}_I - d\ln\bar{\ell})$ into account. In fact, the main argument for production function estimation as opposed to growth accounting via (6) is precisely that it offers a way of finding an econometric model which will pick up varying utilization rates and systematic biases in capital and output measurement. Once one has found such a model one can measure changes in productivity correcting for changes in rates of utilization.[iii]

3 Utilization and the Measurement of Labour Input

Most writers on productivity discuss the measurement of quality and compositional changes in the labour force. I have nothing new to say here. Effective hours of work per unit time (e.g. one quarter), distinguished from paid for hours, consist of effective hours per week h and working weeks per quarter WW, the latter depending on paid holiday arrangements. Write labour input in (1) as $WW_{it}h_{it}\ell_{it}$.[2]

---

1 An approximation to the aggregation bias for labour can be obtained by taking

$$\Sigma w_I^q \alpha_I (\Delta\ln\bar{\ell}_I - \Delta\ln\bar{\ell})$$

where w_I^q and α_I are averages of the beginning of period and end of period weights. This gives less than one half of one percent respectively over the entire periods of 1955–70 and 1970–83.

2 It should be noted that there is another aspect of production, multiple shift working, which (1) thus amended may not represent well. One can argue that adding a night shift to an existing day shift is like replicating the plant without altering the capital stock. Where d and n superscripts refer to day and night shift magnitudes, this suggests in place of (1),

Labour utilization, measured as the proportional deviation of effective weekly hours from normal hours,[1] is defined by

$$u_{it} = \ln h_{it} - \ln \bar{h}_{it} \tag{8}$$

Various ways of proxying utilization have been used in the past. Denison (1979) used the cyclical deviation in the share of profits but this is sensitive to short run movements in input and output prices which may not be reflected in utilization rates. Sometimes unemployment rates of workers have been used, see e.g. Baily (1981), but this seems inappropriate for various reasons including the confusion of supply side effects arising from demographics and supply incentives and the difficulties in defining a sectoral unemployment rate. The Wharton approach of measuring deviations from trend output is little help since it begs the question of what determines the trend. The Jorgenson and Griliches (1967) technique of taking electricity use as a proxy seems sensible but was heavily criticised by Denison (1969) though it has been used with some success by Heathfield (1972, 1983) in a study of the British engineering industry[3]. Baily (1981, 1983) also tried lay offs, deviations of the rate of change of employment from trend and survey based indices of capacity utilization. Another technique, see Chatterji and Wickens (1982), is to estimate production or employment functions where distributed lags pick

---

$$q_{it} = F_i(WW^d_{it} h^d_{it} \ell^d_{it}, K_{it}, t) + \lambda_i F_i(WW^n_{it} h^n_{it} \ell^n_{it}, K_{it}, t)$$

and λ_i takes account of the possibility that night workers are less productive. For British manufacturing, observations on the proportion of shift workers ranging from 13% to 25% are available only for three dates pre–1973, and there is no option but to ignore this complication. With better data, labour utilization could be defined separately for each shift and aggregated by the proportion of workers in each.

3 Electricity consumption as a percentage of installed wattage is the utilization variable here. One problem with it is that it may be sensitive to variations in electricity prices relative to other inputs.

up utilization effects, and use the steady state solution of such an equation to measure cyclically corrected productivity growth. However, my experience is that the steady state solution is typically not precisely enough determined to detect trend shifts before 3 or 4 years have elapsed.

The utilization measure proposed in this paper has a stronger theoretical foundation than most of the above. The method involves an old idea, that of "smoothing by aggregation". In the aggregate production function we need a concept of aggregate labour utilization, u_t so that:

$$\alpha u_t = \Sigma w_i^q \alpha_i u_{it} \tag{9}$$

where w_i^q is the share of the ith plant in reference year output. With w_i the wage, $\alpha \approx \Sigma w_i \ell_i / \Sigma p_i q_i$, $\alpha_i \approx w_i \ell_i / p_i q_i$, $w_i^q = p_i q_i / \Sigma p_i q_i$ it follows that

$$u_t \approx \Sigma w_i \ell_i u_{it} / \Sigma w_i \ell_i$$

Since $w_i \ell_i / \Sigma\, w_i \ell_i$ is a close approximation to the share of the ith plant in the total number of workers, u_t is a close approximation to the average utilization rate averaging over all workers.

Let u_j now refer to the jth worker. If such a worker is working overtime $u_j = \ln(\bar{h}_j + \text{overtime hours}_j) - \ln\bar{h}_j$. What cannot usually be observed is below normal utilization when "undertime" is being worked but the worker is still paid for a normal week.[vi] Thus we observe u_j when $u_j \geq 0$ but not when $u_j < 0$. Approximating the distribution of u_j with a continuous density function $\phi(u)$, define

$$u^* = \int_{u>0} u\phi(u)du \tag{10}$$

Since u_j for $u_j \geq 0$ is overtime proportional to normal hours for the jth worker, think of u^* as proportional overtime averaging over all workers whether working overtime or not.

Analogously, define

$$\hat{u} = \int_{u<0} (-u)\phi(u)du \tag{11}$$

and think of $\hat{u}$ as unobserved proportional "undertime" again averaging over all workers. One can also think of u^* and $\hat{u}$ as truncated means weighted by Prob $(u > 0)$ and Prob $(u < 0)$. Figure 1 illustrates the distribution of u.[vii] By definition, mean utilization

$$E(u) = u^* - \hat{u}. \tag{12}$$

Now imagine the distribution shifting horizontally but with its spread constant. As it moves to the right, u_t^* increases and $\hat{u}_t$ declines. This traces out a smooth trade off between u_t^* and $\hat{u}_t$. For each type of distribution a particular type of trade will be associated.[iv]

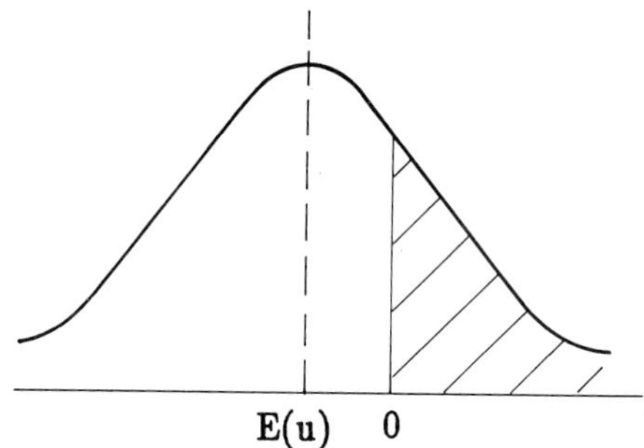

Figure 1: the distribution of proportional deviations of utilization rates

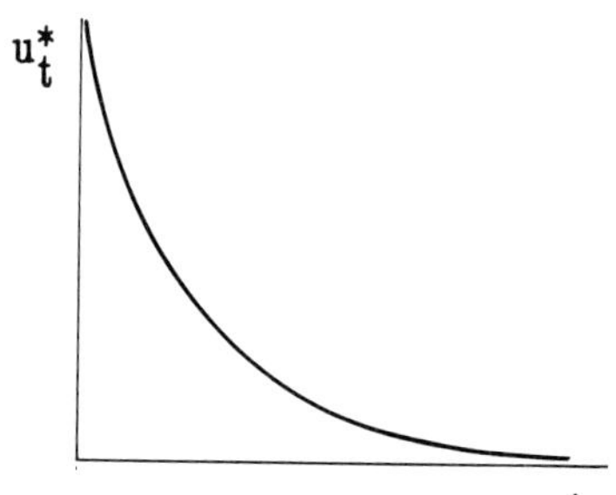

Figure 2: the u_t^*, $\hat{u}_t$ trade off

For example, a symmetric distribution always implies a symmetric trade off. Since the cost per effective hour of paying for a normal week tends to infinity as h_i tends to zero and since h_i has a finite upper limit, it seems likely that the distribution is bounded. A simple functional form which implies both a bounded distribution and which can allow for potential asymmetry is $(u_t^* + c_1)(\hat{u}_t + c_2) = c$ where $c_1 > 0$, $c_2 > 0$ and defined for $u_t^* \geq 0$, $\hat{u}_t \geq 0$. Then, applying (12), mean utilization is[4]

$$E(u_t) = c_2 + u_t^* - \frac{c}{u_t^*+c_1} \tag{13}$$

c reflects the spread of the distribution in Figure 1. If this is constant over time, we can "observe" $E(u_t)$ from (13) even though "undertime" $\hat{u}_t$ is not observable.

In practice, there is another complication. It appears that for some firms a part of overtime is regarded by the workers as normal or systematic. Indeed, the reductions in normal hours in Britain in the late 1950s and the early 1960s seem to have been accompanied by an increase in systematic overtime.[v] This suggests replacing u_t^* by the "true" concept $u_t^* - s_t$ where s_t refers to the systematic overtime component and where s_t increases as normal hours fall.[ix] Thus

$$E(u_t) = u_t^* + c_2 - s_t - \frac{c}{u_t^*+c_1-s_t} \tag{14}$$

Appendix 1 discusses how mean utilization can be derived when u_t follows a Normal distribution and shows that the empirical results are remarkably similar to those discussed in the body of the paper below.

---

4 Similar ideas have been used in the context of measuring aggregate excess supply Z from data on unemployment U but with data on vacancies V missing. Assuming UV = c, see Hansen (1970), $Z=U-V=U-cU^{-1}$.

It is an interesting curiosity demonstrated to me in a 1980 correspondence by Angus Deaton that, if $c_1 = c_2 = 0$, this trade–off corresponds to a t–distribution with two degrees of freedom. Such a distribution has tails so fat that the variance is unbounded and is thus the assumption implicit in Hansen (1970).

4 Capital Measurement, Capacity Utilization and Unobserved Scrapping

It can be argued that a separate concept of capital utilization has cutting power; for example, the tonnage of shipping laid up ought to be subtracted from the total tonnage in a shipping production function. In manufacturing, were the data available, one might want to subtract unused buildings from the total stock of buildings. There is a widespread view, however, see Kendrick (1973), Denison (1974) and Gollop and Jorgenson (1980), in which it is argued that capital is a kind of overhead concept and that a separate concept of capital utilization has no place in a production function, given that the other inputs are correctly measured.

One might ask then what surveys of firms' capacity utilization measure. Price (1977) reports that in a special inquiry into answering practices in the CBI survey[5] two thirds of respondents measured current output against the capacity only of buildings and plant while one fifth included, in addition, the availability of labour. This sort of information is not decisive on the issue since a relatively high level of output may simply reflect a relatively high level of employment and overtime work. But since one might want to entertain the hypothesis that something special to capital is being measured by such surveys, Appendix 2 below is devoted to the construction of a capacity utilization variable based on the proportion of firms (weighted by size) who report below capacity operation. Section 6 contains an empirical test of whether such a variable has any explanatory power given that the labour utilization measure is incorporated in the production function.

As in most countries, the CSO figures for gross capital stock are derived by a perpetual inventory method from gross investment flows and

---

[5] The Confederation of British Industry (CBI) Industrial Trends Survey is similar to surveys in many other countries and asks: "Is your present level of output below capacity (i.e. are you working below a satisfactory or full rate of operation)?"

assumed retirements. The figures on retirements are based on assumptions about service lives of different kinds of assets which have remained unchanged over a long period but were revised in 1983. In fact, the assumption is that for each type of asset there is a smooth distribution of service lives around a central estimate, see Griffin (1976). The assumed service lives do not vary cyclically nor respond to relative factor prices.

However, this does not mean that economic scrapping is entirely ignored in the capital stock figures. When a firm purchases new equipment, the second hand market value of equipment it scraps is subtracted from the purchases to give a net investment figure. There is a conceptual problem here since the gross capital stock figures are meant to reflect productive capacity rather than a market asset valuation. Thus, particularly at times of low profitability, the second hand market value may underestimate the productive capacity of equipment being scrapped. Furthermore, when companies become bankrupt and assets are sold abroad or simply taken out of use, it appears that no allowance whatever for scrapping is made by this procedure. Thus it seems likely that scrapping is understated when economic recession or relative price shifts raise economic scrapping. A similar point is made by Baily (1981) and Raasche and Tatom (1981) to help explain the slowdown in productivity growth which occurred in U.S. manufacturing from 1973.

The question is what observables can be taken as proxies for scrapping not directly observed. Economic theory would suggest estimating a vintage production model in which the capital stock as such does not appear but which exploits the optimality conditions that govern which vintages remain in use. Malcolmson and Prior (1979), Malcolmson (1980) and Mizon and Nickell (1983) have estimated such models and report considerable empirical success. However, the results are not informative on questions of productivity change. Shifts in the parameter which measures technical progress in the path–breaking paper by Malcolmson and Prior and in Malcolmson cannot be identified.[6] Mizon and Nickell report some

6 Prior restrictions on several of the other parameters are required for identification.

difficulty in their model in finding sensible estimates of the technical progress parameter. More work is clearly needed on what can be learned about productivity change from vintage models when no observations on scrapping are available.

Baily (1981) favours a stock market valuation of the capital stock and takes a weighted average of it and the gross capital stock as his measure of the true capital stock. As some of his discussants remark, this may not be ideal since other influences such as variations in interest rates that can have an exogenous source will influence the stock market valuation. Scott (1976, 1981) indeed argues that no observable capital stock concept is meaningful though gross investment is. Apart from unobserved scrapping, he notes that maintenance is not properly measured and questions the assumption made in defining the gross capital stock that each asset remains as productive when new until the end of its life.

The practical response taken below to these difficulties is to fit time trends with linear lines allowing slope changes to occur at times when, on a priori grounds, one would expect a great deal of unobserved scrapping.

5 Measurement Biases in Output

In most countries, indices of aggregate output are constructed from a mixture of basic indicators, some of them measuring output in physical units and others in current prices deflated by price indices. Both raise issues of how to incorporate quality changes. A great deal has been written on the theory and practice of quality measurement and the quality correction of price indices, see for example, Deaton and Muellbauer (1980), ch.8 and Triplett (1983) for recent overviews.[7] This paper does not contribute to this literature. As far as estimating aggregate production functions is concerned, measurement biases in output quality are likely to be rather trend like and will show up in the trend coefficients. This

---

7 Though curiously little applied research on the topic has been undertaken for British output statistics.

paper is concerned with other sources of bias. Although the context of the discussion below is British manufacturing, there are parallels in other countries for all four of the biases to be discussed.

The index of output at constant factor cost aims to measure movements in real value added i.e., to separate out the contribution to final output of labour and capital from that of other inputs such as raw materials and imported intermediate manufactured goods. This separation is not easy to make, especially at an industry level. To do it properly requires frequent input–output tables derived from censuses of production whose costs have been thought prohibitive compared with the value of the information yielded. Instead, the CSO approximates changes in value added for each of a quite disaggregated list of goods by changes in gross output. These are weighted by value added weights from a quinquennial census of production. This is fine as long as value added and gross output change in the same proportion. However, when raw material prices of imported intermediate manufactured goods change relatively to those of labour and capital, firms have an incentive to substitute.

Substitution can be of two types, the first and probably the larger results in changes in the pattern of output towards goods less intensive in the inputs whose relative prices have increased. This results in no bias since the weights applied to the gross output changes are fixed and so unaffected by relative input price changes. However, the second, substitution for each type of good between labour and capital on the one hand and other inputs on the other does give a bias. Value added increases faster (slower) than gross output when the relative prices of raw materials or imported intermediates increases (decreases) so that the output index understates (overstates) the true increase. Since substitution takes time, one would expect this measurement bias to be negatively correlated with the lagged (log) ratio of raw material prices to domestic wholesale prices of output PR and with a (log) index of foreign competitors' wholesale output prices in Sterling relative to domestic ones, PW.

The theory can be more formally explained as follows. Given substitution possibilities in production between raw materials and other inputs, the analogy of (2), again abstracting from utilization, is

$$\mathrm{dlnq_i}^* = \alpha_i \mathrm{dln}\ell_i + \beta_i \mathrm{dlnK_i} + (1 - \alpha_i - \beta_i)\mathrm{dlnm_i} + \theta_i \mathrm{dt} \qquad (15)$$

Here q_i^* is gross output. Having assumed constant returns, the derived demand for raw material input takes the form

$$\mathrm{dlnm_i} = \mathrm{dlnq_i}^* + \mathrm{dlng_i}\ (\text{relative factor prices}) \qquad (16)$$

Substituting (16) into (15), gives

$$\mathrm{dlnq_i}^* = \frac{\alpha_i}{\alpha_i+\beta_i}\mathrm{dln}\ell_i + \frac{\beta_i}{\alpha_i+\beta_i}\mathrm{dlnK_i} + \left[\frac{1-\alpha_i-\beta_i}{\alpha_i+\beta_i}\right]\mathrm{dln\ g_i} + \frac{\theta_i}{\alpha_i+\beta_i}\mathrm{dt} \quad (17)$$

Note that the weights on dln ℓ_i, dlnK_i sum to unity and that since $g_i(\cdot)$ is a decreasing function of PR given some substitution possibilities, a negative response of gross output to PR is implied. These points generalize easily when there is a vector of other inputs.

Bruno (1984) and Bruno and Sachs (1982) suggest that this bias, the 'gross output bias', is an important part of the explanation of the slowdown in productivity growth in industrial countries after 1973. However, as Grubb (1984) has argued, the empirical magnitudes are unlikely to make this a major part of the story.

About two thirds of the British manufactured output index in recent years has been based on value deflated data. A further bias to be considered arises because no satisfactory export price deflators exist for most of output which is exported. The CSO therefore uses domestic wholesale price indices instead. However, because of exchange rate movements, it is likely that in the short run there can be significant divergencies in these domestic prices from the unobserved export prices so that a 'domestic price index bias' in output results. An observable indicator of this bias is the ratio PW of foreign to domestic wholesale prices. Much of this effect operates immediately when exchange rates change but then tends to unwind as competitive and cost pressures act on export prices. This predicts a positive coefficient on PW_t in an equation for measured output with a smaller or zero coefficient in the long run. It is likely that movements in the unobserved export prices are somewhat more attenuated than in foreign wholesale prices but this will be reflected in the estimated coefficients on PW.

A third source of bias arises from another problem with the deflators. Although these aim to capture transactions prices, it is probable that they are partly based on list prices, hence giving rise to a 'list price bias'. Discounts measure the gap between list and transactions prices and are likely to be sensitive to changes in competitive pressure and to changes in underlying costs. An increase in PW reflects a reduction in competitive pressure and so a reduced gap between list and transaction prices. Then measured price indices will tend to understate true price increases and measured output increases overstate true ones. This would imply a positive PW effect, though one that eventually unwinds at least partially, in an equation for measured output. One expects similar effects for increases in cost pressures. Since transactions prices are likely to be more flexible than list prices, an increase in costs will be associated in the short run with measured price increases understating true ones. This would imply positive short run effects for PR and PW, the latter representing imported intermediates, in an equation for measured output. In the long run, these effects should be zero, as list prices adjust fully.

Finally, consider the effects of price controls. Darby (1984) argues that price controls instituted in the U.S. in 1971 were widely evaded, for example, by firms claiming spurious quality improvements or simply relabelling goods. The increase in the official price indices was therefore understated and in output overstated. These biases reversed in 1974 as price controls were taken off and reported output and so productivity fell by more than the true figures. In Britain, price controls were introduced in April 1973, slightly relaxed in December 1974 and August 1976 and replaced in August 1977 by the much weaker Price Code. The Price Commission which operated these policies was finally abolished in 1979. The

Price Commission's quarterly reports give figures which measure intervention both in number of cases and by the money value of sales affected. These make it possible to measure roughly the intensity of the controls. The hypothesis is that the more intense the controls, the greater the incentive of firms to evade them and the greater the bias in the official price indices and so in measured output.

6 Empirical Results

(a) The Data

I begin by briefly[8] describing the data and their sources, whose names are abbreviated as follows: ET is Economic Trends, HABLS is the Historical Abstract of British Labour Statistics, DE is the Department of Employment and DEG is the DE Gazette. Manufacturing is defined by the 1986 SIC.

q_t = index of manufacturing output at constant factor prices, seasonally adjusted and stock adjusted from 1970. This is the only seasonally adjusted variable in the data set. Source: ET, CSO. Range: 65.3 in 1955.1, 115.3 in 1974.2.

ℓ_t = employment in manufacturing, an average of 3 monthly figures and refers to all employees, part time and full time. Source: HABLS, DEG and DE. Range: 5.347 million in 1983.4, 8.491 million in 1965.4.

K_t = gross capital stock in 1980 prices. This includes assets leased to the manufacturing sector and is based on service life assumptions newly introduced in 1983. Source: 1983 National Income and Expenditure and CSO. Range: 88.4 billion in 1955.1, 211.1 billion in 1983.4.

---

8 A fuller description is in Mendis and Muellbauer (1984).

$\bar{h}_t =$ normal hours = 0.4425 x NH_t where NH_t = index of normal hours per week. Source: HABLS, DEG. Range of NH: 100 in 1955.1, 88.6 in 1983.4 and 90.4 from 1968 to 1979, ie. 40 hours per week.

$$OH_t = \frac{\text{weekly overtime hours per operative on overtime} \times \text{fraction of operatives on overtime}}{\text{normal hours}}$$

an average of 3 monthly observations from 1961 and a mid quarter observation before 1961. Source: HABLS, DEG. Range: 0.0400 in 1958.3, 0.0876 in 1973.4.

$$PR_t = \ln \frac{\text{wholesale price index for raw materials purchased by manufacturing}}{\text{wholesale price index for home sales of manufacturing}}$$

Source: ET. Range: –0.352 in 1972.2, 0.128 in 1974.1

$PRD_{t-3} = PR_{t-3} - PR_{1969\cdot 2}$ from 1970.1 and 0 before 1970.1.

$$PW_t = \ln \frac{\text{wholesale price index for foreign competitors}}{\text{wholesale price index for home sales of manufacturing}}$$

Source: ET and earlier figures from U.N. Monthly Bulletin of Statistics.

Range: –0.346 in 1981.1, 0.119 in 1976.4

$PWD_t = PW_t - PW_{1970\cdot 1}$ before 1970.1 and 0 from 1970.1.

$$PC_t = \frac{\text{Price Commission intervention in £ terms}}{\text{wholesale price index for home sales of manufacturing}}$$

Source: Price Commission Reports. Range: 0 up to 1973.1 and from 1977.4, 7.74 in 1974.1.

TRJ = 0 before observation J, 1 at J, 2 at J + 1, 3 at J + 2 etc.

Si = 1 for i^{th} quarter, 0 otherwise.

SiTR = Si × trend.

EX1 = Excess of average January and February temperature over 1941–70 mean, in Centigrade and defined for the 1st quarter only. Source: Annual Abstract of Statistics and Monthly Digest of Statistics. Range: –3.0 in 1963.1, 2.35 in 1957.1.

EX2 = Excess of preceding December temperature over 1941–70 mean, in Centigrade and defined for 1st quarter. Source: as for EX1 Range: –3.7 in 1982.1, 3.2 in 1974.1.

π = Proportion of firms operating below full capacity reported by CBI Industrial Trends Survey. Triannual 1958–72, then quarterly. Quarterly interpolation centered at mid–quarter. Source: CBI. Range: 0.38 in 1965.1, 0.84 in 1980.4.

ww = ln(52 – 1.2 – average weeks annual holiday entitlement), linear quarterly interpolation. This is an indicator of the number of weeks in a normal working year, assuming 1.2 weeks of public holidays. The coverage is all manual workers in national collective agreements or Wages Councils orders. Source: HABLS, DEG. Range: 3.887 in 1955.1, 3.827 in 1983.4.

PO = proportion of employees who are operatives, linear quarterly interpolation. Data are biannual from 1963–1974 and otherwise annual. Source: HABLS, DEG. Range: 0.700 in 1980.4, 0.801 in 1955.1.

(b) Derivation of a parsimonious specification

The process which led to the final equation for manufacturing output can be explained as follows. Given the theoretical discussion in

Sections 2–5, write an aggregate Cobb–Douglas production function, imposing constant returns, in the form.

$$\ln(q_t/K_t) = \alpha_0 + \alpha(\ln(\ell_t \bar{h}_t/K_t) + ww_t + u) \\ + \text{output measurement bias effects} + \text{trends effects} \quad (18)$$

where u_t represents the average proportionate deviation from normal of weekly labour utilization. Section 3 suggests how u_t should be measured. In equation (14), take $s_t = s_0 - s_1(NH_t - 90.4)$ where $s_0 > 0$, $s_1 > 0$ and $NH_t - 90.4$ is positive before 1968, zero for 1968–79 and negative after 1979. Since $u_t^* \approx OH_t$, mean utilization is

$$u_t = OH_t + c_2 - s_0 + s_1(NH_t - 90.4) - c(OH_t + c_1 - s_0 + s_1(NH_t - 90.4))^{-1} \quad (19)$$

I approximate (19) by the expression

$$u_t = \text{const.} + OH_t - cOH_t^{-1} - c_0(NH_t - 90.4)OH_t^{-1} \quad (20)$$

which avoids the use of nonlinear estimation.

To allow for the possibility that the proportion π of firms reported in the CBI Industrial Trends Survey to be operating below full capacity contains information additional to that in the overtime data, I define the variable $CU = (\pi/1-\pi)^{0\cdot4}$, see Appendix 2. CU can then be included in (18) as an additional regressor.

The measurement biases in output discussed in Section 5 are proxied through the variables PR, PW and PC. q_t in (18) refers to measured output so when this exceeds true output there are positive measurement biases. Fours biases were considered: the gross output bias, the domestic price index bias, the list price bias, and the price control bias. There are good reasons for allowing for dynamic effects in the variables used to proxy these biases. The intensity of price controls PC almost certainly has a lagged effect. The Price Commission reports that firms were changing prices of individual goods about 2 or 3 times per annum. The PC data refer to quarters which are one month in advance of the conventional definition. Thus PC_{t-1} implies an average age of 2 months and PC_{t-2} one of 5 months. For reasons discussed in Section 5, the lag responses to PR and PW are likely to be more complicated. In practice, rate of change effects ΔPR_{t-j}, ΔPW_{t-j} with $j \leq 4$, were included in addition to the levels effects.

The trend effects are considered to be of two kinds. The first represents slow changes in technology and work practices, in aggregation biases and in unmeasured changes in output quality, labour force composition, shift work and paid holidays. The second represents the unmeasured scrapping (or loss of productivity) of capital that would have followed the 1973 oil crisis and the collapse of manufacturing output in 1979-80. Both types are represented by linear splines of the form $\Sigma\beta_i TRJ_i$. This yields a continuous line made of straight line segments which change slope at observations J_1, J_2 etc. To select dates for changes in trend slopes corresponding to the first type of trend effects, cyclical peaks were chosen as coordinates on the grounds that these were relatively frequent and to reduce the risks of confusing the effects of trend and cycle. Since the share of labour in manufactured value added is of the order of 0.7, $\ln q_t - 0.7\ln \ell_t - 0.3\ln K_t$ was plotted against time to pick out these cyclical peaks. This gave the following dates: 1959.4, 1964.4, 1968.3, 1973.1 and 1979.2. Slope changes reflecting unobserved capital scrapping were investigated between 1974.1 and 1974.4 and between 1979.2 and 1980.3.

The equations estimated also allowed for seasonal dummies and for unusual weather affecting output in quarter 1 through the excess temperature variables EX1 and EX2. Finally, special dummies for strikes and other unusual events were included.

In the initial estimation, the sample was split pre– and post–1970. There were two reasons for this. From 1968 and particularly in 1970–1972 there was a major change in the basic sources of output data with physical measures of production increasingly replaced by deflated values from quarterly sales inquiries carried out by the Business Statistics Office[9]. This should, by the arguments of Section 5, have led to a break in the coefficients of the variables corresponding to the output measurement biases. The second reason is that with the moves to flexible exchange rates in 1971 and 1972, it seems likely that a structural break would have occurred in the relationships between price forecasts and past data. To the extent that price expectations play a role in, for example, the substitution effect entailed in the gross output bias, one might therefore expect parameter shifts in the effects of these price variables on measured output. Given that these arguments suggest parameter shifts between 1968 and 1972, so that 1970.1 is a mid–point and given the shift to a stock–adjusted definition of output from 1970.1, this date was chosen for the sample split.

The CBI capacity utilization variable CU is defined only from 1958.3 so that the two periods initially considered were 1958.3–1969.4 and 1970.1-1983.4. For both periods the coefficient on CU_t was positive, against the prediction of theory, though insignificantly different from zero in both cases. Defining CU for different values of θ in the plausible range made only slight differences to the t–ratios on CU. In contrast, the employment and overtime variables had sensible and significant coefficients in both periods. This suggested that CU is dominated by the overtime hours based concept of utilization. Dropping CU from the equation, it was possible to extend the first period back to 1956.1.

The next step was to search for a parsimonious representation in each period of the general distributed lags in PR and PW. This suggested PW_t, Δ_3PR_t and Δ_4PW_t in the first period and PR_t, Δ_3PR_t and Δ_4PW_t in the second. F–tests for the 7 restrictions respectively gave $F_{7,26} = 0.56$ (2.39) and $F_{7,21} = 0.32$ (2.49), where the critical values at the 5% level are given in parenthesis. Having now more parsimonious equations it was thought necessary to go back and check that CU_t was still insignificant in

9 See CSO (1976). CSO (1959) suggests that about 31% of the data was based on deflated values in the 1954 based index and about 33% for the 1958 based index. CSO (1970) suggests a figure of 40% for the 1963 based index while CSO (1976) suggests 66% for the 1970 based index.

case the earlier finding had been due to overfitting. CU_t proved insignificant again in both periods with the sign in the latter period still positive.

The two equations for 1956.1–1969.4 and 1970.1–1983.4 were now simplified further by omitting three of the shifting trends, two trending seasonals, PC_{t-1} (t ratio = 0.6) and refining slightly the (0,1) dummies. With 6 restrictions for 1956.1–1969.4 and 4 for 1970.1–1983.4, the F–tests are $F_{6,33} = 1.89$ (2.40) and $F_{4,28} = 2.27$ (2.70). The resulting equations are shown as R.1(a) and (b) in which PR_t and PW_t have been replaced by PRD_{t-3} and PWD_t from which the 1970.1 values of PR_t and PW_t have been subtracted.

These equations suggest that pooling might well be an acceptable restriction. There are 13 restrictions: on the intercept, the trend, the responses of output to employment, the two overtime variables, $\Delta_3 PR_t$, $\Delta_4 PW_t$, three seasonals, one trending seasonal and the two excess temperature variables. The F–test is $F_{13,71} = 1.61$ (1.90).

The resulting equation is R.2.[xi] The obvious naive alternative hypothesis is $\Delta \ln(q_t/\ell_t)$ = constant which has a standard error of 0.01936 compared with R.2's of 0.007457. $\ln(q_t/\ell_t)$ itself is, of course, heavily trended and has a standard deviation of 0.2322. The estimated elasticity of output with respect to employment at 0.681 is plausible. OH^{-1} and its interaction with normal hours are both highly significant with the anticipated signs. The latter term suggests that part of the increased overtime following reductions in normal hours in the 1950s and 1960s itself became normal and thus should not be included in cyclical overtime. The cumulative trend effect, in annualized terms is 1.8% up to 1959.3, then 2.5% up to 1972.4, dropping to 0.8% from 1973.1 to 1979.2, – 2.3% from 1979.3 to 1980.2 and back to 2.5% from 1980.3. This is consistent with prior expectations of higher rates of unobserved scrapping in the periods of 1973.1 to 1972.2 and 1979.3 to 1980.2.

The price control variable is significant, with a plausible lag and size of coefficient. This result parallels that of Darby (1984) for the U.S. The relative price of raw materials to output has a significant post–1970 levels effect consistent with a rather small gross output bias.[10] The rate of change effect like that of the relative price of foreign manufactures is consistent with the "domestic price bias" and the "list price bias" discussed in Section 5. The pre–1970 levels effect of the relative price of foreign manufactures suggests that export prices were able to diverge more permanently from foreign wholesale prices in that period. Finally, the excess temperature variables suggest a significant first quarter output effect as the result of unusual weather in December, January and February.

(c) Further tests and comparisons

The main features of the extensive tests and comparisons of R.2 with alternative specifications are discussed next. Since there is inevitably some arbitrariness about some of the 0,1 dummies which were included, R.3 shows the effect of excluding all except for the 1972.1 miners' strike and the 1974.1 "three–day week" energy crisis dummies over which there can be no argument. The results show that no coefficient in R.3 differs by more than one estimated standard error from the value estimated in R.2 which is reassuring for R.2.

Next consider the results of tests of structural stability and lack of residual autocorrelation. A Lagrange multiplier test of this hypothesis against the alternative of up to fourth order residual autocorrelation and shifts in the parameters between 1956.1 to 1969.4 and 1970.1 to 1983.4 gives $F_{21,63} = 1.38$ (1.74) while the test of structural stability alone had given $F_{13,71} = 1.53$ (1.90). A structural stability or forecast test for 1980.1 to 1983.4 compared with 1956.1 to 1979.4 gives $F_{16,69} = 1.33$ (1.73). A structural stability test for 1973.2 to 1980.2 compared with the pooled sample 1956.1 to 1973.1, 1980.3 to 1983.4 gives $F_{12,71} = 0.85$ (1.89). This is interesting because the intervening period is that of the two oil shocks.

Let us now turn to the question of whether this estimated production function suffers from bias because of the possible endogeneity of

---

10 At its peak in 1974 the bias is about 2% on a base of 1970. This supports Grubb's (1984) argument that the effect posited by Bruno (1984) and Bruno and Sachs (1982) should not be exaggerated.

employment and overtime hours. The first step is to re–estimate R.2 by instrumental variables. To do this ($\ln\ell_t + OH_t$) and OH_t^{-1} are the two variables to be instrumented, though the instrumented value of OH_t^{-1} also enters in interaction with normal hours. The instrumenting equations were estimated for 1955.3 to 1969.4 and 1970.1 to 1983.4. The list of instruments includes lags of the following variables: $\ln\ell$, OH, lnq, lnNH, PR, PW, the national vacancy rate, ln(world industrial production), ln(world exports of manufactures) and a real interest rate term. Relatively parsimonious forms of these equations with sensible long run values of the coefficients were selected. The results of thus instrumenting the production function as specified in R.2 are shown in R.4. The standard error of the equation increases from 0.007457 to 0.008305 and the parameter estimates are all very close to those in R.2. This provides informal support for the proposition that the endogeneity bias can be ignored.

The validity of the over–identifying restrictions entailed in the instrumentation is tested by comparing the likelihoods of the unrestricted and restricted reduced forms. This gives a chi–squared statistic of 19.4. With 30 degrees of freedom the critical value is 43.8 at the 5% level. To test the hypothesis of zero endogeneity bias the Revankar and Hartley (1973) test was used, this being also interpretable on the Lagrange multiplier test principle as discussed, e.g. by Engle (1982). Under the null hypothesis, the residuals from regressions of the potentially endogenous variables on instruments independent of the disturbance in R.2 should have zero coefficients when included as additional regressors in R.2. The resulting F–test gives $F_{3,81} = 1.44$ (2.73) where the variables and instruments are as described in connection with R.4. It seems, therefore, that we need not worry about endogeneity bias.[x]

In contrast, omitting labour utilization as represented by the overtime variable OH_t from the regression produces all the symptoms of gross mis-specification. As can be seen from R.5, the residual standard error doubles, the Durbin Watson statistic falls to 0.87 and the elasticity of output with respect to employment is estimated at 1.70. Such absurd returns to labour are sometimes associated with "Verdoorn's Law". I would interpret this "Law" merely as a cyclical measurement error phenomenon: the result of omitting the utilization of labour.[11] As remarked in the introduction, there are standard cost of adjustment arguments to explain why output expands faster than employment in the upswing and contracts faster than employment in the downswing, thus giving rise to apparently large returns to labour. Further evidence of the misspecification resulting from the omission of labour utilization can be found in the substantial alterations in many of the other coefficients in R.5 compared with R.2.

The nonlinearity entering through OH_t^{-1} seems to be important. Omitting the OH_t^{-1} terms, unrestricting the OH_t coefficient and including an interaction between NH_t and OH_t raises the standard error to 0.008143 compared with R.2's 0.007457. On the other hand, (19) above suggested a more sophisticated non–linear specification. Estimating (19) by least squares gives a t–ratio for $c_1 – s_0$ of 1.0. Setting $c_1 - s_0 = 0$ gives $\hat{c} = 0.01226$ (9.3), $\hat{s}_1 = 0.002015$ (6.2), where t–ratios are in parenthesis. But, although the other parameter estimates are very close, the standard error is slightly higher at 0.007799 compared with 0.007457 in R.2 which is meant to be an approximation to this specification. The Durbin–Watson statistic is 1.97 compared with 2.09 in R.2. This supports the more convenient R.2 specification.

(d) The Feldstein–Craine Result

In more standard specifications of the production function, investigators sometimes define labour input as the number of employees and the paid hours per employee h^0.[12] Sometimes a test is carried out of the hypothesis that in a Cobb–Douglas context the elasticity of output with

---

11 Verdoorn's Law says that the rate of growth of output per head is an increasing function of the rate of growth of employment. Chatterji and Wickens (1982) also provide evidence consistent with the interpretation of Verdoorn's law as a cyclical phenomenon.

12 This is observed average hours as conventionally understood rather than effective hours and will be referred to as 'average hours' or simply 'hours' in what follows.

respect to hours is the same as that with respect to employment. As noted in the Introduction above, Feldstein (1967) argued that the hours elasticity should exceed the employment elasticity. Averaging his cross-section estimates for British manufacturing gives $\ln q$ = const. + 0.773 $\ln\ell$ + 2.046 $\ln h^0$ + .210 $\ln K$. Craine (1973) examined Feldstein's hypothesis for time-series data on U.S. manufacturing. Imposing constant returns to scale he obtained for 1949.2 to 1967.4:

$$\ln(q/K) = \text{const.} + \underset{(23.2)}{0.007t} + \underset{(17.3)}{0.789} \ln(\ell/K) + \underset{(14.8)}{2.177} \ln h^0 \tag{21}$$

s.e. = 0.012, DW = 0.87, d.f. = 71.

In the context of the current paper, average hours per operative h^0 is normal hours plus the difference between average overtime hours and average short time. But short–time is quantitatively unimportant being typically less than 10% of overtime. Thus we can approximate $\ln h$ by $\ln(\bar{h}\,(1 + OH))$. Re–specifying R.2 in the Craine manner leads to the fourth to seventh terms in R.2, again estimating for 1956.1–1983.4, being replaced by

$$\underset{(23.2)}{1.207} \ln(\ell/K) + \underset{(14.3)}{2.176} \ln(\bar{h}(1+OH)) + \underset{(14.6)}{0.00934t} \tag{22}$$

and s.e. = 0.00967, DW = 1.47. All other variables are as in R.2. There is a remarkable similarity in the elasticities of output with respect to average hours from these three very different data sets.

My interpretation of these results, given the theory presented in Section 3 above, is in terms of the correlation between ln(average hours) and the overtime based utilization measure. However, since this correlation is disturbed by variations in normal hours, we ought to find that entering normal hours as an additional variable to remove this source of variation ought to improve the results. This indeed is what happens. Instead of (22), again for 1956.1–1983.4, we now find

$$0.697\ \ln(\ell/K) + 3.479\ \ln(\bar{h}(1+OH)) - 1.852\ \ln\bar{h} + 0.00449t \quad (23)$$
$$(7.9) \qquad (14.8) \qquad (6.6) \qquad (5.0)$$

s.e. = 0.007910, DW = 2.14. The fit is much improved; there is now no sign of first order residual autocorrelation and the elasticity of output with respect to employment is much more satisfactory.[xii] The nonlinearity in response implied by the theory suggests that adding a quadratic term in $\ln\bar{h}(1+OH)$ might improve the fit further. The s.e. now is 0.00765, DW = 2.19 and the elasticity of output with respect to employment is estimated at 0.775 (8.5).

Equation (23) suggests that one can obtain results almost as good as R.2 without the effort of constructing a special overtime series. Instead, readily available average hours data can be used. However, if normal hours went through substantial changes, as they did in the 1950s and 1960s, it is essential that a normal hours variable be included.

Overall, these results strongly support the hypothesis that the Feldstein-Craine finding of an elasticity of output with respect to average hours considerable in excess of the elasticity with respect to employment is the result of the correlation between average hours and an omitted labour utilization variable. Leslie and Wise (1980), who call this the labour hoarding explanation, reject this hypothesis on the basis of a time-series/cross-section study on annual British data for 28 industries for 1948-1968. They find that the inclusion of industry specific dummies and trends in a pooled cross–section reduces the hours elasticity to a value close to the employment elasticity. Hence they argue that there is an upward bias in the hours elasticity caused by omitted industry–specific efficiency effects. This is a serious challenge to the interpretation of the Feldstein-Craine result given here and deserves comment.

One problem with the study is that the hours data, though used to explain annual output, are based on hours observed over only two weeks. Thus there is likely to be quite a serious random measurement error in this data which will bias downwards the hours coefficient. Note that the cross-section variation in hours is already being picked up by the industry specific coefficients and it is the measurement error in hours relative to the cyclical variation in annual hours which matters. Secondly, a great deal of the variation in average hours over 1948–1968 is due to the considerable reduction in normal hours in the 1950s and early 1960s. As we have seen, this tends to reduce the correlation between average hours and labour utilization.[13] In this respect, it is noteworthy that Leslie's (1984) study of annual data for 20 U.S. industries for 1948–1976 shows the hours elasticities to be significantly higher than the employment elasticities despite the inclusion of both industry specific dummies and a Wharton capacity utilization index.

(e) Further aspects of labour input

Another aspect of hours is paid holiday entitlements. Unfortunately the Department of Employment does not publish information which relates to manufacturing but gives annual figures on holiday entitlements in all national agreements covering manual workers. Of these workers, the proportion in manufacturing is probably now a little under one half. Assuming that the national figures are representative of manufacturing permits the construction of the logarithmic weeks worked measure ww described in the data section above. This assumes 5 working days per week so that 6 public holidays per annum corresponds to 1.2 weeks and shows a fall of 6% from 1955 to 1983. Specifying labour input as ($\ln \ell_t \bar{h}_t + ww_t + u_t$) gives an equation standard error of 0.007549 and the other coefficients virtually unaltered even for the trend coefficients. The largest changes for these are 8% reductions in the absolute size of the coefficients for TR79.3 and TR80.3 compared with R.2 and a 16% increase in that for TR59.4.

---

[13] My explanation of the Feldstein–Craine result thus predicts three conditions under each of which the hours elasticity would increase in a study of the Leslie and Wise type: firstly, use hours data which are more annually representative, secondly, include industry–specific normal hours as regressors or thirdly, estimate over a period in which there is little variation in normal hours.

Another trend like change which has occurred is in the incidence of shift work and is no doubt reflected in the estimated trend effects. Before 1973 the only official figures are very sparse. These show a percentage of workers in manufacturing on some kind of shift work of 12.5% in 1954, 20% in 1964 and 24.9% in 1968, see National Board for Prices and Incomes (1970), p.65.[14]

Yet another aspect of labour input is in PO the proportion of employees who are operatives. Data on this are biannual from 1963 to 1974 and otherwise annual. Using a linear quarterly interpolation of PO, one might replace the labour input term in the production function (18) by

$$\alpha_1 \ln(\ell_t PO_t \bar{h}_t / K_t) + \alpha_2 \ln\{\ell_t(1 - PO_t)\bar{h}_t / K_t\} + (\alpha_1 + \alpha_2)u_t \quad (24)$$

The assumption here is that the same utilization factor applies to non-operatives as to operatives. Although the equation s.e. improves to 0.007346, the estimated $\alpha_2 = 0.044$ (0.6) is implausibly low. Applying the utilization term to operatives only raises the equation standard error marginally. Given a share of operatives in the wage bill of roughly 0.7, it is interesting to test the hypothesis $\alpha_2/\alpha_1 = 3/7$. This is easily rejected, the t-test being 2.42. But clearly the implausible hypothesis that only the input of operatives matters is easily acceptable. Perhaps this is not altogether surprising, given the crudity of the data on the proportion of operatives but leaves one in a quandary over which definition of labour input to accept. Since the differences in fit are slight and the quarterly data on PO suspect, my own preference remains for R.2.

(f) The Role of Capital

Finally, the hypothesis of constant returns to scale was tested by adding $\ln K_t$ to the list of regressors in R.2. This produces a coefficient of

---

[14] An estimate by Fishwick (1980) for 1979 of 26% suggests a much slower rate of increase since 1968 and this is consistent with annual New Earnings Survey figures from 1973. Bosworth and Dawkins (1981) discuss comparability problems of these data.

–1.216 (3.2) and very little change in the other parameters except for the trends.[xiii] The hypothesis that the capital stock plays no role in the model versus the unrestricted alternative is just rejected at $t = 2.33$ and gives an equation s.e. of 0.007258. These results should be taken not so much as a rejection of constant returns to scale but as a symptom of the measurement errors in the gross capital stock data, perhaps because unobserved scrapping is highest in downturns and lowest in upturns. An alternative hypothesis is that R.2 is cyclically mis–specified in some way: in cyclical behaviour, lnK lags a little over one year behind overtime hours and is negatively correlated with current overtime hours. However, one can easily accept the hypothesis of zero coefficients on $u_{t-1},\ldots,u_{t-6}$ added as regressors to R.2 and, it should be recalled, for R.2 the Lagrange multiplier tests for residual auto–correlation are insignificant. It should also be noted that Craine (1973) too reports a better fit for U.S. manufacturing when capital is omitted from the production function.

The model selection question here is more one of the economic interpretation one wishes to put on the model than one of goodness of fit. If one wishes to interpret some of the trend shift effects as correction factors to apply to the observed capital stock, R.2. is clearly preferable to a specification where the capital stock is omitted entirely.

(g) The Estimated Equations

R.1 (a): 1956.1–1969.4

$$\ln(q_t/K_t) = \underset{(5.1)}{-7.113} + \underset{(5.0)}{0.752}\,(\ln(\ell_t/K_t) + \ln NH_t + OH_t)$$

$$-\underset{(7.5)}{0.01318}\,OH_t^{-1} + \underset{(7.1)}{0.06062}\left[\frac{NH_t-90.4}{100}\right]OH_t^{-1} + \underset{(3.6)}{0.00536t}$$

$$+\underset{(2.9)}{0.00178}\,TR\,59.4 + \underset{(2.5)}{0.0929}\,PWD_t + \underset{(1.5)}{0.0744}\Delta_4 PW_t + \underset{(1.9)}{0.0799}\Delta_3 PR_t$$

$$+ \underset{(1.0)}{0.0020}EX1_t + \underset{(2.2)}{0.0058}\ EX2_t + \text{ terms in S1, S2, S3, S1TR and 2 (0, 1)}$$

dummies.

s.e.= 0.007447,SSE = 0.002163,$\overline{R}^2$= 0.9453,DW = 2.38, n = 56, d.f. = 39

R.1 (b): 1970.1–1983.4

$$\ln(q_t/K_t) = \underset{(6.4)}{-7.444} + \underset{(6.4)}{0.768}(\ln(\ell_t/K_t) + \ln NH_t + OH_t) - \underset{(5.0)}{0.01035}\ OH_t^{-1}$$

$$+ \underset{(0.9)}{0.05998}\left[\frac{NH_t-90.4}{100}\right]OH_t^{-1} + \underset{(4.8)}{0.00883}t - \underset{(5.2)}{0.00596}\ TR\ 73.1 - \underset{(4.2)}{0.00846}$$

$$TR\ 79.3 + \underset{(6.0)}{0.01403}\ TR\ 80.3 + \underset{(3.3)}{0.00434}\ PC_{t-2} - \underset{(3.4)}{0.0732}\ PRD_{t-3}$$

$$+ \underset{(2.1)}{0.0540}\ \Delta_4 PW_t + \underset{(2.6)}{0.0602}\Delta_3 PR_t + \underset{(2.3)}{0.0019}EX1_t + \underset{(1.0)}{0.0019}EX2_t + \text{ terms}$$

in S1, S2, S3, SITR and 6 (0,1) dummies

s.e.= 0.006817,SSE = 0.001487,$\overline{R}^2$= 0.9974, DW = 1.82, n = 56, d.f.= 32

R.2 1956.1–1983.4

$$\ln(q_t/K_t) = \underset{(8.3)}{-6.457} + \underset{(8.2)}{0.681}(\ln(\ell_t/K_t) + \ln NH_t + OH_t) - \underset{(10.2)}{0.01207}\ OH_t^{-1}$$

$$+ \underset{(8.2)}{0.05140}\left[\frac{NH_t-90.4}{100}\right]OH_t^{-1} + \underset{(4.2)}{0.0050}\ PC_{t-2} + \underset{(3.6)}{0.1015}\ PWD_t$$

$$- \underset{(3.1)}{0.0668}\ PRD_{t-3} + \underset{(2.7)}{0.0591}\ \Delta_4 PW_t + \underset{(3.7)}{0.0776}\ \Delta_3 PR_t + \underset{(5.2)}{0.00451}\ t$$

$$+ \underset{(4.0)}{0.00172}\,TR\,59.4 - \underset{(8.2)}{0.00412}\,TR\,73.1 - \underset{(4.4)}{0.00788}\,TR\,79.3 + \underset{(6.0)}{0.01193}$$

$$TR\,80.3 + \underset{(2.4)}{0.0040}\,EX1_t + \underset{(2.7)}{0.0035}\,EX2_t$$

+ terms in S1, S2, S3, SITR and 8 (0.1) dummies.

s.e.= 0.007457, SSE = .004672, $\overline{R}^2$ = 0.9972, DW = 2.09, n = 112, d.f. = 84.

R.3 1956.1–1983.4

$$\ln(q_t/K_t) = -\underset{(6.9)}{6.200} + \underset{(6.9)}{0.653}(\ln(\ell_t/K_t) + \ln NH_t + OH_t) - \underset{(9.0)}{0.01250}OH_t^{-1}$$

$$+ \underset{(7.4)}{0.05489}\left[\frac{NH_t - 90.4}{100}\right]OH_t^{-1} + \underset{(3.4)}{0.00472}PC_{t-2} + \underset{(2.3)}{0.0772}PWD_t - \underset{(2.2)}{0.0537}$$

$$PRD_{t-3} + \underset{(2.4)}{0.0592}\Delta_4 PW_t + \underset{(3.3)}{0.0783}\Delta_3 PR_t + \underset{(4.1)}{0.00414}t$$

$$+ \underset{(3.4)}{0.00192}TR\,59.4 - \underset{(7.8)}{0.00434}TR\,73.1 - \underset{(3.5)}{0.00727}\,TR\,79.3$$

$$+ \underset{(4.7)}{0.01104}\,TR\,80.3 + \underset{(4.1)}{0.0068}\,EX1_t + \underset{(2.1)}{0.0028}\,EX2_t + \text{terms in S1, S2, S3,}$$

SITR and 1972.1, 1974.1 dummies.

s.e.= 0.008863, SSE = 0.007070, $\overline{R}^2$ = 0.9960, DW = 1.83, n = 112, d.f. = 90.

R.4 1956.1–1983.4 estimation by instrumental variables

$$\ln(q_t/K_t) = -\underset{(7.5)}{6.750} + \underset{(7.5)}{0.711}(\ln(\ell_t/K_t) + \ln NH_t + OH_t) - \underset{(8.1)}{0.01124}\,OH_t^{-1}$$

$$+ \underset{(5.6)}{0.04964}\left[\frac{NH_t-90.4}{100}\right]OH_t^{-1} + \underset{(2.8)}{0.00466}\, PC_{t-2} + \underset{(1.9)}{0.0574}\, PWD_t$$

$$- \underset{(3.5)}{0.0834}\, PRD_{t-3} + \underset{(2.8)}{0.0700}\Delta_4 PW_t + \underset{(3.6)}{0.0837}\Delta_3 PR_t + \underset{(4.7)}{0.00463}\, t$$

$$+ \underset{(3.8)}{0.00203}\, TR\ 59.4 - \underset{(6.8)}{0.00372}\, TR\ 73.1 - \underset{(5.6)}{0.01077}\, TR\ 79.3$$

$$+ \underset{(6.7)}{0.01484}\, TR\ 80.3 + \underset{(2.6)}{0.0047}\, EX1_t + \underset{(2.4)}{0.0034}\, EX2_t + \text{terms in S1, S2, S3,}$$

SITR and 8 dummies.

s.e.= 0.008305,SSE = 0.005800,$\overline{R}^2$= 0.9954,DW = 2.02,n = 112,d.f. = 84.

R.5 1956.1–1983.4

$$\ln(q_t/K_t) = \underset{(17.3)}{-15.909} + \underset{(16.9)}{1.702}\,(\ln(\ell_t/K_t) + \ln NH_t) + \underset{(1.3)}{0.00298}\, PC_{t-2}$$

$$+ \underset{(1.8)}{0.0846}\, PWD_t - \underset{(1.2)}{0.0530}\, PRD_{t-3} + \underset{(5.3)}{0.2039}\Delta_4 PW_t + \underset{(2.5)}{0.1028}\Delta_3 PR_t$$

$$+ \underset{(11.5)}{0.01340}\, t + \underset{(5.5)}{0.00474}\, TR\ 59.4 - \underset{(7.0)}{.00527}\, TR73.1 - \underset{(3.7)}{0.01180}\, TR79.3$$

$$- \underset{(8.0)}{0.0270}\, TR80.3 + \underset{(1.2)}{0.0040}\, EX1_t + \underset{(1.3)}{0.0033}\, EX2_{t-1} + \text{terms in S1, S2, S3,}$$

SITR and 8 (0,1) dummies.

s.e.= 0.01488,SSE = 0.01905,$\overline{R}^2$= 0.9888, DW = 0.87, n = 112, d.f. = 86.

7. Conclusions

This paper has referred to three classic problems in estimating an aggregate production function on aggregate time series data. The first is the unobservables problem, particularly in regard to utilization and the measurement of capital and to a lesser extent the measurement of output and labour input. The second is the aggregation problem and the third the simultaneous equation bias problem. It was the hypothesis of this paper that the first of these problems is the most serious.

A new proposal for the measurement of labour input was put forward. This argued that high rates of labour utilization relative to a norm would be reflected in high aggregate weekly overtime hours being reported. However, since most workers are paid for a standard week even when being under–utilized, there is no corresponding direct measure of below normal rates of labour utilization. This is a situation where we can observe the mean of the truncated upper tail of the distribution of utilization rates over firms. Given a reasonably constant spread the proposal was to derive an estimate of the mean of the whole distribution from the information about the truncated upper tail. This implies that the mean labour utilization rate is a nonlinear function of overtime hours as a proportion of normal hours. With an additional allowance for a British institutional feature in which some of what is called overtime is in fact part of the usual work week for some employees, this measure of labour utilization gave excellent empirical results.

In particular, an elasticity of output with respect to employment of 0.681 is very reasonable in contrast to the huge elasticity estimated when labour utilization is omitted. Large elasticities are often interpreted as evidence of "Verdoorn's Law" which says that the rate of growth of output per head is an increasing function of the rate of growth of employment. On the current evidence, this "law" is a cyclical measurement error phenomenon.

The theory put forward provides a convincing explanation of the Feldstein-Craine result that the elasticity of output with respect to average paid for hours of work substantially exceeds the elasticity with respect to employment. Since average paid for hours is approximately normal hours plus overtime hours averaged over operatives whether they work overtime or not, ln(average paid for hours) $\approx$ ln(normal hours) + OH, where OH is overtime hours averaged over all operatives and scaled by normal hours. According to the theory put forward, labour utilization as I have defined it, is a nonlinear function of OH with a derivative greater than unity. Although the derivative of ln(output) with respect to labour utilization is the same as that with respect to ln(employment), the elasticity with respect to average paid for hours will therefore be greater. This gives the Feldstein-Craine result. Moreover, normal hours enters the relationship between average paid for hours and utilization. The theory therefore predicts and the empirical evidence agrees that for periods when normal hours are changing, normal hours should make a significant negative contribution in a production function in which ln(average paid for hours) enters as a regressor. An implication when normal hours is omitted is an unstable coefficient on ln(average paid for hours) for different samples containing variations in normal hours.

The paper also considered biases in the measurement of capital and output. The former are likely to be large because the gross capital stock as usually measured assumes that assets have service lives and yield service flows which are invariant to changes in economic conditions. Although the device of proxying these measurement biases by shifts in the slopes of time trends is crude, the empirical evidence favoured the hypothesis of increased unrecorded scrapping and perhaps reduced service flows from 1973 with a particularly heavy incidence during the 1979.3 to 1980.3 period when British manufacturing experienced a crisis in which output fell by 16%. The finding, like Craine's (1973), that omitting the capital stock altogether from the production function improved the standard error, could be consistent with a badly measured concept. The hypothesis that a measure of capacity utilization based on a survey of firms' opinions on whether they were operating below full capacity contains no information additional to the overtime based labour utilization variable could be accepted. It is conceivable that this result is a consequence of the poor quality of the capital stock series which this concept might be adjusting. However, it is consistent with the widespread idea that the capital stock should not be utilization adjusted.

As far as output is concerned, four potential biases were considered and observable proxies constructed to measure their effects. The evidence is that for British manufacturing the "gross output bias" that stems from the approximation in the construction of the output index of value added changes by gross output changes is small, probably about 2% at the peak of relative raw materials prices. The "domestic price index" bias and the "list price bias" cannot be fully disentangled empirically but have considerable short run significance, particularly when sharp changes in real exchange rates occur. The former comes from the CSO's approximation of unavailable export price deflators by domestic price deflators. The latter arises to the extent that the prices reported by firms are not transactions prices but list prices. Finally, there is strong evidence that, as Darby (1984) reports for the U.S., attempts by governments to control prices have a significant distortionary effect on price deflators and so on measured output as firms attempt to side–step these controls.[xiv]

Appendix 1: An Alternative Measure of Labour Utilization

Apart from the overtime measure OH there are also data on the proportion p of operatives on overtime. If the distribution of $u = \ln h - \ln \bar{h}$ depends on two parameters, it is possible to deduce both parameters from OH and p. Suppose $u \sim N(\mu, \sigma^2)$. Then

$$OH = \int_0^\infty u f\left(\frac{u-\mu}{\sigma}\right)du, \quad p = \int_0^\infty f\left(\frac{u-\mu}{\sigma}\right)du = 1 - F\left(\frac{-\mu}{\sigma}\right)$$

As is well known (see Johnson and Kotz (1972), p.112–113),

$$\int_0^\infty u f\left(\frac{u-\mu}{\sigma}\right) du = \mu p + \sigma f\left(\frac{-\mu}{\sigma}\right) \tag{A.1}$$

Let $\frac{-\mu}{\sigma} = x$ where $x = F^{-1}(1-p)$. Then

$$OH = \mu p - \mu x^{-1} f(x) \tag{A.2}$$

so that $\mu = OH/(p - x^{-1}f(x))$ (A.3)

Thus we can derive an estimate of mean utilization μ from observing p and OH. Changes in systematic overtime as normal hours fell are allowed for by including an interaction term $(100 - NH)\mu$ which is zero in 1955.[viii]

A variation on R.2 is estimated in which the employment term is $\ln(\ell_t/K_t) + \ln NH_t$ and in which OH_t^{-1} and its interaction with normal hours are replaced by μ_t and $(100 - NH_t)\mu_t$. With $\hat{\alpha} = 0.684$ (6.6) and a coefficient on μ of 0.730 (9.0), the hypothesis that these two coefficients are equal is accepted. Imposing this restriction gives $\hat{\alpha} = 0.710$ (22.1) and a coefficient of –0.0158 (2.4) on $(100 - NH)\mu$. However, with s.e. = 0.008733, SSE = 0.006483, $\overline{R}^2 = 0.9962$, DW = 1.39, this equation is clearly inferior to R.2 even though the remaining parameter estimates are quite similar.

Given that both p and OH are subject to sampling variations it may be unrealistic to derive both μ and $\sigma = -\mu/x$ in this way. An obvious alternative is to treat σ as a constant. Then there are two alternative estimates of μ conditional on σ. The one based on OH, μ_{1t} is derived by solving the implicit function based on (A.1)

$$OH_t = \mu_{1t}\left(1 - F\left(\frac{-\mu_{1t}}{\sigma}\right)\right) + \sigma f\left(\frac{-\mu_{1t}}{\sigma}\right) \tag{A.4}$$

The one based on p is

$$\mu_{2t} = -\sigma F^{-1}(1 - p_t)$$

To investigate the empirical implication, R.2 was respecified with the employment term in the form $\ln\ell_t/K_t + \ln NH_t$ and OH_t^{-1} and its interaction with normal hours replaced by μ_{1t}, μ_{2t} and their respective interactions with normal hours. For a grid of values of σ, $\sigma = 0.7$ was found to give the lowest error sum of squares. Moreover, the hypothesis that the coefficients on μ_{2t} and its interaction with normal hours are zero could be easily accepted. Finally, the hypothesis that the coefficient on μ_{1t} is α could also be accepted. The resulting equation reads as follows:

R.6 1956.1–1983.4

$$\ln(q_t/K_t) = \underset{(39.8)}{-6.284} + \underset{(37.8)}{0.682}\,(\ln(\ell_t/K_t) + \ln NH_t + \mu_{1t})$$

$$-\underset{(7.8)}{0.00921}\,(100-NH_t)\,\mu_{1t} + \underset{(4.4)}{0.00508}\,PC_{t-2} + \underset{(3.3)}{0.0947}\,PWD_t$$

$$-\underset{(3.4)}{0.0709}\,PRD_{t-3} + \underset{(3.4)}{0.0637}\Delta_4 PW_t + \underset{(3.3)}{0.0675}\Delta_3 PR_t + \underset{(10.6)}{0.00440}\,t$$

$$+\underset{(4.8)}{0.00206}\,TR59.4 - \underset{(8.8)}{0.00429}\,TR73.1 - \underset{(5.2)}{0.0888}\,TR79.3 + \underset{(7.4)}{0.01341}\,TR80.3$$

$$+\underset{(2.5)}{0.00406}\,EX1_t + \underset{(2.7)}{0.00349}\,EX2_t + \text{terms in S1, S2, S3, and SITR and 8}$$

(0, 1) dummies.

s.e.= 0.007367, SSE = 0.004613, $\overline{R}^2 = 0.9979$, DW = 2.20, n = 112, d.f. = 85

The equation s.e. and the t–ratios are conditional upon $\sigma = 0.7$. Adjusting the equation s.e. as if the equation were linear in σ gives s.e. = 0.007410. This is very slightly better than R.2's s.e. of 0.007457. The similarity in the parameter estimates and fit of R.6 and R.2 is remarkable and reassuring for the robustness of R.2. There is virtually nothing to choose between them except convenience. Both are easily accepted against the maintained hypothesis which nests them both. R.6 requires the solution of a non–linear implicit function to generate estimates of average utilization and estimation for a grid of values of σ. R.2 can be estimated by OLS and linearity in the parameters has other convenient properties such as the ease of estimation by instrumental variables techniques.

Finally, it is worth noting that there are signs that the true underlying distribution of u is positively skewed. The value of μ from R.6 ranges from about –0.66 to –0.92. This is implausibly negative which suggests that the Normal distribution contains more mass in its lower tail than is plausible. Another way of seeing this is to note that using $p = 1 - F((-\mu)/\sigma)$, the values of p implied by these values of μ/σ turn out to be smaller than the observed values. In terms of (13), (14) or (19), positive skewness implies $c_2 > c_1$, though in fact c_2 cannot be separately identified. This is another piece of evidence which favours the representation of mean utilization in R.2 as an accurate as well as convenient approximation to the true relationship of mean utilization with OH_t.

Appendix 2:
A method for measuring capacity utilization from surveys

In the CBI Industrial Trends Survey as in similar surveys, the question is asked "Is your present level of output below capacity?" (defined as satisfactory or full rate of operation). Let q = output, q(max) = maximum output. Let $-u_c = \ln q(\max) - \ln q$ so that u_c is a proportionate deviation measure of capacity utilization. Suppose that different firms share the same view of what constitutes a satisfactory level of operation. It might, for example, be 80% of the maximum. Let

$$z = \ln q\,(\max) - \ln q(\text{satisfactory}) \tag{A.5}$$

Now suppose that there is a distribution across firms of capacity utilization measured by $\ln q(\max) - \ln q$ as illustrated in Figure 3. Imagine the distribution and so its mean, which measures the average degree of utilization that we want,

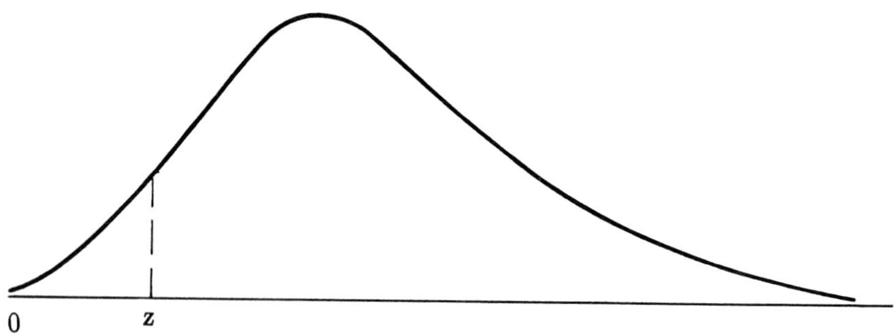

Figure 3: the distribution of (minus) capacity utilization across firms

shifting through time (though always in its lower limit fixed at zero). We will observe corresponding shifts in π, the proportion of firms (weighted by size) to the right of z.

The aim is to make deductions about the unobserved mean of the distribution from observations on π.[15] An obvious candidate for the distribution in Figure 3 is the lognormal, i.e. ln $(-u_c)$ is Normal with mean μ and variance σ^2. Then

$$\pi = 1 - F\left[\frac{\ln z - \mu}{\sigma}\right] \tag{A.6}$$

where $F(\cdot)$ is the distribution function of the standard Normal distribution. Then

$$\mu = \ln z - \sigma F^{-1}(1-\pi) \tag{A.7}$$

μ measures $E(\ln(-u_c))$ but we want $E(-u_c)$. When $\ln(-u_c)$ has a normal distribution,

15 Thiel (1966) has one of the earlier discussions of this type of problem in an applied context.

$$E(-u_c) = \exp\tfrac{1}{2}\sigma^2 \ \exp \mu$$

$$= \exp \tfrac{1}{2}\sigma^2 \exp\{\ln z - \sigma F^{-1}(1 - \pi)\}$$

$$= \text{const.} \exp\{-\sigma F^{-1}(1 - \pi)\} \tag{A.8}$$

there the constant of proportionality is z $\exp\frac{1}{2}\sigma^2$.

The Normal distribution can be quite well approximated by the logistic. As Amemiya (1981), p.1487 points out, if x is a standard Normal variable,

$$F(x) \approx \frac{e^{1.6x}}{1+e^{1.6x}} \tag{A.9}$$

where the RHS is the distribution function of a logistic variable with mean zero and variance $1.6^2/(\pi^2/3)$. Then (A.8) becomes

$$E(-u_c) = \text{const.}(\pi/1{-}\pi)^{\sigma/1.6}$$

Let us illustrate the empirical magnitudes this can take for some plausible parameter values. Suppose $z = 0.2$ which corresponds to satisfactory output being 82% of the maximum. Further, suppose that $\sigma = 0.4$. Then for the minimum observed value of $\pi = 0.38$, $E(-u_c) = 1.083 \times 0.2 \times 0.8848 = 0.192$. For the maximum observed value of $\pi = 0.84$, $E(-u_c) = 1.083 \times 0.2 \times 1.4888 = 0.323$. This would correspond to a range of variation of 0.136 in ln(output) between the highest and lowest observed degrees of capacity utilization and this is not an unreasonable range. The Logistic approximation is very close, also giving 0.192 for $\pi = 0.38$ and 0.323 instead of 0.328 for $\pi = 0.84$.

Some experiments in which the production function was estimated without any overtime variables but with $CU = (\pi/1 - \pi)^{\theta}$ for a range of $\theta = 0.2$ to 0.6 suggested the best fit at $\theta \approx 0.4$ which implies $\sigma = 0.64$ with the corresponding value of z estimated at 0.090. This suggests that "full capacity" is about 91% of the physical maximum. However, with a standard error of 0.0113 and a Durbin–Watson statistic of 1.42, this is a much less satisfactory equation than R.2.

References

Amemiy, T, (1981): "Qualitative Response Models: A Survey", *Journal of Economic Literature*, 19, pp 1483–1536.

Baily, M, (1981): "Productivity and the Services of Capital and Labour", *Brookings Papers on Economic Activity* no 1, pp 1–65.

——— (1982): "The Productivity Growth Slowdown by Industry", *Brookings Papers on Economic Activity* no.2, pp 423–453.

Bosworth, D & P Dawkins, (1981): *Work Patterns: An Economic Analysis*, Aldershot: Gower.

Bruno, M (1984): "Raw Materials, Profits and the Productivity Slowdown", *Quarterly Journal of Economics*, 99, pp 1–30.

Bruno, M & J Sachs, (1982): "Input Price Shocks and the Slowdown in Economic Growth: the Case of U.K. Manufacturing", *Review of Economic Studies*, 49, pp 679–706.

Central Statistical Office (CSO), (1959): *The Index of Industrial Production: Method of Compilation*, Studies in Official Statistics no.7, London: HMSO.

——— (1970): *The Index of Industrial Production and Other Output Measures*, Studies in Official Statistics no 17, London: HMSO.

——— (1976): *The Measurement of Changes in Production*, Studies in Official Statistics no 25, London: HMSO.

——— (1983): *Industry Statistics*, Occasional Paper no 19, London: HMSO.

Chatterji, M & M Wickens (1982): "Productivity, Factor Transfers and Economic Growth in the U.K.", *Economica*, 49, pp 21–38.

Craine, R, (1973): "On the Service Flow from Labour", *Review of Economic Studies*, 40, pp 39–46.

Darby, M (1984): "The U.S. Productivity Slowdown: a Case of Statistical Myopia", *American Economic Review*, 73, pp 301–322.

Deaton, A and J Muellbauer (1980): *Economics and Consumer Behaviour*, New York: Cambridge University Press.

Denison, E, (1969): "Some Major Issues in Productivity Analysis: an Examination of Estimates by Jorgenson and Griliches", *Survey of Current Business*, part II, 49, pp 1–27.

Denison, E, (1974): *Accounting for U.S. Economic Growth, 1929 to 1969*, Washington DC, The Brookings Institution.

——— (1979): *Accounting for Slower Growth: The U.S. in the 1970s*, Washington DC: The Brookings Institution.

Divisia, F (1952): *Exposes d'economique*, Vol.1, Paris, Dunod.

Engle, R (1982): "A General Approach to Lagrange Multiplier Model Diagnostics", *Journal of Econometrics*, 20, pp 83–104.

Feldstein, M (1967): "Specification of the Labour Input in the Aggregate Production Function ", *Review of Economic Studies*, 34, pp 375-86.

Fishwick, F (1980): *The Introduction and Extension of Shiftworking*, London: National Economic Development Council.

Gollop, F & D Jorgenson (1980): "U.S. Productivity Growth by Industry, 1947–73", in *New Developments in Productivity Measurement and Analysis*, NBER Studies in Income and Wealth, Vol.44, (ed.) J Kendrick and B Vaccara, Chicago: University of Chicago Press.

Griffin, T (1976): "The Stock of Fixed Assets in the U.K.: How to Make the Best Use of the Statistics", *Economic Trends*, October, HMSO.

Grubb, D (1986): "Raw Materials and the Productivity Slowdown: Some Doubts", *Quarterly Journal of Economics*, 101, pp 175–184.

Hansen, B D (1970): "Excess Demand, Unemployment, Vacancies and Wages", *Quarterly Journal of Economics*, 84, pp 1–23.

Heathfield, D (1972): "The Measurement of Capital Utilization Using Electricity Consumption Data for the U.K", *Journal of the Royal Statistical Society*, Series A, 135, pp 208–220.

——— (1983): "Productivity in the U.K. Engineering Industry 1956-1976", mimeo, University of Southampton.

Johnson, N & S Kotz (1972): *Distributions in Statistics: Continuous Multivariate Distributions*, New York: J. Wiley.

Jorgenson, D & Z Griliches (1967): "The Explanation of Productivity Growth", *Review of Economic Studies*, 34, pp 249–283.

Kendrick J (1973): *Postwar Productivity Trends in the US, 1948-1969*, New York: NBER.

Leslie, D (1984): "The Productivity of Hours in U.S. Manufacturing Industries", *Review of Economics and Statistics*, 66, pp 486–490.

Leslie, D & J Wise (1980): "The Productivity of Hours in U.K. Manufacturing and Production Industries", *Economic Journal*, 90, pp 74–84.

Malcolmson, J (1980): "The Measurement of Labour Cost in Empirical Models of Production and Employment", *Review of Economics and Statistics*, 62, pp 521–528.

Malcolmson, J & M Prior (1979): "The Estimation of a Vintage Model of Production for U.K. Manufacturing", *Review of Economic Studies*, 46, 719–736.

Mendis, L & J Muellbauer (1984): "British Manufacturing Productivity 1955–1983: Measurement Problems, Oil Shocks and Thatcher Effects"; Center for Economic Policy Research Discussion Paper.

Mizon, G. and S. Nickell (1983): "Vintage Production Models of U.K. Manufacturing Industry"; *Scandinavian Journal of Economics*, 85, pp 295–310.

National Board for Prices and Incomes, (1970), *Report no.161: Hours of Work, Overtime and Shiftworking*, London: HMSO.

Price, R (1977): "The CBI Industrial Trends Survey – An Insight into Answering Practices", *CBI Review*, Summer issue.

Raasche, R A & J A Tatom (1981): "Energy Price Shocks, Aggregate Supply and Monetary Policy – the Theory and the International Evidence", in K Brunner and A Melzer (eds.), *Supply Shocks, Incentives and National Wealth*, Carnegie–Rochester Conference on Public Policy, vol.14, pp 9–93.

Revankar, N S & M J Hartley (1973): "An Independence Test and Conditional Unbiased Predictions in the Context of Simultaneous Equation Systems", *International Economic Review*, 14, pp 625-631.

Scott, M FG (1976): "Investment and Growth", *Oxford Economic Papers*, 28, pp 317–63.

——— (1981), "The contribution of Investment to Growth", *Scottish Journal of Political Economy*, 28, pp 211–116.

Theil, H (1954): *Linear Aggregation of Economic Relations*, Amsterdam: North Holland.

——— (1966), *Applied Economic Forecasting*, Amsterdam: North Holland.

Triplett, J (1983): "Concepts of Quality in Input and Output Price Measures: a Resolution of the User–value Resource–cost Debate"; in M.Foss (ed), *The U.S. National Income and Product Accounts: Selected Topics*, NBER Conference on Research in Income and Wealth, Vol 47.

Conference Discussion

Handout **Aggregate**

Micro: $\ln q_{it} = \text{const.} + \alpha_i(\ln \ell_{it} + \ln WW_{it} + \ln h_{it})$

WW = weeks worked, ℓ = employment, q = value added, h = effective hours per week, K = capital stock.

Define $u_{it} = \ln h_{it} - \ln \bar{h}_{it}$, $\bar{h}$ = normal hours per week.

so that $\ln h_{it} = u_{it} + \ln \bar{h}_{it}$

i **On the Definition of Normal Hours**

AS Normal hours are just paid–for hours?

JM 'Normal hours' is the negotiated normal work week.

TG So it's different from Marty Feldstein's normal hours which included an implicit contract giving an average amount of overtime?

AS You're excluding overtime then?

JM In my theoretical construction, I'm excluding normal overtime or systematic overtime. I'm excluding that part of overtime which is part of an implicit contract.

AS There's data on normal hours as opposed to paid–for hours?

JM The normal hours data get adjusted econometrically to exclude systematic overtime.

Handout **Macro**: $\ln q_t \equiv \Sigma w_i^q \ln q_{it}$

$$= \text{const} + \alpha(\ln \ell_t + \ln WW_t + \ln \bar{h}_t + u_t)$$

$$+ (1 - \alpha)\ln K_t + \theta_t + \text{aggregation biases.}$$

ii Choice of Functional Form

WE Why do you take the Cobb–Douglas of degree one?

JM Because I think it's plausible and I think the measurement errors in the capital stock are so gross, it's a convenient normalisation in a sense. It turns out that if you unrestrict the coefficient on the capital stock you actually get a negative coefficient and you certainly reject constant returns to scale. Does capital make a negative contribution to output? Most people would doubt it.

CB A large amount of that could be explained by what you're analysing really being a value–added production function. It's not surprising that you get a negative coefficient when you free it up because 80 or 90 per cent of what's going on in the economy is being buried. It's hard, even theoretically, to justify looking at a value–added production function.

JM Value–added is meant to measure the contribution of labour or capital to output, it only leaves out raw materials and raw materials ...

CB 80 per cent.

JM Much less than that.

CB A half?

JM 35 per cent.

iii On the Capital Variables

TG I don't personally agree with it, but a lot of people have used capital utilization, in which case maybe the hours variable should be coming in there, and that's what's making a difference.

JM Right. Capital utilization could be part of the effective capital stock.

TG And that may be something to do with the coefficient.

JM I'm perfectly catholic on this, agnostic; catholic and agnostic. K is the capital stock, in principle it's whatever concept of the capital stock you want it to be.

Handout Note: $\alpha u_t = \Sigma w_i^q \alpha_i u_{it}$ where $w_i^q = \frac{p_i q_i}{\Sigma p_j q_j}$

With w_i = wage,

$$\alpha \approx \Sigma w_i \ell_i / \Sigma p_i q_i, \quad \alpha_i \approx w_i \ell_i / p_i q_i$$

$$u_t \approx \Sigma w_i \ell_i u_{it} / \Sigma w_i l_i \approx \Sigma \ell_i u_{it} / \Sigma l_i$$

Thus $u_t \approx$ average over workers of utilization rates..

Let u_j refer to j[th] worker. If overtime is worked by him $u_j = \ln(\bar{h}_j$ + overtime hours$_j) - \ln(\bar{h}_j)$. 'Undertime' typically is unobserved since a normal week is still being paid for. Assume $\tilde{u}_j$ continuous distribution, density = $\Phi(u)$.

Define $u^* = \int_{u>0} u\Phi(u)\, du$ = proportional overtime averaging over <u>all</u> workers

[Proportional meaning relative to normal hours]

$\hat{u} = \int_{u<0} (-u)\Phi(u)\, du$ = unobserved proportional undertime averaging over <u>all</u> workers

$E(u) \equiv u^* - \hat{u}$

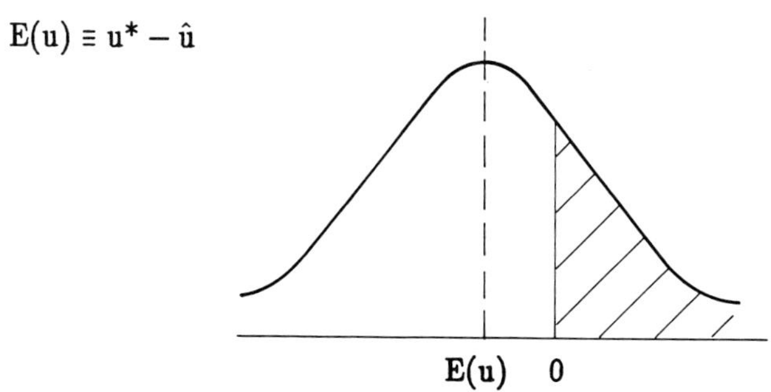

Shift distribution with spread constant to the right then u_t^* increases, $\hat{u}_t$ decreases. Hence u_t^*, $\hat{u}_t$ trade off.

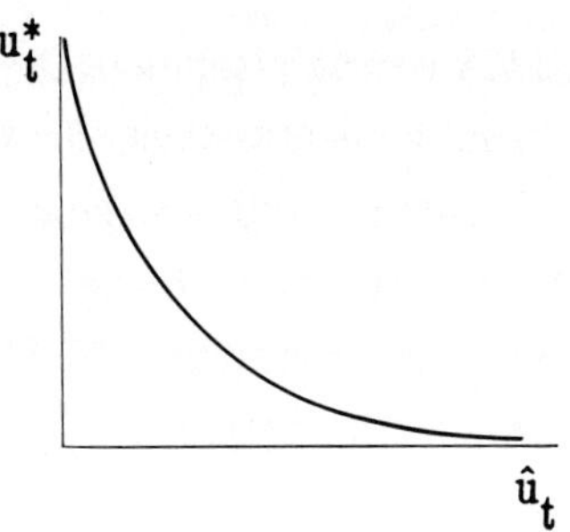

Eg. $(u_t^* + c_1)(\hat{u}_t + c_2) = c$ where $c_1 > 0$, $c_2 > 0$, $c > 0$ defined for u_t^*, $\hat{u}_t \geq 0$

Then $E(u_t) = c_2 + u_t^* - \frac{c}{u_t^*+c_1}$

[Note U,V analogy: if $UV = c$, $z = U - V = U - cU^{-1}$; Hansen (1970)]

iv Trade Off Between u's

AS Presumably, each of these trade–offs correspond to a different density function for the u's?

JM That's right.

AS Is it obvious that by specifying something like that you get a reasonably shaped ϕ function?

JM It turns out that $UV = c$ corresponds to a t–distribution with 2 degrees of freedom.

AS What does $(u_t^* + c_1)(\hat{u}_t + c_2) = c$ correspond to?

JM That, I don't know. It's a generalisation.

AS Is it clear that it's unimodal?

JM Yes, it's unimodal.

Handout But systematic overtime is s_t so "true u_t^*" = observed $u_t^* - s_t$ where s_t depends on normal hours.

v TG In the late 1950s we ran a seminar at Birmingham for managers. We used to berate them for labour hoarding, and they claimed it paid off in good labour relations. So in some sense you want a theory of labour hoarding that runs in terms of good labour relations. Views on labour relations changed a lot in your period, so you might expect some problems with this sort of function. Highly pro–cyclical industries exist too, and trade unions presumably adjust to a considerable extent to that. And, of course, there are white collar workers who used neither to be paid overtime, nor be laid off,...

CJB Presumably in the 1950s good labour relations was a code for meaning that if we try to sack a man everyone comes out on strike.

TG So I'm saying it's changed a great deal since the 1950s, certainly in the last few years, so these trade–offs could be changing.

vi AS Presumably there must be some figures on the distribution of overtime, is that not true?

JM Well, there's sectoral information, but there's no other distribution information.

AS You're just using the average of these u's. Somewhere there ought to be more information.

JM In the New Earnings Survey there is information on overtime but the definition is not quite the same.

AS It does suggest one could make a little more of the functional form of the Φ. Because you should at least have some information to test alternative specifications on that tail.

JM In principle, I suppose what one could do is take the empirical information from the New Earnings Survey, take one particular cross–section, and actually fit an empirical distribution. But I haven't done that.

SP John, you also have information on short–time working and days-off. Is there any way you can use that to tell you about the bottom end of this distribution.

JM Empirical short–time working on the average has been a tenth or less in terms of hours involved of overtime working. It's a curious institution. It's confined to a small group of industries and is tied to subsidies. It's not a very representative indicator.

vii TG Obviously, an alternative people have to consider when overtime becomes in big demand is multiple shift working. I wonder if you've thought of the implication of that for your top tail.

JM Well, again, if you actually look at the number of hours involved in multiple shift working, first of all there's a data problem; there's only a decent source from 1971 onwards. Secondly, there's surprisingly little change in multiple shift working. There's a steady shift upwards, and, of course, in the economic crisis of 1979/81 multiple shift working dropped off. Empirically, I don't think it's of any significance.

Handout Trade–Off in Terms of the Logistic Distribution

The logistic has the neat property

$$1 = \exp\left[\frac{-u^*}{\sigma}\right] + \exp\left[\frac{-\hat{u}}{\sigma}\right],$$

since $\exp\left[\frac{-u^*}{\sigma}\right] = 1 - p$ and $\exp\left[\frac{-\hat{u}}{\sigma}\right] = p$.

where p = proportion of operatives on overtime.
Taking σ as constant we can use

$$\mu = u^* - \hat{u} = u^* - \ln\left[1 - \exp\left[\frac{-u^*}{\sigma}\right]\right]$$

and estimate using a grid for σ.

Alternatively if we really believe the distribution we can deduce

$$\sigma = \frac{-u^*}{\ln(1-p)}$$

Since $\mu = -\sigma \ln p$ we have

$$\mu = u^*\left[1 - \frac{\ln p}{\ln(1-p)}\right]$$

The results should be empirically similar to the normal but the simplicity is preferable.

The range of p is of the order 0.25 to 0.45 so that μ is always negative. This is not entirely plausible. One interpretation is that for institutional or technological reasons some firms do not pay overtime. For example, suppose 'true p' = $p^* = \theta p$ where p = observed proportion on overtime and $\theta > 1$. Then $u^* \approx \theta(OH)$ and

$$\mu = \theta(OH)\left[1 - \frac{\ln \theta p}{\ln(1-\theta p)}\right]$$

where θ could be found by a grid method.

viii Negative μ

AS Why isn't it plausible?

JM Well, because of adjustment costs. Even with asymmetric adjustment costs, it doesn't seem plausible that when you get a positive demand shock, firms on the average mightn't be driven into over–utilitising their workers.

AS What's plausible, that it's zero or positive?

JM Somewhere in the region of zero.

AS Well, this might be in the region of zero.

JM If you take $\theta = 1$ the empirical estimates of μ are substantially negative. That just seems to me implausible. To explain it you have to have extreme risk aversion or some reason why firms systematically under–utilise their workforce.

AS It's actually worse than that because you're assuming that when firms pay overtime everyone is working at full capacity. It may only be that some people in the firm are working overtime.

JM I don't think so. Think of it as aggregation over workers. You're essentially aggregating over those workers working overtime.

AS There's a basic assumption that when you observe a firm paying overtime, some people in that firm are working overtime, you're assuming that all the workers are working up to capacity.

JM You could have a distribution within a plant.

AS Is it the case that maintenance workers are working to capacity if ...

JM Operatives doesn't include maintenance workers.

AS Or the cleaners, or cooks, or office cleaners. One is tending to make the assumption that if one person is working overtime, no–one is working undertime. So in fact your positive observations may themselves be over estimating, so that suggests your μ may be worse than you're estimating.

JM I don't think so.

ix TG I take it that a firm, until quite recently, was offering more or less a guaranteed minimum week which is, so to speak, the 40 hours, and if you take Marty's argument, an expectation that overtime be a certain average. For various reasons, I can well believe that the premium rate for overtime has been going up a lot, there's really no reason why, with risk averse workers, they wouldn't give a great weight to the guaranteed working week, and if the marginal rate on overtime is high, why they wouldn't be willing to shade the bargain towards a higher guaranteed minimum week.

JM The way I handle that empirically is to assume there's some systematic overtime s, and adjust normal hours by systematic overtime.

TG No. I was saying there's a trade–off between the overtime times the overtime rate, which is an average amount they'd hope to have, and the guarantee which, until relatively recently, was normal hours. Now the firm, particularly if it's paying a high overtime rate, may, with risk averse workers, be willing to give a high guarantee with a low expectation of overtime. That seems to me a perfectly rational thing to do and it could give a negative μ.

JM Maybe.

Handout Capital measurement

Is there separate concept of capacity utilization? – see CBI survey and Appendix 2 of main paper.

Gross capital stock figures assume

(1) service lives independent of cycle or relative prices,

(2) constant efficienty of service flow as equipment ages.

Neither seems realistic.

Output Measurement Biases

(1) Bruno type gross output bias

Let q^* = gross output, m = raw materials input. With constant returns,

$$\text{dln } q^* = \alpha \text{ dln } \ell + \beta \text{ dln K} + (1-\alpha-\beta) \text{ dln m} + d\theta$$

(with utilization and weeks worked constant)
Since dln m = dln q* + dln g (relative factor prices), substitution gives

$$\text{dln } q^* = \frac{\alpha}{\alpha+\beta} \text{dln } \ell + \frac{\beta}{\alpha+\beta} \text{dln K}$$

$$+ \left[\frac{1}{\alpha+\beta} - 1\right] \text{dln g} + \frac{d\theta}{\alpha+\beta}$$

Hence negative response of gross output to PR which is ln (rel. price of raw materials to output). But CSO use of fine industrial classification makes probable bias small.

(2) Domestic price index bias: caused by lack of export price indices and use of domestic deflators instead. Gives positive PW, ΔPW effect: PW = ln (rel. foreign/domestic output price).

(3) List price bias: transactions prices more flexible than list prices to pressures of costs and foreign competition.

(4) Price control bias: understated price increases, overstated volume increases when controls introduced (1974) or tightened. Reverse effect when controls removed (1977).

Empirical approximation for mean utilisation

$$\left.\begin{aligned} u_t &= c_2 + u_t^* - s_t - \frac{c}{u_t^* - s_t + c_1} \\ s_t &= s_0 - s_1 (NH_t - 90.4) \end{aligned}\right\} \text{theory}$$

Note: $(NH_t - 90.4)$ was zero for 1968–79, positive before 1968

$$u_t^* \approx OH_t \text{ and}$$

$$u_t = \text{const} + OH_t - c\, OH_t^{-1} + c_0(NH_t - 90.4)\, OH_t$$

is the empirical approximation.

Hence results for R.2 of main paper.

se = 0.007457, $\hat{\alpha} = 0.681$ t = 8.2

DW = 2.09 c has t = 10.2

x Discussion of Estimated Equation

RB On the endogeneity question you test ln $(^{\ell}/k)$ for endogeneity and what happens?

JM I'm sorry, I test OH and $\ln\ell$ for endogeneity. The endogeneity bias is very small, you can accept the hypothesis there's no bias.

RB Using lagged variables as instruments?

JM Yes, and world trade and world production as contemporaneous instruments.

xi TG A Durbin–Watson statistic of 2.09 in equation R2 seems remarkably close to 2.00. You get really surprised when you get χ^2 close to zero, and this suggests that there's been over-mining. Your $\overline{R}^2$ is very close to 1, too, even for time series.

JM Was the model overfitted?

TG Conceivably.

Handout Feldstein–Craine result

Craine uses data on US manufacturing 1949.2–1967.4

$$\ln(q/K) = \text{const.} + \underset{(23.2)}{0.007t} + \underset{(17.3)}{0.789} \ln(\ell/K) + \underset{(14.8)}{2.177} \ln h^0,$$

se = 0.012, D.W. = 0.87, d.f. = 71

h^0 = observed average paid for hours.

$\ln h^0 \approx \ln \bar{h}(1 + OH) = \ln(\text{normal hours} + \text{overtime hours})$

(short–time is negligible) $\approx \ln \bar{h} + OH$

Explanation of Feldstein–Craine result

'True' model is

$\ln (q/K) = \text{const} + \text{trend} + \alpha(\ln(\ell/k) + \ln \bar{h} + u) + \text{error}$

Holding $\bar{h}$ fixed, $\frac{\partial u}{\partial \ln h^0} \approx \frac{\partial u}{\partial OH} = \frac{\partial u}{\partial u^*}$

But from u^*, $\hat{u}$ trade–off and $u = u^* - \hat{u}(u^*)$, $\frac{\partial u}{\partial u^*} > 1$.

Hence $\frac{\partial \ln g}{\partial \ln h^0} > \frac{\partial \ln q}{\partial \ln 1}$.

Craine specification on my data gives

$$\ln(q/K) = \underset{(23.2)}{1.207} \ln(\ell/K) + \underset{(14.3)}{2.176} \ln h^0 + \underset{(14.6)}{0.00934t} + \text{trend shifts etc}$$

se = 0.00967, DW = 1.47.

'True' model further predicts $\frac{\partial \ln q}{\partial \ln \bar{h}} > \frac{\partial \ln q}{\partial \ln 1}$

since 'systematic overtime' implies $\frac{\partial u}{\partial \bar{h}} > 0$ holding OH fixed. Adding $\ln \bar{h}$ as extra regressor to Craine specifications gives:

$$\ln (q/K) = \underset{(7.9)}{0.697} \ln(\ell/K) + \underset{(14.8)}{3.479} \ln h^0 - \underset{(6.6)}{1.852} \ln \bar{h}$$

$$+ \underset{(5.00)}{0.00449t} \quad \text{se} = 0.007910, \ \text{DW} = 2.14$$

ie given OH, $\frac{\partial \ln q}{\partial \ln h} = 1.627$

xii **Test of Craine result**

AS Do you actually release the OH variable as well?

JM It's positive, but not significant.

xiii TS Does this result of a negative capital elasticity occur with all the dummy variables in the equation also?

JM Yes, and the split time trends as well.

TS The only reason I mention that is that there's a study whose author I can't remember, who got exactly this sort of production function estimated for Bell Canada providing telephone services, but basically by splitting the data before and after direct dialling, he cleaned up the results completely. That's why I'm interested in your time trend. The other thing I was wondering about was have you ever included sectoral variables on the composition of industrial output, for instance, into these equations?

JM I haven't tried it but I have done some computations for aggregation bias at the sectoral level. It turns out to be very, very small.

TS To the extent that the sectoral composition is relatively stable there's nothing wrong with using this kind of equation?

JM I think the composition is reasonably stable. It's aggregated into 25 industries. Between 1973 and 1981 the employment of every industry fell. British manufacturing seems to have suffered rather uniformly.

TG You've taken rather a favourable period from just before the first oil shock until after the second oil shock, plus the Thatcher shock. You'd expect almost everything to go down except prices in that period.

JM I did recompute it over two periods and it was also small for the second period.

Handout **Specification and Robustness Tests**

Parameter stability
Lack of residual autocorrelation
Revankar–Hartley test for endogeneity of ln ℓ, OH.
Capacity utilization not significant
Estimates robust to omission of 0,1 special events dummies
Estimates robust to use of Normal distribution for u
Omission of OH $\Rightarrow \hat{\alpha} = \underset{(16.9)}{1.702}$, s e = 0.01488, DW = 0.87
Estimates robust to inclusion of paid holiday proxy
Operatives: data are crude but suggest only operatives matter though parameter estimates are robust
Test of direct Translog given constant returns: accept Cobb–Douglas
Constant returns: rejected; ln K has coefficient = $\underset{(3.21)}{-1.216}$ instead of 0.
Craine finds similar result.

xiv AS What has this to do with the recent productivity trends in the UK? What's the adjusted view of what's been happening? Should we buy gilts or not?

JM The story appears to be that the underlying productivity trend, the composite of all these trend variables, is roughly what it was between 1959 and 1973, from the end of 1980 we return to a trend of greater productivity growth.

AS Is that about 2 per cent?

JM Well, in terms of labour productivity, the trend is as high as about 4 per cent in British manufacturing – about 1 per cent capital deepening, and 3 per cent omitted things. It does look as if this trend is about .3 per cent higher than it was before 1973.

CJB Remind me how that trend is estimated.

JM The underlying trend comes from the trend and the split trends in the model.

AS It's just effectively holding OH constant.

JM Yes, that's right, it abstracts from normal hours as well.

Measurement and Modelling in Economics
G.D. Myles (Editor)

A MODEL OF CAKE DIVISION WITH A BLIND PLAYER

by Christopher Bliss*

A "cake" of random size is divided between two players. If they cannot agree on a division the cake is lost. One player is blind but can weigh the cake he receives. Both players are risk averse. Under the mixed strategy solution (MSS) the sighted player announces the size of the cake and the blind player accepts his allocation with a probability chosen so that the sighted player maximizes by telling the truth. This probability is derived and it is shown that with a sufficiently high degree of risk aversion the MSS produces the outcome that would pertain with sighted players.

1 Introduction

Many writers have considered the solution to a cake division game in which two players must agree on how much cake they will each get, and where the cake is lost to them both if they cannot agree. Sometimes the cake diminishes in size if agreement is delayed. We shall consider the case in which the cake is entirely lost if there is no agreement. This simple game nicely illustrates some fundamental problems of game theory and allows a number of solution concepts to be considered and compared. As is well known, the game may be completely solved only if the specification is enlarged to make it dynamic. When this is done it may be possible to identify a unique solution.

There are two distinct problems which arise with the division of a cake. The first is how to decide on the allocation of benefits between the

* Nuffield College, Oxford.
A simpler version of this paper with the same title was presented to the European meeting of the Econometric Society, Copenhagen, August, 1987.

two players, who will get how much in what circumstances. We call this the problem of determination. The second question is how to implement an agreed division so as to ensure that the players honour the agreement at which they have arrived. We call this the problem of implementation. The problem of implementation only arises when there is scope for dishonesty. The present paper is not concerned with determination. We assume that problem solved. We consider in particular equal division of utility between the players, but the argument would be much the same with any specific division. However our treatment introduces asymmetric information.

Specifically, one player is "blind" and cannot see the cake. The sighted player will be denoted Player I throughout and the blind Player will be Player II. It matters that one player is blind because the size of the cake is a random variable. While he cannot see the cake, the blind player can weigh how much he receives. However this will not inform him how large was the cake from which his portion was cut. Hence, even after they have agreed that the cake will be divided to give them equal utility, and have decided the quantities that will be given to Player II according to the size of the cake, there remains a typical problem of an incentive to cheat. Player I may be tempted to understate the true size of the cake and this will undermine some arrangements that would be possible for two sighted players. The problem with which this paper is concerned is the design of solutions to the game with one blind player.

Given any feasible method of implementing division of the cake when one player is blind, the question of what the division should be is a classic bargaining problem.[1] That is why we assume that part of the

---

[1] Let the size of the cake be as yet unknown to Player I. Then any implementable division rule defines a utility possibility frontier. The players do not need to agree on the method of implementing a division (ie, from which utility possiblity frontier they should choose) as they can bargain over which point to choose on the envelope of the various utility possibility frontiers. An entirely different problem is encountered when the sighted player already knows the size of the cake before the players start to bargain. This gives rise to a distinct version of the bargaining problem with asymmetric information which is not treated here.

problem solved and confine our attention to investigating which divisions can be implemented under asymmetric information.

The basic problem of how to divide a cake when there is no asymmetry of information has been discussed by among others Nash (1953). Rubinstein (1982) showed how expressing the problem as a sequential game could allow a unique solution to be identified, actually the perfect equilibrium of the game. There is a large literature which need not be surveyed here. See Sutton (1985) and Myerson (1986). Several writers have introduced asymmetric information into this model. Fudenberg and Tirole (1983) analyses such an extension and surveys the literature. However the asymmetry of information has concerned the players' preferences, not the scale of the resources to be allocated. To my knowledge this is the first paper to investigate the latter problem.[2] In the following section the formal model is detailed and some examples of a solution are explained. Under the fixed payment solution (FPS) Player II receives a fixed amount of cake from Player I regardless of the size of cake to be shared between them. Player I bears all the risk and the fixed payment reflects this fact. Under another type of solution the players share the risk. Player I announces the size of the cake and there is an agreed division of the cake corresponding to each announcement which we call the division rule. However, Player II adopts a mixed strategy. Following the announcement of the size of the cake he agrees to accept his share with a certain probability, and with the remaining probability he refuses to agree and the cake is lost. The probability is so chosen that is pays Player I to announce the true size of the cake.[3] We call this arrangement the mixed strategy solution (MSS). For the sake of comparison with other models we also consider an extension of our own model in which Player II may shrink the

---

2 But note again that we are concerned with a different question from the one that has engaged the literature: implementation not determination.

3 If this is a one–off game, Player II would need somehow to precommit himself to use the probability values that reveal the true size of the cake despite the fact that after the event he loses by not accepting any positive announcement as true.

cake without discovering its size. We call this the cake shrinking solution (CSS).

These are only examples of solutions. The list could be extended indefinitely. Consider for example the following modification of the FPS.[4] Player I pays a fixed amount to Player II provided that the cake is at least that large, otherwise nothing. To guarantee that Player I will not cheat, Player II always refuses to agree when Player I announces a small cake thus destroying such cakes.

In Section 3 we treat the case in which the size of the cake is a continuous variable. The probability that Player II will agree as a function of cake size which characterizes the MSS is derived for a linear division rule and is shown to depend in a simple way on the utility function and on the division rule. In Section 4 the analysis is extended to cover the case when cake size takes discrete values. This case is much more complicated and we show that different and higher probabilities for accepting announcements may be employed by Player II.

With both continuous and discrete size values it is not possible to say in general which type of solution is best. The FPS suffers from the disadvantage that Player I bears all the risk. However the MSS suffers from the drawback that some cake is wasted, because Player II sometimes refuses to agree and the cake is lost. The same disadvantage, that cake is wasted, is shared by the cake–shrinking solution.[1] Section 5 investigates how the solutions compare when the degree of risk aversion is very high. In Section 6 the principles that have emerged from the model are applied to implicit contract theory, particularly as expounded by Grossman and Hart (1980). Section 7 contains some concluding remarks. The Appendix treats the MSS solutions with a non–linear division rule.

---

[4] I owe this suggestion to Barry Nalebuff.

2 The Model in Continuous Form

In the present section, the size of the cake is a continuous variable, denoted s, with a range $[0,s^{max}]$ and probability density p(s). The utility of each player is the same, that is U(x), where x is the amount of cake received. $U(\cdot)$ is assumed strictly concave. Let h^{max} denote the greatest amount that Player I could receive under the MSS division rule. For reasons that will become clear later we assume that $U(\cdot)$ has been normalised so that $U(0) = 0$, and $U(h^{max}) = 1$. We now examine some solutions.[ii]

The Fixed Payment Solution (FPS)

A constant b is chosen such that:

$$\int_0^{s^{max}} U(s-b).p(s).ds = U(b). \tag{1}$$

Note that if utility is divided, b must exceed values of s to which positive probability attaches. Hence the utility function must be defined for negative arguments but this need not imply negative consumption. The utility function measures the relative benefits of quantities of cake received from the game and the players are assumed to have other resources, including the cake that Player I must hold to pay Player II when s_i is less than b.

The Mixed Strategy Solution (MSS)

Player I announces the size of the cake and Player II agrees with a probability which depends on the announcement. If Player II agrees he receives $b + (1-\theta)s$, where $0.5 \le \theta < 1$.[5] If the announcement is s, the probability that Player II will agree is q(s). When the function q(s) is such that Player I maximises by announcing the true value of s it will be said to have the incentive compatability property (it is IC). So we are assuming that the q(s) function is IC. If utility is divided b and θ are such that the two players do equally well. We do not investigate here the choice of optimal values for b and θ, which is a problem of quite terrible

---

[5] Note that the FPS is a special case of this rule with $\theta = 0$.

complexity. Our results apply to any values of b and θ including the optimal values.

As $U(0) = 0$ the value of the game to the players is simply the integral of the utilities that they obtain if Player I's announcement is accepted multiplied by the probability that the announcement will be accepted. The remaining expected value is terms of the form:

$$\{[1-q(s)]\cdot U(0)\}\cdot p(s); \tag{2}$$

which will all be zero. Hence the values of the two players are respectively:

$$\text{I} \qquad \int_0^{s_{max}} q(s).U[\theta s - b]ds;$$

and (3)

$$\text{II} \qquad \int_0^{s_{max}} q(s).U[(1-\theta)s + b]ds.$$

This specification of the game assumes that it is possible to choose an IC function q(s).[iii,iv] In the following section we show that this is indeed the case and we derive that function.

The Cake Shrinking Solution (CSS)

To assist comparison with other models we extend our own model to allow "cake shrinking". Player II cannot see the cake but he can put into effect an action which is known to shrink the cake to a specific extent. Shrinking the cake may be used to ensure that Player I will reveal its size. The effect is similar to the MSS above in that untruthful announcements are penalized.

Consider again a linear division rule. Let Player I pay to Player II b plus a fraction $(1-\theta)$ of the shrunk cake. A cake of size s will be shrunk to $v(s)s$, where $v(s) \leq 1$. The $v(s)$ function is IC if $s = s_0$ maximizes:

$$U[v(s)\{s_0 - (1-\theta)s\} - b], \tag{4}$$

which is the utility that Player I would obtain from announcing a cake of size s when the true size is s_0. Player I maximizes (4) by choosing s to maximize:

$$v(s)\{s_0-(1-\theta)s\} \tag{5}$$

if

$$v(s) = \left[\frac{s}{s_{max}}\right]^{(1-\theta)/\theta}, \tag{6}$$

then $s = s_0$ solves (5). The first– and second–order conditions for the maximum will be respectively:

$$\left[\frac{1-\theta}{\theta}\right]\cdot\left[\frac{s-(1-\theta)s}{s_{max}}\right]\cdot\left[\frac{s}{s_{max}}\right]^{(1-2\theta)/\theta} - (1-\theta)\cdot\left[\frac{s}{s_{max}}\right]^{(1-\theta)/\theta} = 0; \tag{7}$$

$$(s_0 - (1-\theta)s)\cdot\left[\frac{1-\theta}{\theta}\right]\left[\frac{1-2\theta}{\theta}\right]\left[\frac{1}{s_{max}}\right]^2\left[\frac{1}{s_{max}}\right]^{(1-3\theta)/\theta}$$

$$-\left[\frac{2(1-\theta)}{s_{max}}\right]\left[\frac{1-\theta}{\theta}\right]\left[\frac{s}{s_{max}}\right]^{(1-2\theta)/\theta} < 0 \tag{8}$$

Clearly (7) is satisfied by $s = s_0$, and (8) ensures a regular minimum. Hence the value of the cake shrinking solution to Player I follows as:

$$\int_0^{s_{max}} U[\theta s(s/s_{max})^{(1-\theta)/\theta} - b].p(s).ds\ . \tag{9}$$

The Utopian Solution

Finally for completeness notice the solution that would obtain were both players to be sighted. In that case they would always divide the cake in half and it would never be wasted. This is the utopian solution. The value of the game would be:

$$\int_0^{s^{max}} U(s/2).p(s).ds. \tag{10}$$

3 Solving for an IC q(s) Function in the Mixed Strategy Case

Recall that the utility function $U(\cdot)$ has been normalized so that $U(0) = 0$ and $U(h^{max}) = 1$. We now show that the probabilities of acceptance will be a power of the utilities of the various outcomes. Player II makes the probability that he will agree to accept that the cake is of size s equal to a power of the utility that Player I would obtain from receiving his share of a cake of size s.

Theorem 1 If $q(s) = U(\theta s - b)^{(1-\theta)/\theta}$ Player I maximizes by announcing the true size of the cake.

Proof Suppose the cake is s_0 and Player I announces s. The expected value of this announcement is:

$$U(\theta s - b)^{(1-\theta)/\theta} U[s_0 - b - (1-\theta)s]. \tag{11}$$

We show that $s = s_0$ maximizes (11) which is the result of the theorem. The first order conditions for the maximization of (11) is:

$$(1-\theta)U(s_0 - b - (1-\theta)s) \cdot U_1(\theta s - b) \cdot U(\theta s - b)^{(1-\theta)/\theta}$$

$$- (1-\theta)U_1(s_0 - b - (1-\theta)s) \cdot U(\theta s - b)^{(1-\theta)/\theta} \tag{12}.$$

Note that (12) is satisfied by $s = s_0$. The second–order condition, evaluated at $s = s_0$ is

$$(1-\theta)U_{11}U^{(1-\theta)/\theta} - (1-\theta)U_1^2U)^{(1-2\theta)/\theta} < 0 \tag{13}$$

(13) ensures a regular maximum because $U(\cdot)$ is strictly concave. This completes the proof.$^\nabla$

It is interesting to compare the incentive compatible MSS with the utopian solution for the simple case $b = 0$; $\theta = 0.5$. In this case $q(s) = U(s/2)$. The utopian solution is:

$$\int_0^{s_{max}} U(s/2).p(s).ds, \tag{14}$$

which compares with the value of the MSS which is:

$$\int_0^{s_{max}} U(s/2)^2.p(s).ds, \tag{15}$$

As $U(s/2) \leq 1$, the value (15) falls short of the value of (14) which reflects the loss of cake inherent in an IC solution. The only consolation for the fact of this loss is that large cakes are less likely to be lost, as $U(s/2)$, the probability that the cake will be consumed, increases with s.

4 The Model in Discrete Form

We now consider the case in which the cake will take one of N sizes $s_1, s_2, \ldots, s_N$ arranged in increasing order with probabilities respectively $p_1, p_2, \ldots p_N$. Otherwise the model will be as in Section 3 and the basic types of solution will be the same. Thus the fixed payment solution will be defined by a constant b such that:

$$\sum_{i=1}^{N} p_i.U(s_i - b) = U(b). \tag{16}$$

The mixed strategy solution will be defined for probability values q_i which will be IC. The value of the game to Player I will be:

$$\sum_{i=1}^{N} p_i \cdot q_i \cdot U(\theta s_i - b). \tag{17}$$

We show that the solution which worked for the continuous case, in which the probability that Player II would accept an announcement that the cake is of size s was $U(\theta s - b)^{(1-\theta)/\theta}$, works again for the discrete case. However we then show that in the discrete case it is always possible to improve on those probability values.

Theorem 2 If $q_i = U(\theta s_i - b)^{(1-\theta)/\theta}$ Player I maximizes by announcing the true size of the cake.

Proof Suppose the cake is s_i and Player I announces s_j, where j and i are distinct. The expected value of this false announcement is:

$$U(s_i - b - (1-\theta)s_j) \cdot U(\theta s_j - b)^{(1-\theta)/\theta}. \tag{18}$$

If the theorem does not hold there will exist distinct i and j such that:

$$U[s_i - b - (1-\theta)s_j] \cdot U(\theta s_j - b)^{(1-\theta)/\theta}$$

$$> U(\theta s_i - b)U(\theta s_i - b)^{(1-\theta)/\theta} = U(\theta s_i - b)^{1/\theta} \tag{19}$$

Hence:

$$[(1-\theta)/\theta]\log U(\theta s_j - b) + \log U[s_i - b - (1-\theta)s_j] > [1/\theta]\log U(\theta s_i - b). \tag{20}$$

However, since $U(\cdot)$ is strictly concave, so is $\log U(\cdot)$ strictly concave. Therefore:

$$(1-\theta)\log U[\theta s_j - b] + \theta\log U[s_i - b - (1-\theta)s_j]$$

$$< \log U[(1-\theta)(\theta s_j - b) + \theta(s_i - b - (1-\theta)s_j)] = \log U[\theta s_i - b], \quad (21)$$

which contradicts (20).

Thus we see that the same probability values as functioned to give an IC solution in the continuous case also work in the discontinuous case. However, there is an important difference between the two cases. In the continuous case the <u>only</u> values that will do the job are given by $q(s) = U(\theta s - b)^{(1-\theta)/\theta}$. As they are derived from a tangency condition it is clear that we could not alter them even sightly without altering the response of Player I and upsetting the solution. In the discrete case this is not so and we may always improve upon the $q_i = U(\theta s_i - b)^{(1-\theta)/\theta}$ values.

That this is so is most easily seen by considering the case of $N = 2$.[6] Then $U(\theta s_2 - b)$ will equal 1 and q_1 will be $U(\theta s_1 - b)^{(1-\theta)/\theta}$. However, the largest value of q_1 that does not induce Player I to pretend that the cake is s_1 when it is really s_2, q_1^*, satisfies:

$$1 = q_1^* U[s_2 - b - (1-\theta)s_1]; \quad (22)$$

that is:

$$q_1^* = 1/U[s_2 - b - (1-\theta)s_1]. \quad (23)$$

However, $1/U[s_2 - b - (1-\theta)s_1]$ is greater than $U(\theta s_1 - b)^{(1-\theta)/\theta}$ as may be seen by noting that, as $\log U(x)$ is a strictly concave function, we have:

6 I am grateful to John Broome for pointing out to me that the case of two sizes shows at once that it is sometimes possible to do better than the solution of Theorem 2.

$$\log U(\theta s_1 - b) > (1-\theta)\log U(\theta s_2 - b) + \theta \log U[s_2 - b - (1-\theta)s_1]. \quad (24)$$

However, $U(\theta s_2 - b) = 1$, so that $\log U(\theta s_2 - b) = 0$, and dividing by θ and taking the antilogarithm of both sides of (24) yields:

$$1 > U(s_2 - b - (1-\theta)s_1]^{(1-\theta)/\theta} \cdot U(\theta s_1 - b), \quad (25)$$

as required.

In general with cake sizes s_N down to s_1 the maximum values that the probabilities q_i may take are defined recursively as follows:

$$1 = q_{N-1}U[s_N - b - (1-\theta)s_{N-1}];$$

$$q_{N-1}U(\theta s_{N-1} - b) = q_{N-2}U[s_{N-1} - b - (1-\theta)s_{N-2}]; \quad \ldots(26)$$

$$q_{N-k}U(\theta_{N-k} - b) = q_{N-k-1}U[s_{N-k} - b - (1-\theta)s_{N-k-1}]; \text{ etc.}$$

These q_i values are so constructed that it is immediate that Player I will never want to claim that the cake is of size s_{k-1} when really it is of size s_k. However we have to show that this property holds in the large; i.e. that Player I will never want to claim that the cake is of size $s_{k-m}(m > 1)$ when it is really of size s_k.[7] This is the result of the following theorem.

Theorem 3 When the probability values are those defined recursively in (26) Player I will reveal the true size of the cake.

Proof Assume that Player I falsely announces the size of a certain cake. We may assume without loss of generality that the cake whose size is not truly announced is the largest cake. This is so because should false

---

7 It is assumed that where Player I is indifferent between telling the truth and claiming that the cake is smaller than it actually is he will tell the truth.

revelation happen with a cake smaller than the largest it would also happen if all cakes larger than the one which triggers a false announcement were no longer possible outcomes. The utility values would be renormalized and new probabilities would be defined by (26). However the relative merits of true and false announcements to do with remaining cakes would be unaffected, because all utility values would undergo the same linear transformation, as would remaining q values.

We next show that if one can construct a case of false revelation with the q_i values now under consideration then one may construct a case with only three different sizes of cake. This is so because the operation of removing sizes from the menu of possibilities can only increase the q_i values that attach to those which remain. To see this consider removing the size s_{k-1} from the list which defined the sequence of (26). Then q_{k-2} would have to be recomputed, as would all q values for lesser sizes. We would have:

$$q_{k-2} = q_k \cdot \frac{U(\theta s_k - b)}{U[s_k - b - (1-\theta)s_{k-2}]} \tag{27}$$

where previously

$$q_{k-2} = q_k \cdot \frac{U(\theta s_{k-1} - b)U(\theta s_{k-1} - b)}{U[s_k - b - (1-\theta)s_{k-1}]U[s_{k-1} - b - (1-\theta)s_{k2}]} \tag{28}$$

The ratio of the previous value to the new value is:

$$\frac{U(\theta s_{k-1} - b)\ U[s_k - b - (1-\theta)s_{k-2}]}{U[s_k - b - (1-\theta)s_{k-1}]\ U[s_{k-1} - b - (1-\theta)s_{k-2}]} \tag{29}$$

and the logarithm of this ratio is:

$$\log U(\theta s_{k-1} - b) + \log U[s_k - b - (1-\theta)s_{k-2}]$$

$$- \log U[s_k - b - (1-\theta)s_{k-1}] - \log U[s_{k-1} - b - (1-\theta)s_{k-2}]. \tag{30}$$

Now note that since $\log U(x)$ is a strictly concave function the average of its values for two values of x with a given mean is larger the closer together are the x values. However the first two terms of (30) represent twice the average values of $\log U(x)$ for arguments the mean of which is:

$$-b + \frac{s_k}{2} + \frac{\theta s_{k-1}}{2} - (1-\theta)\frac{s_{k-2}}{2}. \tag{31}$$

The third and fourth terms of (30) subtract twice the average value of $\log U(x)$ for arguments the mean of which is again (31). The first pair of arguments are clearly further apart, as the value $s_k - b - (1-\theta)s_{k-2}$ is larger than any other argument. Hence (30) is negative, which means that it is the logarithm of a value less than 1. Thus (29) is less than 1 which shows that the ratio of the previous value of q_{k-2} to the new value is less than 1, so that deleting a value s has allowed a q value (and all values further on) to increase, as was to be shown.

It only remains to show now that we cannot have a case with three sizes of the cake in which announcing s_1 when s_3 is the true size is better than announcing the true value. Suppose that we have such a case. Then:

$$\begin{aligned} &U(\theta s_3 - b)^{1/\theta} = q_2 U[s_3 - b - (1-\theta)s_2]; \\ &q_2 U(\theta s_2 - b) = q_1 U[s_2 - b - (1-\theta)s_1]; \text{ and} \\ &U(\theta s_3 - b)^{1/\theta} < q_1 U[s_3 - b - (1-\theta)s_1] \end{aligned} \tag{32}$$

However $U(\theta s_3 - b)^{1/\theta} = 1$, and:

$$q_1 = \frac{U(\theta s_2 - b)}{U[s_3 - b - (1-\theta)s_2]\,U[s_3 - b - (1-\theta)s_1]}; \tag{33}$$

so that we must have:

$$1 < \frac{U[s_3-b-(1-\theta)s_1]\ U[\theta s_2-b]}{U[s_3-b-(1-\theta)s_2]U[s_3-b-(1-\theta)s_1]}; \tag{34}$$

which implies:

$$\log U[s_3 - b - (1-\theta)s_1] + \log U(\theta s_2 - b)q$$

$$> \log U[s_3 - b - (1-\theta)s_2] + \log U[s_3 - b - (1-\theta)s_1] \tag{35}$$

The same argument that was used above to show (30) to be negative also serves to show that (35) is a contradiction. Thus the arguments on the left– and right–hand sides of the inequality both have the same mean but the right–hand side must have the larger spread as it includes the values $s_3 - b - (1-\theta)s_1$. This completes the proof.

The conclusion at which we arrive is that the discrete case is importantly different from the continuous case in that different and larger q-values can be employed than would be obtained by applying the formula derived from the continuous case. That formula will work but it can always be improved upon.

5 A Limit Theorem

In the following discussion we return to the case of Section 2 and 3 in which s is a continuous variable. All the arguments to follow apply equally and immediately to both cases. It is convenient to refer to the FPS as something distinct from the MSS. Hence an MSS is taken to imply $\theta > 0$.

It is clear that the utopian solution is unattainable and that one cannot say in general which of the FPS and the MSS will be better for the players. Suppose, for example, that neither player is risk averse. Then U will be linear and the FPS will be as good as the utopian solution. It will not matter that Player I is asked to bear all the risk as he will be risk neutral. In particular no cake will be lost. With risk neutrality there is no point in considering the MSS. However for the sake of completeness note that although $U(\cdot)$ is linear the proof of Theorem 1 goes through because $\log U(\cdot)$ is strictly concave. Hence a MSS solution is feasible in which some cake will be pointlessly lost and the outcome will be inferior to the FPS. Moreover, the loss of cake will impose a heavy cost because the probability that a cake will be consumed decreases linearly with the size of the cake measuring from the largest cake, whereas if $U(\cdot)$ is concave the decrease in the probability is initially less.

The above suggests that there will be a connection between the relative performance of the FPS and the MSS and the degree of risk aversion of the players. It is after all the merit of the MSS that it transfers some of the risk from Player I to Player II, so one would expect the extent of risk aversion to play an important role. As both players are assumed to have the same utility function, both are equally risk averse. Hence Player I's gain in shedding risk must be weighted against Player II's loss in taking it on. However, the gain will exceed the loss as long as Player I bears more risk.

The extent of risk aversion plays an additional role in that it influences the amount of penalty in terms of lost cake that must be imposed under the MSS to ensure that Player I reveals the true size of the cake. The more risk averse is Player I the higher can be the probabilities that his announcements will be credited without inducing him to dissemble and announce a false value. This is because false revelation increases the risk that Player I undertakes. A heavily risk averse sighted player is more likely to play safe and tell the truth.

To see exactly how this works out we consider the following utility function which has a constant value of Arrow's relative risk aversion (see Arrow [1970]).

$$U(x) = \left[\frac{1}{1-\alpha}\right] x^{1-\alpha}, \tag{36}$$

where α is the coefficient of risk aversion. The larger is α the higher is risk aversion. To use the function (36) for our problem we need to interpret x as total consumption, that is cake consumed plus additional consumption, a constant. Otherwise, if α is large the function is not defined for the negative x which would arise with the FPS or with small cakes in an MSS with b positive. Also the function has not been normalized, but as this consists of a linear transformation it does not affect the coefficient of relative risk aversion.[8]

First we add a constant $c > b$ to cake consumed. The level of c may be taken to represent the consumption which an agent enjoys independently of the cake and its division. So our function becomes:

$$U(x + c) = \left[\frac{1}{1-\alpha}\right](x + c)^{1-\alpha}.^{9} \tag{37}$$

As c is a constant we may continue to denote $U(x + c)$ simply by $U(x)$. However, the reader should bear in mind that our function is now (37) and not (36).

From (37) we subtract:

$$\left[\frac{1}{1-\alpha}\right]\cdot(c - b)^{1-\alpha}, \tag{38}$$

so that $U(0)$ which corresponds to a consumption of c will be 0, and we divide the result by:

$$\left[\frac{1}{1-\alpha}\right](\theta s^{max} + c - b)^{1-\alpha} - \left[\frac{1}{1-\alpha}\right]\cdot(c - b)^{1-\alpha}, \tag{39}$$

so that $U(s^{max})$ which corresponds to a consumption of $c - b + \theta s^{max}$ will be 1.

The consequence of these transformations is that our utility function becomes:

---

8 The coefficient of relative risk aversion is: $\frac{-x \cdot U_{11}}{U_1}$, where subscripts denote the order of differentiation.

9 Now the coefficient of relative risk aversion is: $\frac{-(x+c)\cdot U_{11}}{U_1} = \alpha$.

$$U(s) = \frac{\left[\frac{1}{1-\alpha}\right](\theta s+c-b)^{1-\alpha} - \left[\frac{1}{1-\alpha}\right](c-b)^{1-\alpha}}{\left[\frac{1}{1-\alpha}\right](\theta s^{max}+c-b)^{1-\alpha} - \left[\frac{1}{1-\alpha}\right]\cdot(c-b)^{1-\alpha}}$$

$$= \frac{(\theta s+c-b)^{1-\alpha} - (c-b)^{1-\alpha}}{(\theta s^{max}+c-b)^{1-\alpha} - (c-b)^{1-\alpha}} \quad (40)$$

Writing the utility function in this form leads directly to the next result.

Theorem 4 As the coefficient of relative risk aversion α increases without limit, utility levels corresponding to positive quantities of cake all approach 1, and the MSS solution tends to the utopian solution.

Proof That utility levels for positive s tend to 1 as α tends to infinity follows immediately from an inspection of (40). Next note that as the U(s) values tend to 1, so do $U^{(1-\theta)/\theta}$ values so that the probabilities that announcements are accepted tend to 1 and the players obtain their allocations with probability 1. As this is the case there is no longer any need to tailor the division rule to discourage misrevelation, and $b = 0$, $\theta = 0.5$, as in the utopian outcome, does very well. It remains to notice that with a positive coefficient of risk aversion the FPS is inferior to the utopian solution. That this must be so is clear when one considers that with a strictly concave utility function the average of two distinct allocations which give equal utility gives more utility. However the average of what the two players get in the FPS is the utopian solution.

The relative ranking of the CSS solution and the MSS solution as α tends to infinity exhibits a curious feature. Let the cake have size s. Under cake shrinking Player I will receive $(\theta s - b)[s/s^{max}]^{(1-\theta)/\theta}$ and Player II will receive payments of equal expected utility. Under the MSS the expected value of the cake that Player I will receive is $(\theta s - b)$ $U(\theta s - b)^{(1-\theta)/\theta}$. As $U(\theta s - b)^{(1-\theta)/\theta}$ tends to 1 this last value will clearly exceed $(\theta s - b)[s/s^{max}]^{(1-\theta)/\theta}$. However the MSS will retain uncertainty of outcome even subject to the cake being of size s, and as α

increases the cost of this uncertainty will also rise. These two effects do battle with each other as α increases. Must the MSS eventually do better than the CSS? It might seem so because the actual allocations under the MSS are tending in the limit to the utopian solution and under the CSS they tend to outcomes strictly worse than the utopian outcomes. However note that as α tends to infinity the players in the limit only care about avoiding the worse outcome, $s = 0$. All other outcomes are ranked equally in the limit, $U(s) = 1$. Furthermore the probability of the worst outcome is independent of the use of the MSS against the CSS. We thus arrive at a paradox. It takes a high degree of risk aversion to rid the MSS solution of waste but when risk aversion is high enough waste ceases to matter. The players become obsessed by the avoidance of the worst outcome.[10]

6 Interpretations of the Model and Implicit Contract Theory

Several features of the foregoing model will be seen to be inessential to the result obtained. Thus it is not necessary that the players divide utility equally but only that it be divided according to some predetermined principle. Similarly it does not matter whether both players have the same utility function but only that the sighted player should be risk averse.

An interesting feature of the MSS solution is that it seems to explain apparently irrational and inefficient behaviour. The players knowingly waste cake, because cakes are sometimes destroyed when Player II refuses to accept an announcement. According to one interpretation the game is a repeated game. Player II would benefit in the short run if he set $q(s) = 1$ for s, i.e, accepted any announcement. However Player I would eventually come to realise what was happening and this would disturb the equilibrium under which truthful revelation is optimal for Player I and

10 This last result is evocative of Pascal's wager. A sufficiently risk averse agent will care nothing for pleasures on Earth if he or she can reduce, no matter how little, the probability of eternal damnation. What we have shown in fact is that Hell need not even be such a bad place if it is worse than any state on Earth and humans are highly risk averse.

would drive the parties back to the FPS. Knowing this the players may rationally tolerate waste.

We are not however inclined to press the interpretation of the game as one round of a repeated game because, as is well known, the analysis of repeated games raises complicated issues which we have not addressed. It is better to consider the game as a one–round game in which Player II can precommit himself to a certain strategy. Then our argument shows that this strategy may rationally commit Player II to cause waste. Can the model be extended to explain real life examples of apparently wasteful behaviour such as strikes and involuntary unemployment?

A model of the strike emerges naturally from cake division. Suppose that the firm negotiates with labour and agrees to give the workers an expected utility level of U_0. However it is proposed that the operating surplus of the firm, which corresponds to the cake, should be divided in such a way that both workers and the employer share the risk of the uncertainty concerning the size of the operating surplus. The only problem is that workers are "blind" – they cannot accurately measure the true surplus. This point has been argued by among others Grossman and Hart (1983) (see also Hart (1983)) whose solution is equivalent to cake-shrinking. However we consider a different approach.

Let the employer announce the surplus that is available to be divided between profits and wages. If the workers accept the announcement as true, the surplus is divided according to a predetermined agreement. This agreement has the effect of making the workers bear some of the risk which the firm faces. This is necessary for an efficient arrangement because the employer is assumed to be risk averse. However sometimes with a carefully chosen probability the workers will call a strike and the surplus, whatever it may be, will be lost. Because of the strike threat the employer will find it optimal to announce the true profitability. However, although the employer tells the truth, strikes sometimes occur and must occur to sustain the solution. Notice that strikes, which are a characteristic of the MSS will be encountered when the MSS is better than the FPS, which will be when the employer is risk averse. With an extremely risk averse employer on the other hand few strikes will be generated because q(s) will always be close to 1.

The foregoing argument alerts us to look for commitment mechanisms when the model is applied to observations of reality. Suppose that the truth revealing strategy of the union is to call a strike with 20 per cent probability if the employer announces a profitability which would imply a wage of 100. The union faces the difficulty that a threat to call a strike with a certain probability may not be credible, as the employer knows that it will harm the union to go on strike. Nevertheless, a skillful union leader may be able to create genuine randomness beyond his control by "putting it to his members", with the outcome weighted by the way he puts it, of course. The strange conclusion that the autonomous player may be weaker than the constrained player is perfectly natural within the logic of mixed strategy solutions.

Now consider the Grossman–Hart model. The players are a sighted employer and a group of blind workers. The employer is risk averse and the cake is net revenue when workers are fully employed. The action which corresponds to what we have called cake–shrinking is laying-off workers.[11] The fact that he must lay–off workers to an extent determined by his announcement ensures that the employer will truly reveal his profitability.

Note however an important difference of principle between the mechanism of laying–off workers and the MSS solution. Given an announcement, the extent to which workers must be laid off depends only on the levels of payments to workers in various states of the world and the true values of net revenue. It is independent of the employer's utility function and of his risk aversion. Hence the Grossman–Hart solution, although its justification is the employer's risk aversion, is unaffected by an

---

[11] Our cake shrinking is not identical to the Grossman–Hart solution because its extent has been chosen only to ensure incentive compatability and not to given an optimal outcome. However this point does not affect the validity of the argument.

increase in risk aversion. There is no limit theorem and no tendency to approach the utopian solution as risk aversion increases. It follows therefore that if the coefficient of risk aversion is sufficiently large the Grossman-Hart solution can be dominated in outcome space by a MSS. However, as was demonstrated above, dominance in outcome space does not imply dominance in terms of expected utility.

7 Concluding Remarks

It is a well–known result of game theory that introducing randomness to a game with asymmetric information may increase the value of the game even if players are risk averse. The model of this paper has provided another example where randomness can help. Uncertainty affects not only the value of the pay–offs from a game but also the incentive structure under which the agents choose their strategies. In particular, it allows true revelation of information and mutual insurance. We have seen this in our cake division game. It is not the only means to true revelation, as the Grossman-Hart solution illustrates, but when the players are very risk averse it may well be the best.

APPENDIX

In this appendix we look at the continuous case with a non–linear division rule and we derive the q(s) function that should apply to that case. Specifically we allow the amount of cake which Player I may retain to be a general non–linear function of the cake size. A well–known problem of non–linear taxation (see Mirrlees (1986), p.1203 et passim) applies equally to the present problem. The agent, in this case Player I, may face a maximization problem in which his pay–off is not a concave function of his choice. The satisfaction of first–order conditions for a maximum may not imply even a local maximum. We ignore this difficulty in what follows. In other words we assume that the departure from linearity (in which case there is no problem of ensuring a maximum) is not so considerable as to give rise to the problem of a non–concave objective function.

The division rule now takes the following form. If Player I announces s, and Player II agrees, Player I gives s – h(s) to Player I and retains the remainder of the cake. If Player I was telling the truth he will therefore retain h(s). Subject to the point about concavity already discussed, h(s) may be any increasing function.

To state the central result we need to introduce some new functions. The function R(s) is defined as follows:

$$\frac{d}{ds}[R(s)] = U_1[h(s)]/U[h(s)]; \qquad (A.1)$$

$$R[s^{max}] = 0. \qquad (A.2)$$

Note that R(s) is not proportional to log[U(x)], the derivative of which would be:

$$U_1[h(x)]h_1(x)/U[h(x)]. \qquad (A.3)$$

As R[s] is zero when s takes its maximum value and as (A.1) implies that R(s) is an increasing function of s, R(s) is generally negative. Now T(s) is simply the exponential of R(s).

$$T(s) = \exp\{R(s)\}. \qquad (A.4)$$

T(s) is positive and increases with s. Also, from (A.2):

$$T(s^{max}) = \exp\{R(s^{max})\} = 1. \qquad (A.5)$$

Theorem 6 If q(s) = T(s)/U[h(s)], Player I maximizes by announcing the true size of the cake.

Proof We only look at the first order conditions for a maximum as we are assuming that the second order conditions will be satisfied. When the cake is of size s_0 Player I solves:

$$\underset{s}{\text{Max}} \frac{T[(s)]}{U[h(s)]} U[s_0 - s + h(s)]; \tag{A.6}$$

which is equivalent to:

$$\underset{s}{\text{Max}}\ R(s) + \log U[s_0 - s + h(s)] - \log U[h(s)], \tag{A.7}$$

where log is the logarithm to base e. Differentiating (A.7) with respect to s and equating the result to zero yields:

$$\frac{U_1[h(s)]}{U[h(s)]} - \frac{U_1[s_0-s+h(s)]}{U[s_0-s+h(s)]}[1-h_1(s)] - \frac{U_1[h(s)]}{U[h(s)]}h_1(s) = 0 \tag{A.8}$$

which is satisfied at $s = s_0$. This completes the proof.

Notice that $q(s) = U[-b + \theta s]^{(1-\theta)/\theta}$ is a special case of the theorem. As $h_1(s) = 0$, a constant, (A.1) is satisfied by:

$$R(s) = (1/\theta)\log[h(s)]. \tag{A.9}$$

Then:

$$T(s) = U[h(s)]^{(1/\theta)}; \tag{A.10}$$

and $T(s)/U[h(s)]$ becomes $U[h(s)]^{(1-\theta)\theta}$, as required.

Finally note that the limit theorem still implies that q(s) values tend to 1 as α becomes very large. As α tends to ∞, $U_1[\cdot]$ tends to zero, $U[\cdot]$ tends to 1 and R(s) tends to a constant, which must be zero by (A.2). Hence T(s) tends to 1, as required.

References

Arrow, K J (1970): *Essays in the Theory of Risk Bearing*, Amsterdam, North Holland.

Fudenberg, D & J Tirole (1983): "Sequential Bargaining and Incomplete Information", *Review of Economic Studies*, 50, pp 221–247.

Grossman, S J & O Hart (1980): "Implicit Contracts With Imperfect Information", *Quarterly Journal of Economics* (Suppliment), 98, pp 123–156.

Hart, O D (1983): "Optimal Labour Contracts under Asymmetric Information: An introduction", *Review of Economic Studies*, 50, pp 3–35.

Mirrlees, J A (1986): "The Theory of Optimal Taxation", Ch. 24, in Vol III of *Handbook of Mathematical Economics*, edited by K J Arrow and M Intriligator, North–Holland.

Myerson, R B (1986): "Negotiation in Games: a Theoretical Overview", in W.P. Heller et als., eds., *Uncertainty, Information and Communication*, Vol III of Essays in Honour of K J Arrow, Cambridge University Press.

Nash, J F (1953): "Two Person Cooperative Games", *Econometrica*, 21, pp 128–140.

Rubinstein, A (1982): "Perfect Equilibrium in a Bargaining Model", *Econometrica*, Vol 50, pp 97–109.

Sutton, J (1985): "Non–Cooperative Bargaining Theory: An Introduction", Review of Economics Studies Lecture, Oxford, reprinted as ICERD Discussion Paper 85/125, London School of Economics.

GAMBLERS AND LIARS

by Christopher Bliss

1 Introduction

Imagine that a group of agents may enter into an agreement of some kind. Examples would include an employment contract between an employer and his workers (see Hart (1983)), and an arms reduction agreement between great powers. The framework is that of Nash's bargaining problem, Nash (1950). Following Myerson (1979), imagine that the form of the agreement will be determined by an arbitrator. On the same lines assume that the arbitrator does not possess all the information that he would need to implement the ideal agreement. We then encounter the familiar problem of incentive compatibility. If the agents are asked to reveal their private information, it may not be in their interests to reveal it accurately. This fact constrains what the arbitrator can achieve. Indeed, from the consideration of a bilateral bargaining problem (but the conclusion is much more general), Myerson and Satterthwaite (1983) show an impossibility result. Put simply this says that full efficiency of arbitrated solutions is impossible within the constraint of incentive compatible arrangements. Along similar lines, Guesnerie and Laffont (1984) provide a complete solution to a class of principle agent problems in which the agent's private information plays an essential role. A model closer to pure bargaining but with incomplete information is provided in Fudenberg and Tirole (1983).

That there is an excess burden imposed by the constraint of incentive compatibility might suggest the conclusion that it would be a good idea if as much private information as possible could be made known to the arbitrator. However we shall soon see that the issue is not how much private information is made known to the arbitrator, for all will be. The real issue is the design of the mechanism that elicits that information and the way in which it will make use of the information. Hence the use of the term *mechanism design* to describe the study of the design of optimal incentive compatible arrangements.

The well–known public–good free–rider problem will help to illustrate the point. A public good is to be provided, or not provided, according to the willingness to pay of potential beneficiaries. However, how much the provision of the public good will augment the welfare of an individual is information private to that individual. Efficiency requires that the public good be provided if some combination of charges will both make all who participate better off and add up to at least the cost of provision. The requirement that individuals reveal the maximum amount that they are willing to pay to enjoy the public good might be inconsistent with incentive compatibility if individual charges are known to be related to such announcements. Then agents will not reveal true information and the maximum demand prices will not be revealed. Yet provided that nothing depends on the announcement, the maximum demand prices can be elicited, simply by asking the agents to announce them. If nothing depends on the announcement of maximum demand prices, then all must be charged alike for the provision of the public good, and that may be inconsistent with efficiency.

The example is instructive. Maximum demand prices are either revealed, in which case they must play no role in the mechanism, or not revealed, when again they can play no part in the mechanism. However if they play no role in the mechanism, there is no difficulty in the way of their accurate revelation. A general principle, the *revelation principle*, shows that any mechanism that uses certain information may be implemented through a scheme in which that information, at least, is truthfully revealed. The result of course will typically be inefficient. Suppose that information I will be communicated whenever the true information is J, where J need not be single–valued. Let the consequence to the agent of communicating I be C(I). Note that C(I) is what the mechanism does for the agent and can only be single–valued. It is not the ultimate implication for the agent, which will not be single valued in the case that J is not single-valued. Now if the mechanism is altered to associate C(I) with the communication of all information – that is any element of J – which previously induced the agent to communicate I, then no essential difference is made, and the revelation of true information, at least as regards elements of J, has been made incentive compatible. The general application of this principle leads to the conclusion that the only mechanisms which the arbitrator need consider are those in which the revelation of true information is incentive compatible. As it is always incentive compatible to reveal accurately information of which the arbitrator makes no use, and as the revelation principle says that the mechanism can always be designed so that the accurate revelation of information which is used will be incentive compatible, it follows that we need consider only mechanisms in which all information is accurately revealed.

From the above general and well–known points we may conclude that the problem of mechanism design is to do with the comparison of different incentive compatible mechanisms all of which will fully and accurately reveal private information. The revelation of private information and mechanism are entwined with each other. It would usually be wrong to think of the revelation of information as a separate or independent process.

2 Random Exclusion.

In this paper I look at at a scheme in which the arbitrator discovers private information by means of a procedure which may be called *Random Exclusion.* Random exclusion means what its name suggests. Each agent communicates his private information to the arbitrator. The communication may consist of a declaration or of an action. Now the arbitrator excludes an agent from the benefits of the agreement which he is organizing with a probability which depends functionally on the private information communicated. The idea is to have the arbitrator choose the functional relationship between an agent's announcement of what he claims to be his private information and the probability of exclusion so that it will be incentive compatible for agents to truthfully communicate their private information.

Notice that from the formal point of view exclusion is an unclear notion. In actual cases it will be represented by such outcomes as a strike (in an employer–labour negotiation) or signing no agreement (in arms negotiations). However formally these are outcomes, if unpleasant outcomes. What matters is that there should be available a bad outcome which may be imposed randomly on agents to induce them to accurately reveal their private information. It is clear that a bad outcome is a relative notion and such will be the case here. Hence the cost of not revealing private information accurately can only be assessed in terms of its implications for the agent and for others of possibly being excluded.

The arbitrator will choose the probability that an agent will be excluded to be a function of the content of the private information communicated by the agent. He will choose that functional relationship to be such that the agent will maximize by accurately revealing the true content of his private information. At first sight it may appear that the problem of inefficiency inherent in the incentive compatible mechanism with private information has disappeared. Once the arbitrator has collected the private information accurately communicated by the agents he may organize an efficient outcome. In a sense this is true. However the Myerson and Satterthwaite result points us to the conclusion that some inefficiency must remain. We find it when we take note that the arbitrator may organize an outcome efficient only for those agents who have not been excluded in the first stage. Some agents will have been excluded and already assigned to a bad and non–participating state by the chance operation of the probability parameters. This will be so despite the fact that they will have told the precise truth concerning the content of their private information. The excluded agent is not being punished for mis–representing the truth. As telling the truth is incentive compatible, there will be no mis-representation. Rather, the excluded agent is an innocent victim of a process which cannot function as intended unless it includes, and sometimes puts into effect, a threat to exclude.

As exclusion is inefficient, can the problem be circumvented by substituting an efficient but unattractive outcome for exclusion as such? It seems that it can be. Suppose that arbitrator proceeds as follows. He assigns a utility indicator to each agent, the ordering being the same as the agent's own. Then he announces that he will decide on the form of an agreement in which all will be participants by maximizing a weighted sum of the utilities of the various agents. For each agent the weight will be either a large weight or a small one. Then the probability that a small weight will be assigned to a particular agent is made to depend on what he claims to be his private information in such a way that it will be incentive compatible for the agent to maximize by accurately revealing the true content of his private information. Because the arbitrator is maximizing a weighted sum of welfares, the outcome must be efficient. However the whole procedure is surely incentive compatible as care has been taken to design it to be incentive compatible.

One difficulty is that efficiency has been purchased at the price of surrendering full control over the distribution of the benefits of the agreement, which have indeed become random. More importantly the arrangement produces efficiency only in a restricted sense. It shares this feature with the random exclusion method. The exclusion method was efficient only for those agents who had not been excluded in the first stage. The latest method described, with its randomly determined weights, is efficient for all, but only in an *ex post* sense; that is after the weights have been determined. Meanwhile it enforces on the agents an *ex ante* uncertainty as to what weights will be assigned to their utility and this reduces expected welfare after all, and again we have the typical inefficiency of an incentive compatible arrangement in accord with Myerson and Satterthwaite.

Returning to the method of random exclusion, we may ask what the point is of examining it further. We have seen that it is inefficient, a deficiency which it shares with all incentive compatible arrangements where private information is involved, but a deficiency from which it suffers nonetheless. There are no general rankings of inefficient methods. Usually one will suit a particular case better than another. On a different occasion the ordering will be reversed. Hence the fact that the method of random exclusion may sometimes serve well is not a reason for attaching a special importance to it. It is of course interesting to solve for the functional relationship between announcement and probability of exclusion which will elicit true revelation of private information, especially if one is informed in advance that the solution is simple and convincing. Yet unless the method could be supported by something more than taxonomic appeal, even that exercise would seem to be sterile.

The reason for persisting with an interest in the method of random exclusion is that it has a fascinating limiting property as the agents involved become progressively more and more risk averse. This property may be demonstrated for a special case and probably holds more generally. It is reflected in the title of the paper, which may be explained as follows. Supposing that one is an individual who may enter into arbitrated bargaining with others. What kind of company should one seek? Many considerations could be taken into account in that decision, and one consideration would surely be the choice of partners with whom cooperation or exchange was judged to offer particularly attractive gains. However, lacking information on the last range of questions, there is another principle that applies when the method of random exclusion is to be used. Choose the highly risk averse as partners. Relatively low average probabilities of being excluded are needed to induce them to tell the truth about their private information. Hence the excess burden of ensuring incentive compatibility will be relatively low. Avoid gamblers – agents with a low level of risk aversion. They tend to be liars, in the sense that relatively high probabilities of being excluded are needed to induce them to tell the truth about their private information. Of course the moralistic overtones of a term like liar are out of place here. Mechanism design theory starts from the presumption that all agents will lie if it suits them. None is in a good position to cast a stone.[vi]

3 An Information–Revealing Probability Function

The method of random exclusion depends upon choosing the probability of exclusion as a function of the agent's specification of his private information so as to to induce the agent to tell the truth. The time has come to examine the form of the function that will achieve this end. I first develop a general treatment and then discuss particular cases in the following section. So far as the relationship between risk aversion and the size of the probability of exclusion mentioned above is concerned, it can only be established firmly for a special case. However I hope to convince the reader that something like it will apply rather generally. In any case, enough solid results will be established to allow the reader to form an independent judgment of the question.

A special assumption that will be employed throughout is only for convenience. It will be assumed that the agent's private information concerns the value of a scalar quantity denoted by β. We take β to range from zero to infinity, but this is only for the sake of having some definite assumption. Any connected range for β would give us similar results and even a disconnected range for β can be covered by a simple extension of the method. We assume that $\beta = \infty$ corresponds to the state in which the agent makes the greatest demands on the system relative to his own contribution. Hence $\beta = \infty$, and by extension high values of β, are particularly attractive to the agent tempted to dissemble. For this reason the announcement or indication of a high value of β has to be penalized by means of a high probability of exclusion (a low probability of staying in). It is convenient to suppose that $\beta = 0$ is the converse case to $\beta = \infty$. An individual who claims that $\beta = 0$ cannot be dissembling to his own advantage and may be credited with probability 1 (excluded with probability zero).

In a situation when the agent has to reflect his private information in an action, say by declaring it, as with an income tax return, he may "declare" any value he pleases. He declares, or acts as if he has, a value α. In general, without the implementation of an incentive compatible mechanism, α will not correspond to β. The idea is to select a function of

α, $\pi(\alpha)$, such that an agent who will be excluded with probability $1 - \pi(\alpha)$ will maximize by choosing to declare, or act as if his value was, the true value β. If the agent is excluded from the agreement he will enjoy a reservation level of utility denoted V_0. In general V_0 may depend on β, but I assume it to be the same for all types of agents, as this simplifies the analysis. How is $V(0,0)$ related to V_0? Suppose that $V_0 < V(0,0)$. Then the agent affecting to have $\beta = 0$ is excluded with probability zero and does strictly better than if he were excluded. This is entirely possible. Indeed it will be necessary later to require $V(0,0)$ to be positive to ensure that a definite integral may be evaluated. [See equation (20) below.] The opposite strict inequality, $V_0 > V(0,0)$, is more awkward to dispose of. If $V_0 > V(0,0)$, an agent owning a value of β close to zero would prefer to be excluded. If he could ensure his exclusion, the accurate revelation of low values of β would not be incentive compatible. It may make sense to assume that an agent can always withdraw (exclude himself), in which case $V_0 \leq V(0,0)$ must hold. However the mechanism may not allow it. Furthermore, it need not be inefficient for a mechanism to prohibit the own-welfare–improving withdrawal of individual parties, because it may to the benefit of the welfare of others to keep a certain agent in.

Suppose for the moment that the agent may declare, or act as if he has, any value of β he pleases and not be excluded from the benefits of the agreement. So he acts as if his true value were α, although really his value is β. The mechanism treats him accordingly, bestowing on him costs and benefits appropriate to what we may call an α–individual, although really he is a β–individual. As a result the agent will enjoy a utility level which will depend upon both α and β. We denote it:

$$V[\alpha,\beta]. \tag{1}$$

We assume $V[\alpha,\beta]$ to be a twice differentiable function of both its arguments.

Sometimes the mechanism is such that, for all or some values of β, $V[\alpha,\beta]$ is maximized by setting α equal to β. We can call this case *perfect self-selection.* It is the social economist's ideal of a policy for which the intended candidates select themselves. The problem of course is implicit in the immovable Myerson and Satterthwaite result. Arrangements which satisfy the perfect self–selection property may do so only because they achieve rather little. The provision late at night of free hot soup at a depressing inner–city site to anyone dressed like a drop–out is not a policy that needs to concern itself with the problem of mis–representation. On the other hand its coverage is necessarily extremely limited in scope.

If the indirect utility function (1) satisfies perfect self–selection, then $\pi(\alpha)$ may equal 1 for all values of α. More generally consider the following proposed solution:

$$\pi(\alpha) = \exp\left[-\int_0^\alpha \frac{V_1(\theta,\theta)}{V(\theta,\theta)}\cdot d\theta\right]. \tag{2}$$

Notice that $\pi(0) = 1$, as required. The equation (2) is obtained by integrating the first–order conditions obtained from the agent's maximization problem. That is not a valid method in general unless the satisfaction of second–order conditions can be assured. I return to that point below. Allowed to choose which type of individual to represent himself as, the agent solves:

$$\operatorname*{Max}_{\alpha}\ \pi(\alpha)\cdot V[\alpha,\beta] + \{1-\pi(\alpha)\}\cdot V_0 = \pi(\alpha)\cdot V[\alpha,\beta]; \tag{3}$$

where (3) takes into account $V_0 = 0$. For an incentive compatible mechanism, $\alpha = \beta$ must solve (3). Therefore first–order conditions will be satisfied at $\alpha = \beta$. Differentiating the maximand of (3) with respect to α, setting the resulting derivative to zero, substituting a common value θ for both α and β, and rearranging, yields:

$$\frac{\pi'(\theta)}{\pi(\theta)} = -\frac{V_1(\theta,\theta)}{V(\theta,\theta)}; \tag{4}$$

where $V_1(\theta,\theta)$ indicates the partial derivative of V with respect to its first argument. Integrating (4) from $\theta = 0$ to α gives:

$$\log_e \pi(\alpha) - \log_e \pi(0) = \log_e \pi(\alpha) = -\int_0^{\alpha} \frac{V_1(\theta,\theta)}{V(\theta,\theta)} \cdot d\theta. \tag{5}$$

Taking the exponents of both sides of (5) yields (2), as required.

We may check that the solution makes sense by substituting (2) into (3) and deriving a first–order condition directly. So the agent solves:

$$\underset{\alpha}{\text{Max}} \exp\left[-\int_0^{\alpha} \frac{V_1(\theta,\theta)}{V(\theta,\theta)} \cdot d\theta\right] \cdot V[\alpha,\beta]. \tag{6}$$

However this is equivalent to maximizing the logarithm of the maximand of (6), hence equivalent to:

$$\underset{\alpha}{\text{Max}} -\int_0^{\alpha} \frac{V_1(\theta,\theta)}{V(\theta,\theta)} \cdot d\theta + \log(V[\alpha,\beta]). \tag{7}$$

The first order condition for (7) is:

$$-\frac{V_1(\alpha,\alpha)}{V(\alpha,\alpha)} + \frac{V_1(\alpha,\beta)}{V(\alpha,\beta)} = 0. \tag{8}$$

When $\alpha = \beta$ (8) is satisfied, as required.

As is often the case when optimization defines a constraint, the typical situation with incentive compatible mechanisms, the second–order condition is problematic. See Mirrlees (1986). That the problem is encountered in the present case may be seen most easily by differentiating the maximand of (3) twice with respect to α and setting β equal to α to obtain:

$$\pi''(\alpha) \cdot V[\alpha,\alpha] + 2 \cdot \pi'(\alpha) \cdot V_1[\alpha,\alpha] + \pi(\alpha) \cdot V_{11}[\alpha,\alpha]. \tag{9}$$

Such a term is impossible to sign. This problem is a constraint on mechanism design. The appropriately designed mechanism must make sure that it takes care of second–order conditions. For this reason I shall not concern myself further with second order conditions.

Let us look more closely at the solution for the $\pi(\alpha)$ function, given in (2), reproduced for the sake of convenience as (10):

$$\pi(\alpha) = \exp\left[-\int_0^\alpha \frac{V_1(\theta,\theta)}{V(\theta,\theta)}\cdot d\theta\right]; \tag{10}$$

At a quick glance it appears that the function under the integral is an exact derivative. It looks like the derivative of $\log V[\theta,\theta]$ with respect to θ. However that would be:

$$\frac{V_1(\theta,\theta) + V_2(\theta,\theta)}{V(\theta,\theta)}. \tag{11}$$

The integrand of (10) includes only the first term of the numerator of (11). To discover the implications of variations in the degree of risk aversion we need to obtain a closed form expression for the probability function $\pi(\alpha)$. To this end I examine a special case in the next section.

4 A Special Case

In a special, but nevertheless important, case, the features that differentiate various agents, some of which constitute the content of agents' private information, are of the same kind as the transactions between the agent and the mechanism. Therefore the welfare of the agent depends on the sum of two vectors, one given and not observed, except by the agent, and denoted $y(\beta)$; the other transferred to the agent by the operation of the mechanism, and denoted $x(\alpha)$. In general $x(\alpha)$ may have negative components. Furthermore, for any θ, $x'(\theta)$ is proportional to $y'(\theta)$, where the superscript $'$, as in $x'(\theta)$, denotes the vector of derivatives of components of x with respect to θ. This implies of course that:

$$x(\theta) = x_0 + \varphi\cdot y(\theta); \tag{12}$$

where x_0 is a vector of constants, and φ is a scalar constant. For ease of reference, I call this case *linear additivity*, although plainly more than just linear additivity as such is involved. Under linear additivity and assuming that the agent truly reveals β, as the incentive compatibility will ensure, $V[\alpha,\beta]$ takes the special form:

$$V[\alpha,\beta] = U[x_0 + \varphi\cdot y(\alpha) + y(\beta)]. \tag{13}$$

Notice that the welfare level in (13) is represented by $U[\cdot]$, reminding us that this is now the direct utility function and not the indirect form denoted by $V[\cdot]$.

From (13) it is immediate that:

$$V_1[\theta,\theta] = u'[x_0+(1+\varphi)\cdot y(\theta)]\cdot x'(\theta); \tag{14}$$

where u' is the vector of marginal utilities of the various components of the vector sum $x_0+(1+\varphi)\cdot y(\theta)$. Similarly:

$$V_2[\theta,\theta] = u'[x_0+(1+\varphi)\cdot y(\theta)]\cdot y'(\theta). \tag{15}$$

However from (12):

$$x'(\theta) = \varphi\cdot y'(\theta). \tag{16}$$

Therefore, from (14) to (16):

$$V_2[\theta,\theta] = \left[\frac{1}{\varphi}\right].V_1[\theta,\theta]. \tag{17}$$

Therefore:

$$V_1[\theta,\theta] + V_2[\theta,\theta] = \left[\frac{\varphi+1}{\varphi}\right].V_1[\theta,\theta]; \tag{18}$$

and:

$$V_1[\theta,\theta] = \left[\frac{\varphi}{\varphi+1}\right]\cdot\left[V_1[\theta,\theta] + V_2[\theta,\theta]\right]. \tag{19}$$

We may substitute (19) into (2) and obtain a closed form expression for $\pi(\alpha)$, as:

$$\pi(\alpha) = \exp\left[-\int_0^{\alpha}\frac{V_1(\theta,\theta)}{V(\theta,\theta)}\cdot d\theta\right] = \exp\left[-\left[\frac{\varphi}{\varphi+1}\right]\int_0^{\alpha}\frac{V_1(\theta,\theta)+V_2(\theta,\theta)}{V(\theta,\theta)}\cdot d\theta\right]$$

$$= \exp\left[\left[\frac{\varphi}{\varphi+1}\right]\cdot\{\log V(\alpha,\alpha) - \log V(0,0)\}\right] = \left[\frac{V(\alpha,\alpha)}{V(0,0)}\right]^{\frac{\varphi}{\varphi+1}}. \tag{20}$$

To summarize (20), we have:

$$\pi(\alpha) = \left[\frac{V(\alpha,\alpha)}{V(0,0)}\right]^{\frac{\varphi}{\varphi+1}}. \tag{21}$$

The probability of not being excluded is a power of the normalized utility function. Clearly $\pi(0) = 1$, as always required.

Although it is plainly a special case, linear additivity covers some important examples. For instance, the income tax game may be characterized as follows. The agent makes return in which he may or may not communicate all his private information concerning his actual income and expenses. Then, unless he is excluded (which here would mean the imposition of an audit conceived as a severe penalty) he pays out money in the form of taxes and/or receives money in the form of refunds. If his utility depends only on the net income after his tax is settled, then linear additivity is satisfied. From among many other examples, consider the optimal labour contracts under asymmetric information reviewed by Hart (1983). In an important special case, when no disutility attaches to work, the transactions between employer and worker concern only sums of money. However the private information is with the employer, and it concerns essentially a sum of money, specifically a point on the revenue function. We have linear additivity.

In the next section I examine the relation between risk aversion and the function $\pi(\alpha)$, making use of the closed form solution (21).

5 Risk Aversion and the Probability of Exclusion

Consider the solution for $\pi(\alpha)$ when we can obtain it explicitly in the case of linear additivity. This is:

$$\pi(\alpha) = \left[\frac{V(\alpha,\alpha)}{V(0,0)}\right]^{\frac{\varphi}{\varphi+1}}. \tag{22}$$

In essence $\pi(\alpha)$ is utility raised to a power less than 1. Notice that utility here is cardinal because it is the von Neumann–Morgenstern measure used to maximize expected utility in (3) above. For this reason we should think of there being a best and worst outcome and of the utility function normalized so that the worst outcome has utility 0 and the best utility 1. We would expect risk–aversion to affect the shape of the utility function and hence of $\pi(\alpha)$. In Bliss (1987) I showed that the limiting form of the utility function as the coefficient of relative risk aversion goes to infinity concentrates all the weight of utility maximization on the avoidance of the worst outcome. In other words, in the limit the utility of the worst outcome may be zero and all outcomes significantly better have weights close to 1.

Here I establish a similar but more general result by a more informal and intuitive route. In Arrow (1970) it is shown that the extent of risk aversion may be measured by a parameter which increases with the curvature, or the degree of concavity, of the utility function. This parameter is the Arrow–Pratt index of absolute risk aversion, see Machina (1987). Suppose that our utility function ranges from the worst to the best outcome, and let the outcomes be arranged in ascending order of desirability.

Suppose that the consumer is indifferent between, on the one hand, the best outcome with probability π and the worst outcome with probability $1 - \pi$, and, on the other hand, outcome X (between the best and worst outcomes) with certainty. Then:

$$U(X) = \pi \cdot U(B) + (1 - \pi) \cdot U(W); \tag{23}$$

where U(B) and U(W) have their obvious meanings as the utilities of the best and worst outcomes. Figure 1 illustrates. Outcomes are shown in increasing order of preference along the horizontal axis. A divides the line the line WB in the ratio WA:AB = π:1$-\pi$. Outcome X has the same expected utility as B with chance π and W with chance 1$-\pi$. Now take the limit of the expression (23), holding U(B), U(W) and π constant, as the agent becomes progressively more and more risk averse. Increased risk aversion means that for given U(B), U(W) and π, X can be worse and the original U(X) smaller. This is the price which the agent is willing to pay to avoid risk. Hence, for given U(B), U(W) and π, the X in (23) must become a worse and worse outcome as we take the limit of increasing risk aversion described.

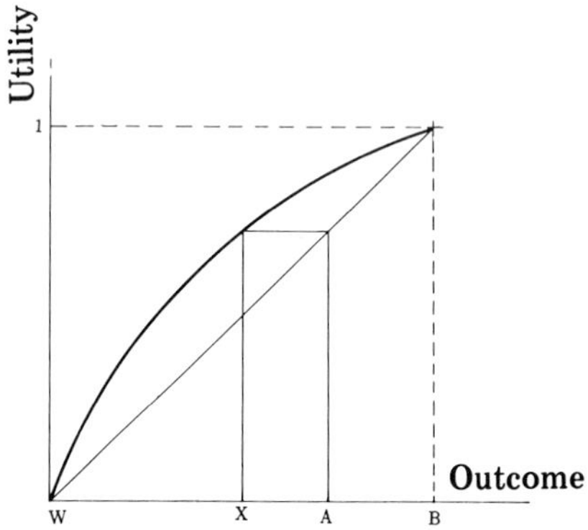

Figure 1

Figure 1 shows the shifting of X to the left, to X′. Notice however that this change requires that U(X) relative to given utilities for W and B must increase. In consequence, for any consumer indifferent between X with certainty and a chance π of B combined with a chance 1$-\pi$ of W, there is always a more risk averse type for whom X with certainty is strictly preferable and whose point of indifference is at something worse than X. The limit of such more risk averse types is X = W in (23) for all

$\pi < 1$. Of course the rate at which X goes to W will be slower for high values of π. As the utility curve shifts left everywhere between its end-points, the utility of every given outcome intermediate between the best and the worst increases. In fact each goes towards U(B), although at different rates for different outcomes. However the conclusion that interests us is now clear. As risk aversion increases without limit the agent will reject the substitution of a better for a good outcome when the former is purchased at the price of an increased probability of the worst outcome.

A striking implication of the result just derived may be noted when utility is normalized so that $U(B) = 1$ and $U(W) = 0$. In that case, $\pi(\alpha)$ in (22) is simply a power of the utility function, and as all utilities tend to 1 as risk aversion becomes extreme. It follows that the probabilities of exclusion needed to provide incentive compatibility decline towards zero. As previously suggested, the highly risk averse, the very opposite of gamblers, may be induced to tell the truth, not to lie, at a cost which approaches zero.

The result has been derived under the assumption of linear additivity but is surely more general. Recall that the general formula for $\pi(\alpha)$ is given by (10), reproduced here for convenience.

$$\pi(\alpha) = \exp\left[-\int_0^\alpha \frac{V_1(\theta,\theta)}{V(\theta,\theta)}\cdot d\theta\right]. \tag{10}$$

Extreme risk aversion should ensure that $V_1(\theta,\theta)$ will be small nearly everywhere, and that will imply $\pi(\alpha)$ close to 1 above values of θ close to the worst outcome.

It is important to note a final caveat. Not for the first time in this argument we seem to have stumbled across a way around the Myerson and Satterthwaite impossibility result, and once again it is not so. It is true that with extreme risk aversion the burden of exclusion as a mechanism is reduced to negligible proportions. However, first, a very special case does not make much a dent in a general result; and, secondly, it is not even necessarily the case that extreme risk aversion reduces the costs, as opposed to the quantitative scale, of the burden of exclusion. Although the point is subtle, it is in fact not difficult to comprehend. As risk aversion increases, the probabilities of exclusion go towards zero and the real outcomes converge on what the mechanism would dictate if all information were freely available to the arbitrator. The uncertainty of outcome which exclusion imposes declines in scale as risk aversion increases but the cost of the uncertainty that remains grows ever larger, because the more risk averse agent detests uncertainty to an increasing extent. For these reasons it is impossible to establish a general limiting result on the comparison of the exclusion mechanism with other, less random, mechanisms.

6 Concluding Remarks

The basic idea of this paper is an example of the theory of mechanism design which is both powerfully illustrative of the general principles involved and which also serves as a vehicle to demonstrate how randomization of the outcome may be usefully incorporated into a mechanism, as a means to ensure the incentive compatibility of the revelation of private information. As the Myerson and Satterthwaite impossibility result leads us to expect, randomization is at best a neat device which may do well in certain situations but is not and cannot be without cost. The costs of randomization vary with risk aversion. A tantalizing possibility, explored above, is that with very high risk aversion small probabilities of exclusion might achieve the desired result, and the revelation of private information would become incentive compatible at what looks like a low cost. Unfortunately, when risk aversion is high small probabilities of exclusion are not a low cost. The high risk aversion that enables them to take low values also implies that the agent comes to care only about avoiding the worst, or very poor, outcomes, and improvements in the average worth other outcomes pales into insignificance.

It has been suggested that negotiation with gamblers (meaning agents with a low degree of risk aversion) is to be avoided. However the agent will be best served by some taste for gambling (in the form of a not-excessive degree of risk aversion) on his own part. If the agent's own risk aversion is very high along with that of his fellow parties to the agreement, then the benefit he enjoys from low probabilities of exclusion for himself and others will do him no good when he cares only about the probability of bad outcomes.

References

Arrow, K J (1970): *Essays in the Theory of Risk Bearing*, North-Holland.

Bliss, C (1987): "A Model of Cake Division with a Blind Player", presented at the European Meeting of the Econometric Society, Copenhagen, 1984.

Fudenberg, D & J Tirole (1983): "Sequential Bargaining and Incomplete Information", *Review of Economic Studies*, 50, pp 221–247.

Guesnerie, R & J–J Laffont (1984): "A Complete Solution to a Class of Principle Agent Problems with Application to the Control of a Self-Managed Firm", *Journal of Public Economics*, Vol. 25, pp 329-369.

Hart, O D (1983): "Optimal Labour Contracts under Asymmetric Information: An Introduction", *Review of Economic Studies*, 50, pp 3–35.

Machina, M (1987): "Expected Utility Hypothesis", in J Eatwell, M Milgate and P Newman, eds, *The New Palgrave: A Dictionary of Economics*. London: The Macmillan Press.

Mirrlees, J A (1986): "The Theory of Optimal Income Taxation", Ch 24, Vol. III, of K J Arrow and M D Intriligator, eds, *Handbook of Mathematical Economics*, Amsterdam: Elsevier.

Myerson, R B (1979): "Incentive Compatibility and the Bargaining Problem", *Econometrica*, 47, pp 61–73.

Myerson, R B & M A Satterthwaite, (1983), "Efficient Mechanisms for Bilateral Trading", *Journal of Economic Theory*, 29, pp 265-281.

Nash, J F (1950): "The bargaining problem", *Econometrica*, 18, pp 155-162.

Conference Discussion

i GM This cake shrinking and the inefficiency: in what sense is there really an inefficiency? When you work back to the solution, the players agree in the first period, and the cake never actually shrinks.

AS It does in the case of incomplete information.

CJB Yes, with the blindness, the cake is always shrunk. My intuition, which I can't actually justify, that cake shrinking is bad relative to randomisation, particularly for very risk averse players, is that the cake has to be shrunk every time.

GM I was thinking of the Rubinstein solution where, although the cake does shrink in each period, you don't agree, because they do agree in the...

AS There's no incomplete information there. As soon as you introduce incomplete information...

CJB There have been extensions of the Rubinstein model with a different kind of incomplete information, incomplete information about preferences.

TG But Gareth is surely right about the standard model.

CJB Gareth is right about the standard model where you foreclose at once and it's mechanical, but this is a different question.

ii **On Game in 'Cake'**

AS I don't think it's completely clear. Is it a one–period game in which one person makes an offer and the other accepts or rejects, or randomises? Is that the game as specified; there must be a game there. Is it something where you randomise, and if you miss out on the first time, you have another go?

CJB Let's explain it this way: you and I have to play a cake dividing game and in order to join with me you have to pay a bus–fare or something to get to the game, so you want to be sure that you have a certain expected value, and we agree in advance how the cake is divided. For simplicity, let's say we agree to divide it down the middle. So I say to you; "okay, Tony, if you come and play this game with me you'll always get half the cake"; and you calculate the pay-off from that. The problem is that since you can't see the cake you don't really trust me, so you want to join a game where I will be locked into telling the truth by my own maximisation.

TG So it is a one–move game?

AS I don't think it can be a one move game. What you're saying is: I'm the blind player; in the first move of this game I announce some rule which is how we must play this game; at the second stage you announce how big the cake is, and that determines the outcome.

CJB That will do, yes.

AS It's very close to a principal agent problem I take it.

CJB It's exactly a principal agent problem.

iii Equation (1)

AS I think I have a little bit of a problem here. This person's objective function is an expected utility function?

CJB Yes.

TG So θ is an exogenous constant for this?

CJB Yes, as far as this problem is concerned.

IJ Some θ's would be better than others?

CJB The θ is a standard part of any bargaining game and determines how much benefit a player gets. If we increase θ we give more to the blind player and less to the sighted player, it's a trade–off, there's no Pareto improvement.

IJ Except the blind player is adopting some probability of accepting each size of cake and that depends on θ.

AS Yes, I think that's my problem. This seems to presume that the opponent's going to tell the truth.

CJB No, it's not, I'm going to demonstrate that. I will derive the q(s) that gives us that property.

IJ That q(s) is going to depend on θ, isn't it?

CJB I'm sorry; this is an incentive compatible q(s). The existence of an incentive compatible q(s) is what both papers are about.

iv Again on (3)

AS But that's not the objective function of the sighted player which is what I though you said.

CJB When he chooses what to announce he already knows the size of the cake, it's no longer a variable, we can forget about the integral, and he has to decide.

TG There should be an argument θ in the function q(s) and that really answers the question, but θ is fixed exogenously presumably in terms of alternative possibilities of using the time so I think that fixes Tony's problem.

AB So what you're looking at is a given rule to split up the cake ex-post, where if you wanted an optimal rule to split the cake 50–50 ex–ante it probably wouldn't be the case in every state of the world that you would want to split the cake $\frac{1}{2}$–$\frac{1}{2}$ ex–post.

CJB That is correct.

v Theorem 1

AS Is it the blind player's objective to get the other player to tell the truth or to maximise his expected utility?

CJB Could you repeat that?

AS The blind player is the one who chooses the incentive structure here.

CJB In your interpretation, yes.

AS This theorem just says something about getting the other player to tell the truth, but that's not the objective, the objective of the blind player is to maximise his utility rather than simply to persuade the sighted to tell the truth.

CJB Well, let's go back to my simple story. You're paying a bus fare to play with me. Your objective is not that I tell the truth, but it certainly is your objective that I shouldn't rip you off after you've come a long way.

JF But there may be another q that leads me to get a higher utility even if you lie. I don't care.

CB The revelation principle surely applies here.

TG You see, there's a theorem that you do no harm by requiring incentive compatibility.

vi Risk Aversion

TG A lot of the papers have dealt in recent years with the problem "I wouldn't tangle with x". If somebody is willing to bluff you and be awfully annoying, you're willing to give in, on the whole. I remember this as an argument by David Kreps in favour of Mrs Thatcher as a good Prime Minister; but people say "I wouldn't do business with x" too. The disadvantage of pretending not to be risk averse in this case is that they may be shunned, sent to Coventry or even boycotted. I think yours, in a way, is the interpretation " wouldn't do business with that character".

CJB It would be absurd for me to claim that I've got a model of truthfulness and good character. All the players in this game have bad characters and will lie if it's in their interest.

Measurement and Modelling in Economics
G.D. Myles (Editor)

THE ESTIMATION OF ENGEL CURVES

by

Stephen Pudney*

This study examines the problems of dealing with zero expenditures in the estimation of Engel curves. Generalised forms of the Deaton–Irish P–Tobit model are proposed and estimated by least–squares and maximum–likelihood methods. These techniques are used in a preliminary application to three expenditure categories from the UK Family Expenditure Survey.

1 Introduction

A major difficulty in applied cross–section demand analysis lies in the fact that observed data usually comes from short–duration expenditure surveys, while the economic theory of the consumer runs in terms of long-term average rates of consumption. The relationship between these two is not simple: if a household is not observed to purchase a particular good in the survey period, it may nevertheless be a consumer of that good on average in the longer run; moreover, even if a purchase is observed, the underlying rate of consumption is not necessarily equal to the observed purchase rate, particularly for goods with a storage life greater than the length of the survey period. This difficulty has been well known to practitioners since the earliest work in the field, but there has so far been no convincing theoretical framework for the analysis of consumption behaviour from short–duration expenditure data. It is the aim of this paper to

---

* London School of Economics

I am grateful to the participants of seminars at LSE, Bristol and Cambridge for helpful comments on earlier versions of this paper. The empirical results reported here are incomplete and preliminary. Please do not quote.

propose, and to examine the estimation problems associated with, some specific econometric models which provide such a framework.

The fundamental distinction we maintain throughout is between consumption and expenditure. The underlying rate of consumption of the good is denoted c_n, where $n = 1....N$ indexes the households in our sample. The variable c_n is a purely theoretical concept: it is the choice variable in a static utility maximisation problem solved by the consumer. It is to be interpreted as the average (weekly) rate of consumption that we would arrive at by observing the household over a very long period during which all external conditions (prices, income, family composition, etc.) remain unchanged. The form of c_n is irrelevant here: it may be measured in quantity, expenditure or budget share terms. The major problem of cross-section analysis is that we cannot observe households for long periods under unchanging conditions, so c_n, which is the variable we are trying to explain, is not observable.

What we do observe is a variable e_n, defined as the total expenditure on the good over some short observation period, divided by the number of weeks in that period. Again, e_n may be expressed in quantity, expenditure or budget share form.

Our aim in this paper is to suggest some simple statistical models for the Engel curve underlying c_n and some mechanisms relating the unobservable c_n to the observable e_n. We then use these to derive estimation techniques for the Engel curve.

The paper is organised as follows. Section 2 considers the problem of specifying statistical Engel curves on a single–equation basis, and suggests a classification of goods into types requiring different statistical models. Section 3 discusses the relationship between e_n and c_n, concentrating particularly on a generalisation of the P–Tobit model of Deaton and Irish (1984). Section 4 proposes a simple nonlinear least squares estimator which does not require the specification of a detailed model of purchasing behaviour; some preliminary results are also discussed. Section 5 introduces the problem of specifying and estimating a joint model of consumption and purchasing behaviour.

2 Modelling the Rate of Consumption

Wales and Woodland (1983) and Lee and Pitt (1984) discuss the problem of estimating a full system of demand equations from cross-section data, allowing for the non–consumption of certain goods. The appalling complexity of the statistical models that result is sufficient reason to abandon a full–system approach at this micro level, and to treat demand behaviour on a single–equation basis. As we shall argue below, different goods tend to have distinctive features which require special treatment, and this also makes a full–system specification rather restrictive. Our approach therefore, is a very simple one: we shall concentrate chiefly on the consumption – purchase problem, using convenient Engel curve models.

However, as a pedagogical device to motivate our choice of models for specific goods, it is helpful to interpret alternative statistical models for c_n in terms of a simple two–good utility–maximisation problem. I propose a classification of goods into the following four types.

Type 1: Everyone consumes

For some goods there can be no non–consumption. Everyone wears clothes, and eats food. For goods of this type, we know that all observed zeros must be purely fortuitous: a household may not buy clothes in a particular two–week observation period, but its members do not habitually go naked.

Thus, if we think of the Engel curve for c_n arising from a utility maximisation problem, we have

$$\max \ u(c,C;\theta_n,\epsilon_n) \tag{1}$$

subject to

$$p_n\, c + P_nC = y_n, \tag{2}$$

where C is consumption of a composite 'other goods' category, p_n and P_n are prices, y_n is exogenous total expenditure, θ_n represents observed demographic influences and ϵ_n represents random preference variation. For convenience, assume that the solution to this problem is an Engel curve linear in some transformations of p_n, P_n, y_n and θ_n:

$$c_n = \beta' x_n + \epsilon_n \tag{3}$$

where x_n is an observable vector constructed from p_n, P_n, y_n and θ_n and β is a vector of constant parameters.[vi]

For this class of goods, u(·) is such that the indifference curves never cut the C–axis, and a corner solution for c_n is impossible. Thus we must choose a distributional form for ϵ_n (or equivalently for $c_n|x_n$) which ensures that c_n is strictly positive with probability one. There are many possibilities, and I shall investigate two simple specifications.

(a) The truncated normal model

If ϵ_n has a $N(0, \sigma^2)$ distribution truncated from below at $-\beta' x_n$, then c_n is strictly positive and has a conditional p.d.f.:

$$\text{p.d.f. } (c_n|x_n) = \frac{\sigma^{-1}\phi\left[\frac{c_n-\beta' x_n}{\sigma}\right]}{\Phi\left[\frac{\beta' x_n}{\sigma}\right]}, \tag{4}$$

where $\phi(\cdot)$ and $\Phi(\cdot)$ are the p.d.f. and c.d.f. of the N(0,1) distribution. This has a mean function:

$$E(c_n|x_n) = \beta' x_n + \sigma\, \lambda^*(\beta' x_n/\sigma) \tag{5}$$

where $\lambda^*(\cdot) = \phi(\cdot)/\Phi(\cdot)$ is the complement of the inverse Mills' ratio.

(b) The lognormal model

If ϵ_n has a displaced lognormal distribution, with displacement parameter $-\beta' x_n$, then $c_n | x_n$ has a lognormal distribution, with p.d.f.:

$$\text{p.d.f.}(c_n | x_n) = \sigma^{-1} c_n^{-1} \phi \left[\frac{\log c_n - \beta' x_n}{\sigma} \right] \tag{6}$$

In this case, c_n has conditional mean function:

$$E(c_n | x_n) = \exp\{\beta' x_n + \sigma^2/2\} \tag{7}$$

$$= \exp\{\beta^{*\prime} x_n\}. \tag{8}$$

In (8), we have absorbed the term $\sigma^2/2$ into the intercept term in the linear form $\beta' x_n$. Thus β^* is identical to β except for its first element. Note that, since:

$$\log c_n | x_n \sim N(\beta' x_n, \sigma^2), \tag{9}$$

our interpretation of the Engel curve is somewhat different here: we have a logarithmic, rather than linear relationship.

It should be observed that neither the truncated normal nor the lognormal distribution is as flexible as we might like: although both are skewed to the left, neither incorporates a separate parameter controlling the degree of skewness. However, identification difficulties prevent the use of more heavily parameterised distributions.

Type 2: Some economic non–consumers

There are some goods which are not consumed by everybody at current prices. However, this non–consumption is economic in nature: if the price were reduced (or income increased) sufficiently, any non-consumer could be induced to become a consumer of the good. Most goods that are generally thought of as luxuries probably fall into this category: consumer durables, various types of entertainment, etc.[i]

For goods of this kind, the utility maximisation problem (1)–(2) must be solved subject to an additional non–negativity constraint:

$$c \geq 0. \tag{10}$$

For this two–good case, the Kuhn–Tucker conditions imply the following demand function:

$$c_n = \max \{\tilde{c}_n, 0\}, \tag{11}$$

where $\tilde{c}_n$ is the solution to (1)–(2) with no non–negativity constraint imposed. If we adopt a normal linear regression for $\tilde{c}_n$, then we have:

$$\tilde{c}_n | x_n \sim N(\beta' x_n, \sigma^2), \tag{12}$$

and the effect of (11) is to censor this normal distribution from below at zero. Thus (11)–(12) constitute the censored regression or Tobit model (Tobin (1958)), which underlies the most widely–used technique for coping with zero expenditures in cross–section work. The Tobit model implies the following mixed discrete–continuous distribution for c_n:

$$\Pr(c_n = 0 | x_n) = 1 - \Phi(\beta' x_n / \sigma) \tag{13}$$

$$\text{p.d.f.}(c_n | x_n) = \sigma^{-1} \phi\left[\frac{c_n - \beta' x_n}{\sigma}\right] \quad \text{for } c_n > 0. \tag{14}$$

This distribution has mean function:

$$E(c_n | x_n) = \beta' x_n \Phi(\beta' x_n / \sigma) + \sigma \phi(\beta' x_n / \sigma). \tag{15}$$

Under this interpretation of the Tobit model, use of the maximum likelihood Tobit estimator to estimate an Engel curve is only valid under two special assumptions: e_n and c_n are always identical, and non–consumption is always the result of a strictly economic decision.

Type 3: Conscientious abstention

Most vegetarians do not abstain from meat because it is too expensive, or because they are too poor to afford it. Similarly, a large reduction in the price of tobacco will induce very few non–smokers to adopt the habit. The same applies to many non–consumers of alcoholic drink.

In all of these cases, non–consumption is the result of a conscientious rather than economic decision: it reflects the fact that the population can be divided into distinct groups of abstainers and non–abstainers, characterised by essentially different preferences. Cragg (1971) proposed a statistical structure, known as the double–hurdle model, which can be interpreted as incorporating this distinction (although Cragg was not very specific about his own interpretation of the double–hurdle model, and as a result there is some ambiguity about the way it is specified in practice: see Atkinson, Gomulka and Stern (1984)).

Suppose that abstainers (non–smokers, say) have no use for the good c: if given a free supply of it, they will simply throw it away. Thus, their preferences are represented by a utility function of the form $u^*(C;\theta_n,\epsilon_n)$. Non–abstainers, however, have preferences representable by the full utility function (1), and we assume also that all non–abstainers consume a positive amount of the good: smokers cannot do without tobacco, no matter how expensive it becomes.

The simplest way of modelling the distinction between abstainers and non–abstainers is to use a probit mechanism, assumed independent of ϵ_n. Thus, define an unobservable indicator, v_n, which is such that:

$$v_n \,|\, \theta_n \sim N(\gamma' \theta_n, 1). \tag{16}$$

Then c_n is generated as follows:

If $v_n > 0$, household n is a consumer, and:

c_n is drawn from a conditional truncated normal (4) or lognormal distribution, (6)

If $v_n \leq 0$, household n is an abstainer, and:

$$c_n = 0.$$

The truncated normal and lognormal versions of this model are Cragg's (1971) models (9) and (11) respectively.

In each case, c_n has a mixed discrete–continuous distribution:

(a) Truncated normal case

$$\Pr(c_n = 0 | x_n) = \Pr(v_n \leq 0 | x_n)$$
$$= 1 - \Phi(\gamma' \theta_n) \tag{17}$$

$$\text{p.d.f.}\ (c_n | x_n) = \sigma^{-1}\phi\left[\frac{c_n - \beta' x_n}{\sigma}\right] \frac{\Phi(\gamma' \theta_n)}{\Phi(\beta' x_n / \sigma)} \quad \text{for } c_n > 0. \tag{18}$$

This has mean function:

$$E(c_n | x_n) = \Phi(\gamma' \theta_n)\ [\beta' x_n + \sigma\lambda^*(\beta' x_n / \sigma)]. \tag{19}$$

This is a rather complicated expression. However, note that if we can discard all households for whom $c_n = 0$, it simplifies to:

$$E(c_n | c_n > 0,\ x_n) = \beta' x_n + \sigma\ \lambda^*(\beta' x_n / \sigma) \tag{20}$$

(b) Lognormal case

$$\Pr(c_n = 0 | x_n) = 1 - \Phi(\gamma' \theta_n) \tag{21}$$

$$\text{p.d.f.}(c_n | x_n) = \sigma^{-1} c_n^{-1} \phi\left[\frac{\log\ c_n - \beta' x_n}{\sigma}\right] \frac{\Phi(\gamma' \theta_n)}{\Phi(\beta' x_n / \sigma)} \tag{22}$$

This has mean function:

$$E(c_n | x_n) = \Phi(\gamma' \theta_n)\ \exp\{\beta^{*\prime} x_n\}, \tag{23}$$

where $\beta^{*\prime}x_n$ is again $\beta' x_n + \sigma^2/2$. If it is possible to discard non-consumers, this becomes:

$$E(c_n | c_n > 0, x_n) = \exp\{\beta^{*\prime} x_n\}. \tag{24}$$

Type 4: Conscientious abstention and economic non–consumption

For some goods there may be a mixture of reasons for non-consumption: some households may be conscientious abstainers, and others may not consume because the good is currently too expensive. There is perhaps a case for classifying both tobacco and alcohol as type 4 rather than type 3 goods.

Again, a double–hurdle model is appropriate here, but with a Tobit mechanism generating zeros for some potential consumers. Thus, with $v_n | \theta_n$ distributed as $N(\gamma' \theta_n, 1)$:

If $v_n > 0$, household n is a potential consumer, and:

$$c_n = 0 \text{ if } \tilde{c}_n \leq 0$$
$$c_n = \tilde{c}_n \text{ if } \tilde{c}_n > 0, \text{ where } \tilde{c}_n | x_n \sim N(\beta' x_n, \sigma^2)$$

If $v_n \leq 0$, household n is an abstainer, and:

$$c_n = 0.$$

This leads to a distribution for c_n of the form:

$$\begin{aligned} \Pr(c_n = 0 | x_n) &= 1 - \Pr(c_n > 0 | x_n) \\ &= 1 - \Pr(v_n > 0 | z_n)\Pr(\tilde{c}_n > 0 | x_n) \\ &= 1 - \Phi(\gamma' \theta_n)\Phi(\beta' x_n/\sigma) \end{aligned} \tag{25}$$

$$\text{p.d.f. } (c_n | x_n) = \Phi(\gamma' \theta_n)\, \sigma^{-1}\phi\left[\frac{c_n - \beta' x_n}{\sigma}\right], \tag{26}$$

which has a mean function:

$$E(c_n | x_n) = \Phi(\gamma' \theta_n)\Phi(\beta' x_n/\sigma)[\beta' x_n + \sigma\lambda^*(\beta' x_n/\sigma)]. \tag{27}$$

In this case, there is no simplification to be gained by conditioning on the event $c_n > 0$.

3 The Relationship Between Purchases and Consumption

The theory of consumer demand is conducted in terms of smooth planned rates of flow of consumption services. What we observe in cross-section data is an aggregate, for each household, of a number of individual purchases of varying amounts over a short observation period. Our task is to provide a statistical mechanism relating these two quantities.

Our interpretation of the fundamental consumption variable, c_n, as a long–term average rate of expenditure, together with our assumption that c_n is predetermined when individual expenditures are made, implies the following fundamental identity[vii]

$$E(e_n | c_n, z_n) = c_n, \tag{28}$$

where z_n is a vector of (observable) variables relevant to the determination of purchasing frequency.[v,ix] Define $P(c_n, z_n)$ as follows:

$$P(c_n, z_n) = \Pr(\text{one or more purchases occur during the observation period} \,|\, c_n, z_n), \quad c_n > 0 \tag{29}$$

$$P(0, z_n) = 0 \tag{30}$$

Using this definition of $P(c_n, z_n)$, we have the following identity:

$$E(e_n | c_n, z_n) = 0 \qquad \text{for } c_n = 0 \tag{31}$$

$$= P(c_n, z_n) E(e_n | e_n > 0, c_n, z_n) \qquad \text{for } c_n > 0 \tag{32}$$

Equations (28), (31) and (32) imply:

$$E(e_n | e_n > 0, c_n, z_n) = 0 \qquad \text{for } c_n = 0 \tag{33}$$

$$= \left[\frac{c_n}{P(c_n, z_n)}\right] \qquad \text{for } c_n > 0. \tag{34}$$

Equations (33)–(34) constitute an important result, since they give a relationship between the mean function of observed expenditure and the unobserved variable to which our theory relates. All that is required to make this operational is a specification for the purchasing probability, $P(c_n,z_n)$.[viii]

A simple example will serve to clarify (33)–(34). Suppose a good is bought regularly once every four weeks, in a quantity q. If the survey lasts one week, there is a probability of observing a purchase, $P(c,z) = 1/4$. The corresponding underlying average rate of consumption is obviously $c = q/4$, and yet any observed positive expenditure will be of size $e = c/P = (q/4)/(1/4) = q$. Thus, the important implications of the distinction between expenditures and consumption are:

(i) observed zeros are not necessarily the result of non–consumption;

(ii) even when expenditure is positive, it is generally a poor measure of consumption.

Author	$P(c_n,z_n)$	Model for c_n	Relation between e_n and $c_n/P(c_n,z_n)$
Deaton and Irish(1984)	constant parameter	Tobit	equality
Kay, Keen and Morris (1984)	constant parameter	linear regression	additive error
Keen (1986)	constant parameter	linear regression	additive error
Blundell and Meghir (1987)	$\Phi(\alpha' z_n)$ where $z_n = x_n$	linear regression lognormal	additive normal error, multiplicative lognormal error

Table 1: Applied models based on (33)–(34),

Deaton and Irish (1984) appear to have been the first to appreciate the importance of identity (33)–(34) and to exploit it in the construction of an applied model. Since then, it has been used extensively. Table 1 classifies the existing applications.

There are two questions to be decided in specifying a model of expenditures: the nature of the conditional purchase probability, $P(c_n,z_n)$, and the relationship between e_n and $c_n/P(c_n,z_n)$ when e_n is positive.

3.1 The purchase probability

How should the function $P(c_n,z_n)$ be specified? As Table 1 shows, existing applications have taken a very simple approach. Deaton and Irish (1984), Kay, Keen and Morris (1984) and Keen (1986) all specify $P(c_n,z_n)$ as a constant parameter, to be estimated jointly with the parameters of the consumption model.[x]

This has not proved very successful. Deaton and Irish, working with a Tobit model for c_n, and using 1973–1974 Family Expenditure Survey (FPS) data, estimate this constant probability, P, as almost exactly unity for tobacco, (implying that their P–Tobit model is identical to a simple Tobit model) and they fail to achieve an estimate at all for two other goods: alcoholic drink and durables. In these latter two cases, the log-likelihood function turns out to be unbounded as P increases above unity, and a Lagrange Multiplier (LM) test of the restriction $P=1$ rejects it in favour of the absurd alternative hypothesis $P > 1$. Thus, Deaton and Irish are left in the unsatisfactory position of having a model that provides significant evidence against the conventional Tobit model, but which is untenable in itself. Our results for these three goods, based on 1983 FES data, confirm Deaton and Irish's conclusion (see Tables 4–11 below).

There are two obvious shortcomings of the Deaton–Irish model. One is that it is unreasonable to treat $P(c_n,z_n)$ as a fixed parameter, and the other is that the Tobit model for c_n classifies all three goods as type 2: with non–consumption possible, and then interpreted as arising from an "economic" corner solution.[xi]

So far, little progress has been made with these shortcomings in applied work. Blundell and Meghir (1987) do present estimates based on alternative models of c_n for a different good, clothing, and they estimate a model with $P(c_n,z_n)$, specified as a simple Probit form, with explanatory variables taken to be identical to those used in the Engel curve itself, x_n. However, this is still restrictive.[xii]

Depending on the good concerned, there are several explanatory variables that are likely to be important in modelling $P(c_n,z_n)$, including the following:

(i) Consumption It is obvious that the rate of consumption, c_n must play a role in determining the purchase probability. If, as in all previous work, $P(c_n,z_n)$ is taken to be independent of c_n, then this carries the absurd implication that we are just as likely to observe a purchase for a household that consumes hardly any of the good as for a heavy consumer. Given that most goods cannot be bought in arbitrarily small units, this is an impossibility. Note that we cannot simply include the determinants, x_n, of c_n rather than c_n itself, since this fails to take correct account of the random component of c_n.

(ii) Survey duration The longer a household is observed, the greater is our chance of observing a purchase (provided $c_n \neq 0$). For a given survey, duration is usually fixed, and this is then only a consideration if we are to make use of evidence from surveys based on observation periods of different lengths. The availability of such surveys would give a valuable source of evidence on the nature of purchasing behaviour.

(iii) Household Consumption Households with a relatively large number of adult members (particularly members who are unoccupied or retired) can be expected to make more purchases of any given good in a fixed period, simply because it has more collective opportunity to do so.

(iv) Durable ownership Households that own cars and freezers have an opportunity to exploit economics of scale in purchasing, since they are able to transport and store larger quantities of goods (foods in particular). These larger quantities imply less frequent purchases, and thus a smaller purchase probability, provided the survey duration is not larger than the re–stocking interval.

If we aim to build a full structural model of the observable, e_n, it is necessary to choose a convenient functional form for P(c,z). The main restrictions on choice are:

(i) $0 \leq P(c,z) \leq 1$ all $c \geq 0$, z

(ii) $\lim_{c \to 0} P(c,z) = 0$ all z

(iii) P(c,z) is monotonically increasing in c, for all z

(iv) $\lim_{c \to \infty} P(c,z) = 1$ all z

Properties (i) and (ii) are uncontentious. Strictly speaking, there is no necessity for property (iii) to hold: we can envisage situations in which an increase in consumption leads to a switch to a new mode of behaviour involving storage of goods and less frequent purchasing. However, most such counter–examples are unconvincing or require a concomitant change in other relevant variables, z. Property (iv) is not particularly compelling, since it implies that very large consumers devote all of their time to making purchases, rather than buying in suitable bulk. However, property (iv) is probably innocuous as an approximation, and we retain it for convenience. It could be relaxed without difficulty.

There are many plausible choices for P(c,z). We could, for instance, use any convenient c.d.f. such as the normal, with c entered in log form:

$$P(c,z) = \Phi(\alpha_1 \log c + \alpha_2' z). \tag{35}$$

This is similar in spirit, but not in its implications, to the model of Blundell and Meghir (1987), but is rather a complicated form to use, when $\alpha_1 \neq 0$. Slightly more convenient for computational purposes is the form:

$$P(c,z) = \frac{c}{\alpha' z + c}, \tag{36}$$

which has the implication that $E(e_n | c_n, e_n > 0, z_n)$ differs from c_n by a constant quantity, $\alpha' z_n$. Pudney (1985) discussed a number of more elaborate specifications.

3.2 The relation between e_n and $c_n/P(c_n,z_n)$

Behaviour is rarely regular. Even if the underlying rate of consumption is unchanged, successive purchases are likely to be of different sizes, unless there is some technical reason, such as indivisibility, dictating otherwise. Thus, it seems wise to assume that e_n is not exactly equal to $c_n/P(c_n,z_n)$, but fluctuates randomly around this value. Since $c_n/P(c_n,z_n)$ is the mean of e_n conditional on the event $e_n > 0$, we must specify a distribution for $e_n | e_n > 0$, c_n, z_n with support only on the positive real line. The practice of adding a regression–style normal disturbance (see Kay, Keen and Morris (1984) and Keen (1986)), is not therefore strictly valid.

Again, there are many convenient possibilities for the distribution of $e_n | e_n > 0$, c_n, z_n; we might use a suitably truncated normal, or alternatively a lognormal distribution parameterised to have mean $c_n/P(c_n,z_n)$, implying that log e_n has a conditional $N(\mu_n, \sigma_e^2)$ distribution where $\mu_n = \log\{c_n/P(c_n,z_n)\} - \sigma_e^2/2$.

An example of a fully–specified model, involving both a consumption-dependent purchase probability and a random element in observed expenditure, is examined in section 5 of the paper.

4 Nonlinear Least–Squares Estimation

Before specifying a particular generalisation of the Deaton–Irish model in section 5, we consider the possibility of constructing a consistent (albeit inefficient) estimator which is limited information in the sense that it does not require any specific assumption about the nature of $P_n = P(c_n,z_n)$. That this is possible is a remarkable feature of the P–Tobit framework.

4.1 The estimator

Consider the expected value of e_n, conditional on c_n and z_n:

$$E(e_n | c_n,z_n) = P(c_n,z_n)E(e_n | e_n > 0, c_n,z_n) + [1 - P(c_n,z_n)] \times 0 = c_n, \tag{37}$$

using (34). Thus the conditional expectation of e_n is independent of the nature of P_n, and hence any technique founded on the expected value of e_n requires no assumptions about the purchase probability.

Since c_n is unobserved, take the further expectation with respect to c_n:

$$\begin{aligned} E(e_n | x_n,z_n) &= E(c_n | x_n,z_n) \\ &= E(c_n | x_n), \end{aligned} \tag{38}$$

since z_n is not involved in the determination of consumption. Equality (38) can be written as a regression equation:

$$e_n = g(\beta' x_n,\sigma) + \nu_n, \tag{39}$$

where $\nu_n = e_n - E(e_n | x_n)$ is a disturbance term with zero mean conditional on x_n, and where $g(\beta' x_n,\sigma)$ is the mean function of c_n. Depending on the good concerned, we might use the expressions (5), (8), (15), (19), (23), or (27) as this mean function.[xiii]

Equation (39) is nonlinear in the parameters β and σ; any further parameters that might be involved in the function $P(c_n,z_n)$ or the p.d.f. of $e_n | c_n/P(c_n,z_n)$ do not appear in (39) and hence cannot be estimated.

However, β and σ determine the Engel curve completely, and we are often not particularly interested in purchasing patterns, so this may not be a serious drawback. More serious is the possibly severe nonlinearity of $g(\beta' x_n, \sigma)$ and the fact that nonlinear least–squares estimation is generally inefficient. There may also be difficulty in identifying both β and σ from the mean of e_n alone.

A further potential difficulty is that the random errors, ν_n, are heteroscedastic, with the precise form of the heteroscedasticity dependent upon the nature of $P(c_n, z_n)$. Therefore, to preserve the limited-information spirit of this approach, we must ignore the heteroscedasticity problem during estimation, but take proper account of it when computing the asymptotic covariance matrix.

Thus, the nonlinear least–squares estimator we propose solves the following problem:

$$\min_{\beta,\sigma} \sum_{n=1}^{N} [e_n - g(\beta' x_n, \sigma)]^2 \tag{40}$$

where N is the sample size. The asymptotic covariance matrix of this estimator can be consistently$_{n=1}$ estimated in the presence of heteroscedasticity of unknown form by the following expression:

$$\text{a. cov}\,(\hat{\beta}, \hat{\sigma}) = \hat{A}^{-1}\,\hat{B}\,\hat{A}^{-1} \tag{41}$$

where:

n=1

$$\hat{A} = \left[\sum_{n=1}^{N} \hat{g}_n^{\delta}\, \hat{g}_n^{\delta\prime}\right] \tag{42}$$

$$\hat{B} = \left[\sum_{n=1}^{N} \hat{\nu}_n^2\, \hat{g}_n^{\delta}\, \hat{g}_n^{\delta\prime}\right], \tag{43}$$

where $\hat{g}_n^{\delta}$ is the vector of derivatives of $g(\hat{\beta}'x_n,\hat{\sigma})$, with respect to $\hat{\delta}' = (\hat{\beta}',\hat{\sigma})$ and $\hat{\nu}_n$ is $e_n - g(\hat{\beta}'x_n, \hat{\sigma})$.

4.2 Nonlinear least squares results

Our applied work uses data from the 1983 UK Family Expenditure Survey, which is based on a two–week observation period.[iii] The full sample available for analysis comprises 6973 households. We have chosen four categories of expenditure for study: tobacco, alcohol, durables and clothing. Our dependent variables, e_n and c_n, are defined as budget shares; Table 2 gives a summary of the properties of e_n. To maintain comparability with the work of Deaton and Irish (1984), we have used the same list of explanatory variables (the elements of x_n), which appear in Table 3.[ii]

(i) Tobacco

For most people, the choice between smoking and non–smoking appears not to be an economic decision: it concerns mainly considerations of health and social opprobrium.[xiv] Thus, as an approximation, it seems reasonable to classify tobacco as a type 3 commodity, for which all non-consumption is conscientious abstention. If we are to apply the nonlinear least squares estimator to the full sample, we must fit either expression (14) (the truncated normal model) or (23) (the lognormal model) to the data. However, it is extremely unlikely that we would be able to obtain good estimates of γ, β and σ in this way, since many variables would appear in both θ_n and x_n.

However, tobacco has one further special feature: it is customarily bought through frequent, small purchase. Thus, except for the few people with very small but positive rates of consumption, observed expenditure and underlying consumption are likely to be almost identical. This approximation has two implications.

The first is that the double–hurdle structure assumed to generate c_n can be fitted directly to expenditure data by maximum likelihood techniques. This was done with some success by Atkinson, Gomulka and

Stern (1984), who, however, classified tobacco as a type 4 good, using a Tobit model for the demands of non–abstainers. We do not repeat their analysis here, since we are concentrating on the simpler nonlinear least-squares techniques.

A second implication of the assumed equality of expenditure and consumption is the $E(e_n | e_n > 0, x_n)$ and $E(c_n | c_n > 0, x_n)$ are identical. The latter expectation is given by expressions (20) and (24) for the truncated normal and lognormal models respectively. These are much simpler than the unconditional regression functions (14) and (23), since γ does not appear. Thus, it is possible to estimate the demand parameters of non-abstainers, β and σ, by fitting a less heavily–parameterised regression equation, provided we truncate the sample by excluding all zero observations. Since this estimator does not require any assumption about the probability of abstention, it is robust against misspecification of this probability, and could be used in a Hausman specification test of the double-hurdle model. Note that abstention can be modelled separately by the Probit technique.

Tables 4 and 5 summarise four different estimates of the demand function for tobacco. These are the Tobit estimate, and three regressions computed from the truncated sample: a linear model and the two non-linear models (20) and (24).

We have also investigated a model based on the interpretation of tobacco as a type 2 good, with non–consumption arising from a corner solution to the utility maximisation problem. This is based on the non-linear regression (15), fitted to the full sample. However, it proved impossible to obtain convergence of the iterative minimisation algorithm applied to the residual sum of squares, and, for this sample, β and σ appear to be unidentifiable by nonlinear least squares. Similar problems were encountered for this model when applied to the Engel curve for alcohol, so this seems likely to be a common problem. Wales and Woodland (1980), simulating the same estimator in a rather different context, also find the technique unreliable.

At first sight, the Tobit results appear very good. The Deaton and Irish LM test against the P–Tobit alternative, distributed as $\chi^2(1)$ under the Tobit null hypothesis, fails to reject. Moreover, the Tobit model's prediction of the proportion of zero observations is very accurate. This prediction is defined as:

$$N^{-1} \sum_{n=1}^{N} [1 - \Phi(\hat{\beta}' x_n / \hat{\sigma})], \tag{44}$$

and is almost exactly equal to the observed proportion of 47.6%.

These predicted zeros must be interpreted as corner solutions, since the estimated "potential" demand, $\hat{\beta}' x_n$, is negative in 46.2% of the sample. The zeros do not arise as an incidental consequence of the stochastic specification and this is the first indication of misspecification, since it is difficult to believe that nearly half the population are non-smokers purely because tobacco is too expensive.

A conclusive rejection of the Tobit model emerges if we compare the Tobit coefficients with those of a simple Probit analysis, which models only the distinction between purchasing and non–purchasing. The Probit coefficients (not reported here) differ significantly from the Tobit coefficients for the variables #AD, #2–5, NOKIDS, CLERK, CAR and 2CARS. These differences suggest that the mechanisms determining the smoking/non-smoking distinction and the quantity purchased are essentially different. A double–hurdle model is thus to be preferred.

The second estimate reported in Tables 4 and 5 is a simple linear regression fitted to the truncated sample. This model implies that consumption may be negative for some non–abstainers, but has the virtue of simplicity. The implied estimate of the proportion of consumers with negative c_n is:

$$\frac{1}{N^+} \sum_{e_n > 0} [1 - \Phi\left[\frac{\hat{\beta}' x_n}{\hat{\sigma}}\right]], \tag{45}$$

where N^+ is the number of households with positive expenditure. This is computed as 5.1% which is probably not large enough to be worrying. Note that these implied negative consumption levels are almost entirely due to the stochastic specification rather than the location of the Engel curve, since very few of the $\hat{\beta}' x_n$ are negative.

The last two sets of results are nonlinear least–squares estimates of the truncated normal and lognormal models. Both fit better than the linear regression, although the heteroscedasticity implied by our model means that the residual sums of squares are not strictly comparable. The truncated normal model displays rather large coefficient standard errors, and the additional parameter σ thus imposes a considerable cost in terms of estimation precision. A clear implication of Table 4 considerably over-estimates the income elasticity of demand for tobacco. The least–squares estimates are unanimous in this conclusion, and they also agree on the general nature of the Engel curve: a significant negative relation between the budget share and log total expenditure.

A final point that is worth making is that there is some support for our use of the truncated sample in estimating these regressions. In contrast with the other three goods studied here, the inclusion of zero observations causes the fit to deteriorate markedly: for the linear regression, for instance, R^2 falls from .236 to .159. Thus a double–hurdle structure seems to be strongly indicated by these results.

(ii) Alcohol

Alcohol is a difficult good to classify. There is certainly some conscientious abstention, and possibly also some economic non–consumption. Moreover, it is a storable good, often bought in units that are large relative to typical rates of consumption: a bottle of whisky may last for months in some households.[iv] This suggests that we classify alcohol as a type 4 good, and attempt to fit a regression of the form (27).

It has proved impossible to compute least–squares estimates of β, γ and σ in this model. Moreover, an attempt to interpret alcohol as a type 2 good and estimate a regression of the form (15) also failed.

Instead, Tables 6 and 7 compare the Tobit estimate with least-squares estimates computed from the truncated sample. These are based on a rather implausible interpretation of alcohol as a type 3 good, with conscientious abstention as the only source of non–consumption, and an assumption that e_n and c_n are always identical. There is little support for these assumptions: for instance, a linear regression on the full sample yields a better fit (R^2 = .135) than the truncated sample regression (R^2 = .119).

Moreover, convergence problems prevented the computation of least–squares estimates of the truncated normal model, so there is little encouragement in these results.

The only definite conclusion we can draw is that the Tobit technique is completely inadequate: it is rejected against the Deaton and Irish P-Tobit model, and it over–predicts the number of zero observations. If we have any confidence in the least–squares estimates, there is also evidence that the Tobit Engel elasticity is too high.

(iii) Durables

The durables category is so broadly defined that it must be interpreted as a type 1 good: everyone consumes some durable goods. This accounts for the surprisingly small frequency of zero budget shares, and the huge range of variation of positive shares. This variability, arising from purely fortuitous deviations of purchases from consumption, means that all our regressions display a rather poor fit.

Again, the Tobit model is strongly rejected: the LM test rejects it against the P–Tobit model, and it grossly over–predicts the number of zero observations.

A simple linear regression fitted to the full sample also fares rather badly: the two nonlinear least–squares estimates fit the data considerably better, and the implied proportion of negative mean consumption levels is nearly 3%.

As in previous cases, there was considerable difficulty in estimating the truncated normal model, with β and σ proving almost impossible to identify separately. As a result, convergence was difficult to obtain, and the estimated coefficients are rather wild, with very large standard errors.

Thus, the lognormal model is the only one for which reasonable results are available, and, if we believe this model, it suggests that the Tobit and linear regression models yield a substantial underestimate of the income elasticity.

(iv) Clothing

Everyone wears clothes, and therefore this category of expenditure must clearly be interpreted as a type 1 good. It also displays a substantial number of zero expenditures in the two–week survey period, and is thus a more interesting modelling problem than the durables category.

Again, the Tobit results are very poor: the Deaton–Irish LM test rejects the model, and there is considerable over–prediction of zeros. A full 10% of households display negative values of $\hat{\beta}' x_n$, and thus the model implies that 10% of the population habitually go naked because clothing is too expensive.

Of the three least–squares estimates, the truncated normal model seems clearly preferable in terms of fit, and produces acceptably precise estimates in this case. The lognormal model fits the data rather poorly in comparison, but agrees quite closely on the shape of the Engel curve. The linear regression seems an adequate approximation in terms of it, but surprisingly yields a considerably lower income elasticity.

On the basis of these very sparse preliminary results, a generalised P-Tobit structure based on a truncated normal model for c_n, seems to be indicated.

5 A Fully–Specified Model

Although the nonlinear least–squares approach of Section 4 is simple and undemanding in terms of model specification, it is inefficient and frequently suffers from severe identification problems. It is therefore of some interest to develop the efficient maximum likelihood estimator for a fully specified model.

Denote the p.d.f. of $c_n|x_n$ by $f_c(c_n|x_n)$, and the p.d.f. of $e_n|e_n > 0$, c_n,z_n by $f_e(e_n|c_n/P(c_n,z_n))$. Then the mixed discrete–continuous distribution for $e_n|x_n,z_n$ is:

$$\Pr(e_n = 0|x_n,z_n) = 1 - \int_0^\infty P(c,z_n)f_c(c|x_n)dc, \tag{46}$$

$$\text{pdf}(e_n|x_n,z_n) = \int_0^\infty P(c,z_n)f_e(e_n|c/P(c,z_n))f_c(c|x_n)dc, \quad e_n > 0 \tag{47}$$

This is a formidably complicated distribution, since, in general, the integrals in (46) and (47) will not be expressible in closed form. As an example, consider the case of type 1 good, with $c_n|x_n$ assumed to have a lognormal $(\beta' x_n,\sigma^2)$ distribution, $e_n|c_n,z_n$ assumed to have a lognormal $([c_n/P(c_n,z_n) - \sigma_e^2/2], \sigma_e^2)$ distribution, and $P(c_n,z_n)$ assumed to be of the form (36). For these relatively simple assumptions, the distribution (46) – (47) is:

$$\Pr(e_n = 0|x_n,z_n) = \frac{\alpha' z_n}{\sigma}\int_0^\infty \frac{1}{c(\alpha' z_n+c)}\,\phi\left[\frac{\log c - \beta' x_n}{\sigma}\right]dc, \tag{48}$$

$$\text{p.d.f.}(e_n|x_n,z_n) = \frac{1}{e_n\sigma\sigma_e}\int_0^\infty \frac{1}{\alpha' z_n+c}$$

$$\phi\left[\frac{\log e_n - \log(\alpha' z_n+c)+\sigma_e^2/2}{\sigma_e}\right]\phi\left[\frac{\log c - \beta' x_n}{\sigma}\right]dc,\ e_n > 0. \tag{49}$$

These expressions can be rewritten:

$$\Pr(e_n = 0 | x_n, z_n) = \int_{-\infty}^{\infty} \psi_{n1}(\zeta) e^{-\zeta^2/2} d\zeta \tag{50}$$

$$\text{p.d.f.}\ (e_n | x_n, z_n) = (e_n \sigma_e)^{-1} \int_{-\infty}^{\infty} \psi_{n2}(\zeta) e^{-\zeta^2/2}\, d\zeta \tag{51}$$

where:

$$\psi_{n1}(\zeta) = \alpha' z_n (\alpha' z_n + c_n(\zeta))^{-1} \tag{52}$$

$$\psi_{n2}(\zeta) = c_n(\zeta)(\alpha' z_n + c_n(\zeta))^{-1} \phi\left[\frac{\log e_n - \log(\alpha' z_n + c_n(\zeta)) + \sigma_e^2/2}{\sigma_e}\right] \tag{53}$$

and:

$$c_n(\zeta) = \exp\{\sigma\zeta + \beta' x_n\}. \tag{54}$$

In this form, these quantities are more suitable for computation, since efficient Gauss–Hermite quadrature algorithms can be used (see Waldman (1985) for an evaluation of Guassian quadrature in a sightly different context).

The corresponding log–likelihood function has the general form.

$$\log L(\alpha, \beta, \sigma, \sigma_e) = \sum_{n: e_n = 0} \log \Pr(e_n = 0 | x_n, z_n) + \sum_{n: e_n > 0} \log \text{p.d.f.}\ (e_n | x_n, z_n). \tag{55}$$

The evaluation of this log–likelihood requires the calculation of N integrals by numerical quadrature; the evaluation of its gradient can be shown by differentiation of (55) to require the calculation of between 3N and 5N still more complicated integrals.

Thus, a model of this type, which recognises the dependence of the purchase probability on the rate of consumption, presents much more substantial computational problems than the simpler models that have appeared in the existing literature.

Work is still proceeding on this approach to estimation. Preliminary trails have established the computation of the log–likelihood (55) and its gradient vector is feasible. However, parameterised in this way, $\log L(\alpha,\beta,\sigma,\sigma_e)$ is poorly scaled and rather difficult to maximise. Alternative parameterisations are being investigated.

For the time being, we consider only a single set of estimates, relating to the clothing category. This model is of the form (48)–(49), but with the linear form $\alpha' z_n$ specified as a simple constant. Thus the purchase probability is taken to depend only on c_n.[xvi]

The model was estimated from a 10% random sample of 697 observations. The parameter estimates, which are presented in Table 12, are broadly similar to the least–squares estimates of the lognormal model in Table 11, and appear quite reasonable. The model under–predicts the proportion of zeros somewhat, since $N^{-1}\Sigma \Pr(e_n = 0 | x_n, z_n)$, evaluated at the ML estimates is 15.5%, compared with an actual 22.8%. The estimated mean of c_n is 0.0817, at which value $P(c, z_n)$ is 0.896. The estimated income elasticity seems plausible, taking the value 1.723 at the sample mean of LNY.

Table 2
Dependent Variables
(6973 households from the 1983 Family Expenditure Survey)

	Mean budget share (full sample)	Mean budget share (truncated sample*)	Mean log per capita expenditure (truncated sample)	Percentage of zero expenditures	Minimum positive budget share	Maximum budget share
Tobacco	.034	.065	7.707	47.6	.0003	.430
Alcohol	.045	.062	7.614	27.7	.0006	.511
Durables	.053	.059	7.154	9.2	.0004	.794
Clothing	.061	.078	7.750	22.7	.0001	.562

*Sample truncation means that zero observations are discarded

Table 3

Explanatory Variables

Variable	Symbol	Mean
log per capita household expenditure	LNY	10.83
log per capita household expenditure squared	LNY2	117.58
dummy: household has no workers	NWRK	0.27
number of adults	AD	1.92
number of children under 2 years	<2	0.08
number of children over 2 and under 5	2-5	0.12
number of children over 5	>5	0.53
dummy: head of household aged 25-35	AGE 25	0.18
head of household aged 35-45	AGE 35	0.19
head of household aged 45-60	AGE 45	0.23
head of household aged 60-65	AGE 60	0.09
head of household aged 65 and over	AGE 65	0.25
Sex of head of household: 1 = male; 2 = female	SEX	0.23
dummy: household has no children	NOKIDS	0.61
dummy: household consists of a single adult	SINGLE	0.23
dummy: occupation is professional class	PROF	0.22
occupation is clerical	CLERK	0.06
occupation is shop assistant	SHOP	0.01
occupation is armed forces	FORCES	0.01
retired or unoccupied	UNOCC	0.38
dummy: household has one or more cars	CAR	0.62
dummy: household has two or more cars	2 CARS	0.17
dummy: household has telephone	TEL	0.77
dummy: household is owner occupier	OWNER	0.59
dummy: household is in Yorkshire	YORKS	0.10
household is in E. Midlands	EMID	0.07
household is in E. Anglia	ANGLIA	0.04
household is in Greater London	LONDON	0.11
household is in South East	SE	0.18
household is in South West	SW	0.07
household is in Wales	WALES	0.05
household is in W. Midlands	WMID	0.10
household is in North West	NW	0.12
household is in Scotland	SCOT	0.09
household is in N. Ireland	ULSTER	0.02

Table 4
Summaries of Tobit and Least–squares Results for Tobacco (47.6% zeros)

	Estimated income elasticity	% of observations in the full sample with $\hat{\beta}' x_n < 0$	$\hat{\sigma}$	Residual sum of squares	R^2
Tobit (LM test against P - Tobit: $\chi_1^2 = 0.001$, predicted zeros = 47.6%)	0.93[a] (.040)	46.2	0.072	-	-
Linear Regression (truncated sample)	0.53[a] (.031)	0.2	0.044	6.8700[c]	.236[c]
Nonlinear Least-squares: truncated normal model (truncated sample)	0.41[a] (.087)	3.0	0.031	6.8435[c]	.239[c]
Nonlinear Least-squares: lognormal model (truncated sample)	0.51[b] (.030)	-	-	6.8435[c]	.239[c]

Notes: (a) Elasticity calculated as $1 + (\hat{\beta}_1 + 2\hat{\beta}_2\overline{y})/\overline{w}$, where $\hat{\beta}_1$ and $\hat{\beta}_2$ are the coefficients of LNY and LNY2, $\overline{y}$ is the mean of LNY in the full sample and $\overline{w}$ is the mean budget share truncated sample; (b) Elasticity calculated as $1 + \hat{\beta}_1 + 2\hat{\beta}_{2\bar{y}}$; (c) Calculated from the truncated sample.

Table 5
Tobin and Least–Squares Coefficient Estimates for Tobacco

	ML Tobit	Linear Regression (truncated sample)	Nonlinear Least-squares: normal model (truncated sample)	Nonlinear Least-squares: lognormal model (truncated sample)
Constant	- .141 (.283)	.424 (.216)	- .321 (.415)	-12.25 (4.84)
LNY	.041 (.052)	- .032 (.039)	.113 (.079)	2.286 (.888)
LNY2	- .003 (.002)	.000 (.002)	- .007 (.004)	- .128 (.041)
NWRK	- .022 (.004)	- .002 (.003)	- .002 (.004)	- .013 (.046)
AD	.007 (.002)	- .002 (.001)	- .003 (.002)	- .030 (.020)
<2	- .005 (.004)	- .005 (.003)	- .004 (.004)	- .033 (.046)
2-5	- .010 (.003)	- .010 (.002)	- .011 (.003)	- .123 (.037)
>5	- .004 (.002)	- .008 (.001)	- .009 (.002)	- .111 (.020)
AGE 25	.020 (.005)	.010 (.004)	.013 (.004)	.152 (.057)
AGE 35	.025 (.005)	.017 (.004)	.019 (.005)	.235 (.061)
AGE 45	.021 (.005)	.013 (.004)	.015 (.005)	.158 (.059)
AGE 60	.020 (.006)	.013 (.004)	.015 (.005)	.182 (.070)
AGE 65	- .002 (.005)	.007 (.004)	.007 (.005)	.087 (.066)
SEX	- .021 (.003)	- .006 (.002)	- .007 (.003)	- .088 (.038)
NOKIDS	.000 (.004)	.006 (.003)	.008 (.004)	.112 (.047)
SINGLE	- .020 (.004)	.010 (.003)	.011 (.004)	.131 (.045)
PROF	- .025 (.003)	- .005 (.002)	- .006 (.003)	- .085 (.036)
CLERK	- .013 (.004)	- .001 (.003)	- .001 (.004)	- .016 (.055)
SHOP	- .007 (.010)	- .009 (.007)	- .012 (.009)	- .160 (.120)
FORCES	- .020 (.019)	- .021 (.009)	- .032 (.012)	- .457 (.143)
UNOCC	.002 (.004)	- .004 (.003)	- .005 (.003)	- .087 (.041)
CAR	- .022 (.003)	- .013 (.002)	- .014 (.003)	- .180 (.030)
2 CARS	- .007 (.003)	.000 (.002)	- .001 (.003)	- .032 (.035)
TEL	- .018 (.002)	- .008 (.002)	- .009 (.003)	- .097 (.031)
OWNER	- .025 (.002)	- .011 (.002)	- .013 (.003)	- .165 (.028)
YORKS	- .003 (.005)	- .001 (.004)	.000 (.004)	- .004 (.058)
EMID	- .017 (.005)	- .010 (.004)	- .012 (.005)	- .163 (.065)
ANGLIA	- .016 (.006)	- .012 (.005)	- .015 (.007)	- .192 (.079)
LONDON	- .011 (.005)	- .007 (.004)	- .008 (.005)	- .098 (.061)
SE	- .014 (.004)	- .007 (.003)	- .008 (.004)	- .111 (.056)
SW	- .011 (.005)	- .005 (.004)	- .005 (.005)	- .048 (.067)
WALES	.006 (.006)	.003 (.004)	.004 (.005)	.064 (.067)
WMID	- .011 (.005)	- .004 (.004)	- .003 (.005)	- .028 (.060)
NW	.000 (.005)	.002 (.003)	.003 (.004)	.041 (.057)
SCOT	.009 (.005)	.009 (.004)	.010 (.005)	.115 (.061)
ULSTER	- .002 (.007)	.006 (.006)	.009 (.007)	.132 (.091)

Standard errors in parenthesis.

Table 6
Summaries of Tobit and Least–squares Results for Alcohol (27.7% zeros)

	Estimated income elasticity[a]	% of observations in the full sample with $\hat{\beta}'x_n < 0$	$\hat{\sigma}$	Residual sum of squares	R^2
Tobit (LM test against P - Tobit: χ_1^2 = 231.7, predicted zeros = 34.7%)	1.589 (.040)	19.6	0.069	-	-
Linear Regression (truncated sample)	1.26 (.038)	0.3	0.057	16.461[c]	.119[c]
Nonlinear Least-squares: truncated normal model (truncated sample)[b]	-	-	-	16.245[c]	.130[c]
Nonlinear Least-squares: lognormal model (truncated sample)	1.17 (.037)	-	-	16.269[c]	.129[c]

Notes: (a) see notes (a) and (b) of Table 4, (b) Non-convergence, (c) calculated from the truncated sample.

Table 7
Tobit and Least–Squares Coefficient Estimates for Alcohol

	ML Tobit	Linear Regression (truncated sample)	Nonlinear Least-squares: lognormal model (truncated sample)
Constant	-2.381 (.270)	-1.469 (.228)	-32.24 (4.02)
LNY	.415 (.049)	.276 (.041)	5.306 (.728)
LNY2	- .017 (.002)	- .012 (.002)	- .237 (.033)
NWRK	- .024 (.003)	- .012 (.004)	- .116 (.057)
AD	.019 (.001)	.011 (.001)	.150 (.020)
<2	.000 (.004)	- .005 (.003)	- .117 (.054)
2-5	- .003 (.003)	- .007 (.003)	- .136 (.042)
>5	.000 (.002)	- .004 (.002)	- .086 (.024)
AGE 25	.003 (.004)	.012 (.004)	.221 (.065)
AGE 35	- .007 (.005)	.010 (.004)	.195 (.070)
AGE 45	- .019 (.004)	.005 (.004)	.134 (.064)
AGE 60	- .025 (.005)	- .001 (.005)	.039 (.072)
AGE 65	- .024 (.005)	.002 (.005)	.077 (.072)
SEX	- .035 (.002)	- .002 (.003)	- .443 (.044)
NOKIDS	.007 (.003)	.007 (.003)	.071 (.051)
SINGLE	.005 (.003)	.021 (.003)	.328 (.049)
PROF	- .010 (.003)	- .008 (.002)	- .122 (.036)
CLERK	- .006 (.004)	- .008 (.003)	- .111 (.063)
SHOP	- .012 (.010)	- .009 (.009)	- .126 (.111)
FORCES	- .023 (.017)	- .024 (.010)	- .474 (.127)
UNOCC	- .004 (.003)	- .002 (.003)	- .050 (.049)
CAR	- .013 (.002)	- .013 (.002)	- .214 (.034)
2 CARS	- .007 (.003)	- .006 (.002)	- .110 (.035)
TEL	- .016 (.002)	- .018 (.002)	- .229 (.033)
OWNER	- .011 (.002)	- .012 (.002)	- .187 (.030)
YORKS	- .006 (.004)	.000 (.004)	- .001 (.060)
EMID	- .010 (.005)	- .004 (.004)	- .034 (.068)
ANGLIA	- .023 (.006)	- .015 (.005)	- .282 (.091)
LONDON	- .022 (.004)	- .010 (.004)	- .136 (.062)
SE	- .019 (.004)	- .009 (.004)	- .127 (.059)
SW	- .019 (.005)	- .009 (.005)	- .122 (.068)
WALES	- .008 (.005)	- .002 (.005)	- .008 (.071)
WMID	- .012 (.004)	- .006 (.004)	- .067 (.063)
NW	- .007 (.004)	- .001 (.004)	- .015 (.059)
SCOT	- .016 (.004)	- .003 (.004)	- .014 (.065)
ULSTER	- .036 (.007)	- .008 (.008)	- .058 (.113)

Table 8
Summaries of Tobit and Least–squares Results for Durables (9.2% zeros)

	Estimated income elasticity[a]	% of observations with $\hat{\beta}'x_n < 0$	$\hat{\sigma}$	Residual sum of squares	R^2
Tobit (LM test against P - Tobit $\chi_1^2 = 2061$, predicted zeros = 31.0%)	2.18 (.056)	8.7	0.088	-	-
Linear Regression (full sample)	2.12 (.045)	2.9	0.083	48.020	.119
Nonlinear Least-squares: truncated normal model (full sample)	5.94 (1.94)	81.4	0.088	46.879	.139
Nonlinear least-squares: lognormal model (full sample)	2.35 (.097)	-	-	46.938	.138

Note: (a) see notes (a) and (b) of Table 4.

Table 9
Tobit and Least–squares Coefficient Estimates for Durables

	ML Tobit	Linear Regression (full sample)	Nonlinear Least-squares: truncated normal model (full sample)	Nonlinear Least-squares: lognormal model (full sample)
Constant	- .221 (.246)	.322 (.270)	-9.49 (4.64)	-44.28 (7.84)
LNY	- .019 (.043)	- .107 (.049)	1.461 (.734)	6.327 (1.38)
LNY2	.004 (.002)	.008 (.002)	- .054 (.029)	- .230 (.061)
NWRK	.010 (.005)	.009 (.004)	.043 (.033)	.200 (.128)
AD	.008 (.002)	.006 (.002)	.004 (.008)	.026 (.037)
<2	.017 (.005)	.014 (.004)	.067 (.034)	.318 (.101)
2-5	.013 (.004)	.010 (.003)	.055 (.025)	.259 (.071)
>5	.010 (.002)	.007 (.002)	.046 (.020)	.225 (.045)
AGE 25	.011 (.005)	.008 (.005)	.051 (.034)	.215 (.130)
AGE 35	.003 (.006)	.001 (.005)	.019 (.031)	.045 (.143)
AGE 45	- .005 (.005)	- .007 (.005)	- .005 (.029)	- .055 (.138)
AGE 60	- .013 (.006)	- .016 (.006)	- .065 (.045)	- .337 (.175)
AGE 65	- .009 (.006)	- .013 (.006)	- .064 (.046)	- .324 (.158)
SEX	.000 (.004)	- .002 (.003)	- .032 (.023)	- .160 (.088)
NOKIDS	- .007 (.005)	- .008 (.004)	.005 (.019)	.005 (.090)
SINGLE	- .032 (.004)	- .024 (.004)	- .115 (.048)	- .574 (.110)
PROF	- .005 (.003)	- .004 (.003)	- .014 (.015)	- .075 (.069)
CLERK	- .002 (.005)	- .001 (.004)	- .001 (.022)	.028 (.105)
SHOP	.002 (.013)	.005 (.011)	.040 (.053)	.178 (.239)
FORCES	.038 (.011)	.037 (.014)	.059 (.044)	.350 (.201)
UNOCC	- .022 (.005)	.001 (.004)	.030 (.026)	.125 (.102)
CAR	- .017 (.003)	- .017 (.003)	- .071 (.030)	- .348 (.070)
2 CARS	- .014 (.003)	- .011 (.003)	- .044 (.022)	- .205 (.071)
TEL	.005 (.003)	- .001 (.003)	- .013 (.019)	- .070 (.083)
OWNER	.004 (.003)	- .001 (.002)	- .003 (.015)	- .016 (.070)
YORKS	- .002 (.006)	.000 (.005)	.038 (.033)	.159 (.142)
EMID	- .008 (.006)	- .005 (.006)	.014 (.031)	.028 (.148)
ANGLIA	- .009 (.007)	- .006 (.007)	- .004 (.038)	- .019 (.190)
LONDON	- .029 (.006)	- .022 (.005)	- .006 (.039)	- .340 (.142)
SE	- .010 (.005)	- .007 (.005)	- .019 (.028)	- .119 (.131)
SW	- .009 (.007)	- .009 (.006)	- .019 (.031)	- .125 (.145)
WALES	- .010 (.007)	- .006 (.006)	- .018 (.036)	- .112 (.164)
WMID	- .012 (.006)	- .008 (.005)	- .013 (.031)	- .090 (.142)
NW	- .012 (.006)	- .008 (.005)	.000 (.029)	- .051 (.245)
SCOT	- .011 (.006)	- .009 (.005)	- .031 (.034)	- .192 (.151)
ULSTER	- .021 (.009)	- .013 (.008)	.006 (.052)	.006 (.248)

Table 10
Summaries of Tobit and Least–squares Results for Clothing (22.7% zeros)

	Estimated income elasticity [a]	% of observations with $\hat{\beta}'x_n < 0$	$\hat{\sigma}$	Residual sum of squares	R^2
Tobit (LM test against P - Tobit $\chi_1^2 = 415.0$, predicted zeros = 30.5%)	1.77 (.035)	10.0	0.084	-	-
Linear Regression (full sample)	1.42 (.028)	0.8	0.070	33.865	.100
Nonlinear least-squares: truncated normal model (full sample)	1.73 (.048)	1.7	0.020	33.754	.103
Nonlinear least-squares: lognormal model (full sample)	1.69 (.041)	-	-	33.939	.098

Note: (a) see notes (a) and (b) of Table 4.

Table 11
Tobit and Least–squares Coefficient Estimates for Clothing

	ML Tobit	Linear Regression (full sample)	Nonlinear Least-squares: truncated normal model (full sample)	Nonlinear Least-squares: lognormal model (full sample)
Constant	-2.936 (.280)	-2.097 (.226)	-2.852 (.334)	-47.61 (4.81)
LNY	.492 (.051)	.358 (.041)	.490 (.059)	7.618 (.869)
LNY2	- .020 (.002)	- .015 (.002)	- .020 (.003)	- .320 (.039)
NWRK	.001 (.004)	.001 (.003)	.000 (.004)	- .042 (.066)
AD	.017 (.002)	.012 (.002)	.013 (.002)	.128 (.023)
<2	.018 (.005)	.011 (.004)	.014 (.004)	.156 (.052)
2-5	.017 (.004)	.011 (.003)	.014 (.003)	.162 (.043)
>5	.019 (.002)	.014 (.002)	.016 (.002)	.192 (.021)
AGE 25	- .013 (.005)	- .010 (.004)	- .011 (.005)	- .124 (.073)
AGE 35	- .017 (.005)	- .013 (.005)	- .014 (.005)	- .172 (.076)
AGE 45	.019 (.005)	- .015 (.004)	- .016 (.005)	- .200 (.073)
AGE 60	- .019 (.006)	- .015 (.005)	- .017 (.006)	- .223 (.087)
AGE 65	- .020 (.006)	- .014 (.005)	- .016 (.006)	- .238 (.086)
SEX	.029 (.003)	.017 (.003)	.019 (.003)	.254 (.047)
NOKIDS	- .010 (.005)	- .005 (.004)	- .004 (.004)	- .080 (.053)
SINGLE	- .038 (.004)	- .016 (.003)	- .018 (.004)	- .284 (.062)
PROF	- .004 (.003)	- .001 (.002)	- .002 (.003)	- .046 (.038)
CLERK	.006 (.005)	.007 (.004)	.006 (.004)	.063 (.060)
SHOP	.019 (.010)	.015 (.009)	.015 (.011)	.168 (.147)
FORCES	.017 (.012)	.015 (.012)	.014 (.017)	.118 (.225)
UNOCC	- .009 (.004)	- .006 (.003)	- .005 (.003)	- .071 (.056)
CAR	- .017 (.003)	- .013 (.002)	- .015 (.003)	- .216 (.042)
2 CARS	- .008 (.003)	- .006 (.003)	- .007 (.003)	- .096 (.039)
TEL	.002 (.003)	.000 (.002)	- .001 (.003)	.014 (.047)
OWNER	- .009 (.003)	- .007 (.002)	- .007 (.002)	- .093 (.037)
YORKS	- .003 (.005)	- .003 (.004)	- .004 (.005)	- .063 (.074)
EMID	- .005 (.006)	- .004 (.005)	- .005 (.005)	- .056 (.080)
ANGLIA	- .009 (.007)	- .006 (.006)	- .008 (.006)	- .089 (.096)
LONDON	- .014 (.005)	- .008 (.004)	- .010 (.005)	- .098 (.073)
SE	- .015 (.005)	- .012 (.004)	- .014 (.004)	- .208 (.067)
SW	- .010 (.006)	- .008 (.005)	- .010 (.005)	- .138 (.084)
WALES	.003 (.006)	.004 (.005)	- .004 (.006)	- .068 (.087)
WMID	- .002 (.006)	- .002 (.004)	- .003 (.005)	- .038 (.073)
NW	- .009 (.006)	- .009 (.006)	- .008 (.005)	- .103 (.070)
SCOT	.003 (.005)	.003 (.005)	.004 (.005)	.058 (.074)
ULSTER	.011 (.005)	.011 (.005)	.011 (.008)	.158 (.110)

Table 12

ML Estimates of Model (48)–(49), from a 10% subsample

Constant	-47.906	(20.20)
LNY	7.611	(3.661)
LNY2	- .318	(.167)
NWRK	- .110	(.278)
AD	.200	(.136)
<2	.361	(.356)
2-5	.241	(.257)
>5	.323	(.161)
AGE 25	- .448	(.330)
AGE 35	- .440	(.353)
AGE 45	- .589	(.334)
AGE 60	- .916	(.393)
AGE 65	- .303	(.345)
SEX	.374	(.196)
NOKIDS	- .139	(.314)
SINGLE	- .789	(.266)
PROF	- .305	(.194)
CLERK	- .092	(.296)
SHOP	.943	(.782)
FORCES	-7.739	(178.6)
UNOCC	- .167	(.265)
CAR	.110	(.184)
2 CARS	- .084	(.223)
TEL	.331	(.173)
OWNER	- .235	(.155)
YORKS	.165	(.328)
EMID	- .024	(.354)
ANGLIA	- .373	(.411)
LONDON	- .259	(.343)
SE	- .312	(.315)
SW	- .472	(.381)
WALES	- .174	(.379)
WMID	- .166	(.323)
NW	- .174	(.325)
SCOT	- .053	(.353)
ULSTER	- .325	(.515)
σ	.847	(.888)
σ_e	1.092	(.124)
α	.0095	(.0046)

REFERENCES

Atkinson, A B, J Gomulka & N H Stern (1984): "Household expenditure on tobacco 1970–1980: evidence from the Family Expenditure Survey", ESRC Programme on Taxation, Incentives and the Distribution of Income, discussion paper 57.

Blundell, R & C Meghir (1987): "Bivariate Alternatives to the Tobit Model", *Journal of Econometrics*, 34, pp 179–200.

Cragg, J G (1971): "Some statistical models for limited dependent variables with applications to the demand for durable goods", *Econometrica*, 39, pp 829–44.

Deaton, A S & M Irish (1984): "A Statistical model for zero expenditures in household budgets", *Journal of Public Economics*, 23, pp 59-80.

Kay, J A, M J Keen & C N Morris (1984): "Estimating Consumption from expenditure data", *Journal of Public Economics*, 23, pp 169-182.

Keen, M J (1986): "Zero expenditures and the estimation of Engel curves", *Journal of Applied Econometrics*, 1, pp 277–286.

Lee, L–F & M M Pitt (1984): "Microeconometric models of consumer and producers demand with limited dependent variables", mimeo.

Marquardt, D W (1963): "An algorithm for least–squares estimation of non-linear parameters", *Journal of the Society for Industrial and Applied Mathematics*, 11, pp 431–441.

Pudney, S E (1985): "Frequency of purchase and Engel curve estimation", discussion paper A56, LSE Econometric Programme.

Tobin, J (1958): "Estimation of relationships for limited dependent variables", *Econometrica*, 26, pp 24–36.

Waldman, D M (1985): "Computation in duration models with heterogeneity", *Journal of Econometrics* (Annals 1985–1), 28, pp 127–134.

Wales, T J & A Woodland (1980): "Sample Selectivity and the estimation of labour supply functions", *International Economic Review*, 21, pp 437–468.

——— & ——— (1983), "Estimation of consumer demand systems with binding non–negativity constraints", *Journal of Econometrics*, 21, pp 263–285.

Conference Discussion

Handout Cross–section expenditure sample

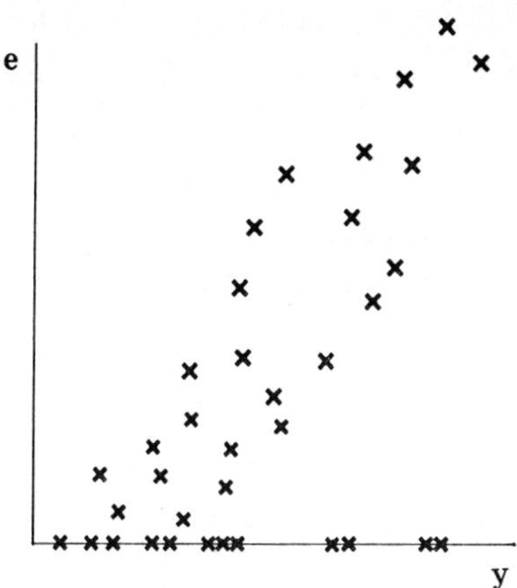

e = household "expenditure" on the good

y = household "income"

Problems

(1) Interpretation of zeros.

(2) Relationship of expenditures to true demand.

(3) Can demand be modelled without modelling purchasing behaviour?

(4) Is the "income" variable reliable?

(5) Can the survey design be improved?

(6) Aggregation over goods.

i Zero expenditures

CJB Why do you get only 30 per cent zeros for refrigerators? I would have thought it was about 90 or 95 per cent.

SP No, it's not for refrigerators. I have four categories of expenditures: tobacco, alcohol, durables which is a very large category and has only 9 per cent zeros, and also clothing.

GM When you say expenditure on durables, does that include hire purchase repayments which would be made weekly, and would make sure there are no zero's?

SP That's what I was afraid of, someone's going to ask me detailed questions about the FES.

AS Repayments on hire purchase wouldn't be counted as purchase of a durable good.

SP I'm trying to remember the exact definition of this durables category. It's a very strange thing, because it's so wide.

TG It includes things like knives and forks, for instance.

SP That's the problem, anything that has any degree of durability is included.

TG The thing about durables is the fantastic difference in size, so you'd expect even the non–zeros to be very peculiarly distributed.

SP This is precisely what I come on to.

ii **On data**

TG I'm not quite clear why you assume the prices don't change over the course of the year.

SP That's the problem.

TG Surely there's data on prices at this level of aggregation from components of the retail price index and so on.

SP The real problem is that as price inflation drops, far more people refuse to take part in the survey. That's why you have such short survey periods.

TG I used to teach survey design, admittedly about 30 years ago. One of the main topics was how to correct for non-participation. One can often make a reasonable stab at it.

iii AS The FES requires an enormous amount of bookkeeping by the household. They actually have to write down everything that they spend their money on. I would imagine that attempts to do that for more than two or three weeks ...

RB It is true that this is an accurate survey for the individual households you actually look at, they actually interview them three times during the period.

SP The FES is quite an impressive piece of work, and it's hard to believe that these one year surveys are going to represent normal behaviour. The mere fact that you keep detailed accounts of everything you do for a year is surely going to change what you do.

TG But again, that's a thing that's subject to interpretation. Perhaps it's confused with a seasonal effect. If seasonal, so you do it for two years, bringing new people in and dropping old out each week or so. It's perfectly susceptible to analysis. Now I'm not claiming whether it's a very good idea or not, but when you consider that we haven't the faintest idea of whether the immense variations about "the Engel curve" are due to intra–household or inter–household effects, and lots of things are based on these variations, it's an important problem.

JM Panel data in the FES would be relatively cheap to introduce and would help. The other thing is that in many surveys in other countries, clothes expenditure is taken over a three month interval and the refrigerators over a one–year recall. The FES could do that.

SP Yes, it depends on the good, a one year or three month recall could be subject to error.

iv AS The FES does it two weeks right the way through the year with different people?

RB Yes.

AS Even alcohol at Christmas?

RB You know the data period.

TG I'm not so sure they do do Christmas. I have a feeling it's left out.

SP The actual Christmas period is left out, but the period before Christmas, when people buy the stuff by the crate, is not.

v **Aggregating Across Goods**

RB If you aggregate across goods with different purchase frequencies, then it's true that the expectation of expenditure equals consumption, presumably, if you're counting the zeros as well, like you would do.

SP Yes.

RB So that if that was your dependent variable, presumably you'd talk about it as your right hand variable, but if it was your left hand variable, it would just be subject to measurement error which wouldn't necessarily cause any problem. Is that right?

SP That's right. The limited information estimators that I will be talking about in a moment will not be affected by that. Unfortunately they're not very successful.

Handout Conventional approach: the Tobit model

Latent Engel curve:

(1) $e_n^* = \beta' x_n + \epsilon_n \quad \epsilon_n | x_n \overset{\sim}{} \text{N.I.D}(0,\sigma^2) \quad n = 1...N$

vi AS You'd actually have to have preferences defined on negative amounts of the goods, I take it.

SP You'd have to set the thing up so that you have a maximisation which could generate negative results.

TG You could have the maximisation generate the $\beta' x$ part, and then something happen afterwards, and the $\beta' x$ should always be positive.

SP But you can't constrain $\beta' x$ to be always positive and that's one of the problems.

Handout Observed expenditure:

(2) $e_n = \max\{e_n^*, 0\}$

Interpretation:

(i) Equation (1) is solution to unconstrained utility maximisation problem.

(ii) Equation (2) is solution to Kuhn–Tucker problem with 2 goods.

(iii) e_n is identical to true consumption

What is true consumption?

Define c_n, $n = 1 \ldots N$

(i) c_n is a choice variable from consumer's utility maximisation problem

(ii) c_n is determined prior to purchasing behaviour

(iii) c_n is implemented by the consumer in the (hypothetical) long run:

$$E(e_n | c_n) = c_n$$

under unchanging conditions

(iv) c_n is stochastic

vii The relation of consumption to expenditure

AS That somehow assumes that people's average stocks are in equilibrium at all times, doesn't it?

SP You're not here holding initial stocks of anything fixed. You're picking people at random. You don't know what their initial stock situation is.

AS Well, just take an example. Suppose you've got a young family that's just bought a house and they're perhaps capital constrained, so their stocks of consumer durables are quite low. You might say they were under their equilibrium and as they moved to their equilibrium...

SP Yes, that would certainly be a reason e_n would depart from c_n in the short run, but again this is a hypothetical long run relationship.

AS Isn't that not so much about the long–run but about a collection of similar individuals. If you take an average over similar households the average should correspond to c_n?

TS But your average households have to be regarded as people at different places in the purchasing cycle.

viii On purchasing frequency

RB There must be some economics behind the purchasing frequency. If you completely disregard the economics of purchasing you might rule out some important determinants of purchasing. For example, different types of working behaviour, different hourly wage rates, would change your optimal purchasing behaviour, which could be built in to the model.

SP Again I think you're asking me to complicate things.

RB Not really because you just put those in the model. You said it's limited information so presumably it's some reduced form idea of determination of purchasing frequency.

SP Well, no, provided you accept all this you need to say nothing about purchasing behaviour. It's all because of this wonderful equation.

ix On $c_n = e_n$

RB If you took people who had a high stock of durables and then some other price fell and they were stuck with too high a rate of durables, would that condition still hold? I could imagine averaging across a group this expenditure and finding them all zero and yet they were still consuming services from a durable as they ran it down. You seem to be hanging quite a lot around that condition.

SP The problem is that you've got to hang quite a lot around something, and it's very difficult to see with what you're going to replace this that's going to be more convincing. I can accept you point but I hope it's going to be a small problem in any given year.

Handout Modelling strategy

(1) Specify demand function generating c_n
(2) Specify distribution for $e_n | c_n$

Alternative strategy

Eg, Poisson purchasing process

m = number of purchases in survey period
e^i = size of i–th purchase

$$Pr(m) = \frac{e^{-\lambda_n} \lambda_n^m}{m!}, \quad \lambda_n > 0$$

Independent purchases:

$$p.d.f.(e^1 \ldots e^m | m) = \prod_{i=1}^{m} f(e^i) \text{ where } Ee^i = \mu_n$$

$$\text{Derive p.d.f. of } e = \sum_{i=1}^{m} e^i\text{: } f^*(e|m)$$

Distribution of observed expenditure:

$$\Pr(e_n = 0 \mid \lambda_n) = e^{-\lambda_n}$$

$$\text{p.d.f.}(e_n \mid \lambda_n) = \sum_{m=1}^{\infty} \frac{e^{-\lambda_n} \lambda_n^m}{m!} f^*(e_n \mid m), \; e_n > 0$$

Implied c_n satisfies:

$$c_n = \sum_{m=1}^{\infty} \frac{e^{-\lambda(c_n)} \lambda(c_n)^m}{m!} m\mu(c_n)$$

x The statisticians' Approach

TG It is one of Ehrenberg's stylised facts that purchasing behaviour is negative binomial which is, of course, a compound Poisson.

SP That's right, it's a Poisson process with a Gamma parameter.

AS You're ruling that people choose their c_n and then go and work out how to achieve it?

SP No, not necessarily. It's just difficult to see how you'd link that with any model of rational choice.

TS Ehrenberg would have the λ follow a certain distribution, the gamma, but the determination of the λ is basically an average rate of purchase.

SP Why λ and why not some parameter of the distribution of the size of the individual purchases?

TG It's just the mean of the Poisson.

CJB It's clear with alcohol one decides how much one wants to drink and then leaves it to a random process.

TG That's the reason λ is the mean of the distribution in question.

Handout The relation between e_n and c_n

Define:

$$P(c_n,z_n) = \Pr(\text{at least one purchase during observational period} \mid c_n, z_n)$$

where

$$P(0,z_n) = 0$$

z_n is vector of exogenous variables relevant to purchasing behaviour.

Identity:

$$\begin{aligned} E(e_n \mid c_n,z_n) &= P(c_n,z_n)\, E(e_n \mid e_n > 0, c_n z_n) \quad \text{for } c_n > 0 \\ &= 0 \qquad\qquad\qquad\qquad\qquad\qquad\quad \text{for } c_n = 0 \end{aligned}$$

Implies:

$$E(e_n \mid e_n > 0, c_n, z_n) = \frac{c_n}{P(c_n, z_n)}$$

The Literature

Author	$P(c_n,z_n)$	Model for c_n	Relation between e_n and $c_n/P(c_n,z_n)$
Almost everyone	1	censored regression	exact equality
Deaton and Irish(1984)	P	censored regression	exact equality
Kay, Keen and Morris (1984)	P	linear regression	additive error
Keen (1986)	P	linear regression	additive error
Blundell and Meghir (1987)	$\Phi(a'x_n)$	linear regression, lognormal	additive normal error, multiplicative lognormal error

xi **The Deaton and Irish model**

RB Isn't there a kind of identification problem in that model. I can't remember off hand, but if you do fix P it can look like a lot of other models. Can't you rephrase it as a double hurdle model?

SP It basically is a double hurdle model with no variables in the first-stage. They also interpret it as a mis–reporting model.

xii TS In the papers that use linear regression for c, how is this P parameter identified? You can scale everything. What you observe is c_n/P and you have this linear model.

SP Oh, no, you don't observe c_n. P comes into the probability of a zero.

TS That's a linear regression model, it's not a censored model any more. If I have $c = X\beta + \epsilon$ and I divide it by P to get e.

RB It's easy to identify P in those models because it's the proportion of buyers. You use the zeros to identify P and then scale it down.

Handout Limited information estimation

$$E(e_n | c_n) = c_n$$

Hence:

$$E(e_n | x_n) = E(c_n | x_n)$$

Regression equation:

$$e_n = g(\beta' x_n, \sigma) + \nu_n$$

where:

$g(\beta' x_n, \sigma)$ is parametric form of $E(c_n | x_n)$
$E(\nu_n | x_n) = 0$

xiii **On g**

RB Presumably you're ruling out corner solutions here?

SP Oh, no, they'd be built into the structure of g. If you had a Tobit model, g would be the mean function coming from the censored normal.

RB How would you know someone was at a corner solution?

SP You wouldn't know if any particular individual had displayed a zero because they were at a corner. You would derive the functional form for g from a specific model for c which allows corner solution to happen.

Handout Parametric estimation

$$\min_{\beta,\sigma} \sum_{n=1}^{N} (e_n - g(\beta' x_n, \sigma))^2$$

Asymptotic covariance matrix:

$$\text{a. cov } (\hat{\delta}) = \hat{A}^{-1}\hat{B}\hat{A}^{-1}$$

where:

$$\delta = \begin{bmatrix} \beta \\ \sigma \end{bmatrix}$$

$$\hat{A} = \sum_{n=1}^{N} \hat{g}_n^\delta \hat{g}_n^{\delta\prime}$$

$$\hat{B} = \sum_{n=1}^{N} \hat{\nu}_n^2 \hat{g}_n^\delta \hat{g}_n^{\delta\prime}$$

$$\hat{\nu}_n = e_n - g(\hat{\beta}' x_n, \hat{\sigma})$$

$$\hat{g}_n^\delta = \frac{\partial g(\hat{\beta}' x_n, \hat{\sigma})}{\partial \hat{\delta}}$$

Inefficient but robust

Forms for $g(\beta' x_n, \sigma)$

Classification of Goods:

Type 1: Everyone consumes (eg clothing, durables)

(a) $c_n | \tilde{x}_n$ truncated $N(\beta' x_n, \sigma)$

$$g(\beta' x_n, \sigma) = \beta' x_n + \sigma \lambda^*(\beta' x_n / \sigma)$$

where $\lambda^*(s) = \phi(s)/\Phi(s)$

$$\phi(s) = (2\pi)^{-1/2} e^{-2^2/2}$$

$$\Phi(s) = \int_{-\infty}^{s} \phi(t)\, dt$$

(b) $c_n | x_n \sim LN(\beta' x_n, \sigma)$

$$g(\beta' x_n, \sigma) = \exp\{\beta' x_n + \frac{\sigma^2}{2}\}$$

$$= \exp\{\beta^{*\prime} x_n\}$$

Type 2: Some corner solutions (most luxuries)

$$c_n | x_n \sim \text{censored } N(\beta' x_n, \sigma)$$

$$g(\beta' x_n, \sigma) = \beta' x_n \Phi\left[\frac{\beta' x_n}{\sigma}\right] + \sigma\, \phi\left[\frac{\beta' x_n}{\sigma}\right]$$

Type 3: Conscientous Abstention (tobacco, alcohol?)

Double hurdle:

$$v_n | \theta_n \sim N(\gamma' \theta_n, 1)$$

is latent preference shifter.

If $v_n \leq 0$ individual n abstains:

$$c_n = 0$$

If $v_n > 0$ individual consumes:

$$c_n | x_n \sim \text{truncated normal or lognormal}$$

$$g(\beta' x_n, \gamma' \theta_n, \sigma) = \Phi(\gamma' \theta_n) \left[\beta' x_n + \sigma \lambda^*(\beta' x_n / \sigma)\right]$$

or

$$g(\beta' x_n, \gamma' \theta_n, \sigma) = \Phi(\gamma' \theta_n) \exp\{\beta^{*\prime} x_n\}$$

xiv On Non–smoking Choice

CJB Don't we conceive of the possibility that someone becomes a non-smoker because they simply get fed–up with what it costs? My father did that annually for years.

SP I consider the case where there are people who don't drink because they consider it the work of the devil and those who don't drink because they can't afford it in the Type 4 classification.

Handout Type 4: Abstentions and corners (tobacco, alcohol?)

If $v_n \leq 0$ then:

$c_n = 0$

If $v_n > 0$ individual is <u>potential</u> consumer

$c_n | x_n \sim$ censored normal

$$g(\beta' x_n, \gamma' \theta_n, \sigma) = \Phi(\gamma' \theta_n) \Phi\left[\frac{\beta' x_n}{\sigma}\right] [\beta' x_n + \sigma\lambda^*(\beta' x_n/\sigma]$$

Least squares results

Alcohol and tobacco

(i) Attempts to fit type 2, 3, 4 form for $g(\cdot)$ failed.

(ii) If $e_n \equiv c_n$, it is valid to fit type 1 or 2 forms to the <u>truncated</u> sample – some success with type 1 lognormal.

Durables and clothing (both type 1)

(i) Lognormal form gives plausible estimates

(ii) Truncated normal "fits" better, but less precision.

Non–parametric estimation?

Problems: x_n is mostly discrete

e_n is censored

e_n is heteroscedastic

Fully–specific models
Form of $P(c_n,z_n)$?

P_n depends on:

(i) c_n
(ii) survey duration
(iii) household size and composition
(iv) ownership of durables

Examples:

$$P_n = \Phi(\alpha_1 \log c_n + \alpha_2 z_n) \quad \text{log–probit}$$

$$P_n = \frac{c_n^{\alpha_1}}{e^{-\alpha_2 z_n} + c_n^{\alpha_1}} \quad \text{log–logistic}$$

xv Random P

RB That is like saying there's an omitted variable in Φ which is correlated with c, and that would be equivalent to a test of the omission of c in Φ.

SP That's right.

TG If P were only a function of c_n and if as you suggest it were strictly monotonic, then actually one can go back to one's Poisson distribution and c_n is a function of λ.

SP But what's λ a function of, how is it related to the sorts of things you write papers about?

TG You shouted me down for daring to suggest the frequency is a function of the c thing.

SP The link is rather complicated and not particularly transparent, to me at least.

Handout Example

Lognormal demand: $c_n \sim LN(\beta' x_n, \sigma^2)$

Simplified log–logistic purchase probability:

$$P_n = \frac{c_n}{\alpha + c_n}$$

Lognormal expenditures:

$$e_n \mid c_n \sim LN(\mu_n, \sigma_e^2)$$

where:

$$\mu_n = \log(\alpha + c_n) - \frac{\sigma_e^2}{2}$$

is such that $E(e_n \mid c_n) = c_n / P_n$

xvi RB If you didn't have your infrequency of purchases problem and it was just zeros, are you saying this is a way of estimating a model with zeros?

SP If there's no infrequency of purchase, c_n and e_n are the same thing, so you would estimate a Tobit model.

RB You seem to be doing something other than that here.

SP Yes, I'm allowing the fact that e_n may depart from c_n by these purely fortuitous random deviations.

RB I agree with that, but doesn't this actually collapse down to a Tobit model then?

SP This is consistent with a Tobit model. There's an identification problem, if you fit this and find that it works there are at least two explanations that you could come up with. One is that c_n and e_n are the same thing and that's the Tobit model; the other is that c_n and e_n are different, you have this purchasing problem, but that the underlying rate of consumption is generated by a Tobit model. Now you can't distinguish between those using this model.

TG In a way it is paradoxical that you welcome σ being completely impossible to estimate, but are really quite worried when it's just rather difficult.

SP No, I'm extremely worried by the whole thing.

TG You did seem rather to go for the earlier one.

SP Well, I have this love of catastrophe you see.

JM If σ's hard to estimate, it doesn't matter what value you assume for it, so you might as well assume a plausible value and away you go.

SP Depending on which value for σ you choose, you get completely different values for certain elements of these. There's a severe identification problem.

TG So you're now worried about the σ because you're assuming these things are homoscedastic and therefore in the previous case it's only affecting constant terms?

SP I'm assuming what's homoscedastic?

TG Look at your (b) in Type 1 and now come down here. You've got your $\sigma^2/2$, which I take is the variance of the log–normal distribution and you're making it's mean depend on x_n. It's the fact that your having the variance not depend on x_n that makes the effect of not knowing σ^2 not muck up any β's except the constant terms. Therefore I'm saying it's the homoscedasticity that's stopping you worrying, otherwise you'd be worried as hell that you haven't the faintest idea what any of the β's were.

SP Oh, yes, but heteroscedaticity in the c_n is just another huge problem to put on top of this. Forget about that, there's a disastrous problem in any case. If you try to fit this to data you find that σ tries to go off to be something very large, and that certain elements of β also do wild things. If you parametrise it and try to estimate β/σ, and $1/\sigma$ as well, you still can't really estimate anything at all precisely. It's almost impossible to make the thing converge.

AS Is it specific x_n's that do that?

SP It's the intercept, certainly; the two income variables; log total expenditure and log(total expenditure)^2, they're the two coefficients you're mainly interested in, and they tend to display this problem.

RB And the reason you're getting this is that you tried to make no assumptions earlier; on because if you'd made some assumptions earlier on, you could have set this up as a generalisation of Tobit.

SP That's right. The σ is normally taken care of by the probability of zeros and that sort of thing. It comes from a different kind of information. Here you're trying to get at it from the mean function.

Handout Implied distribution:

$$\Pr(e_n = 0|x_n) = \int_0^\infty \frac{\alpha}{c(\alpha+c)} \sigma^{-1}\phi\left[\frac{\log c - \beta' x_n}{\sigma}\right] dc$$

p.d.f. $(e_n|x_n) =$

$$(e_n\sigma\sigma_e)^{-1} \int_0^\infty \frac{1}{\alpha+c}\, \phi\left[\frac{\log e_n - \mu_n(c)}{\sigma_e}\right]\, \phi\left[\frac{\log c - \beta' x_n}{\sigma}\right] dc$$

Log–likelihood:

$$L(\beta',\alpha,\sigma) = \sum_{e_n=0} \log \Pr.(e_n = 0|x_n) + \sum_{e_n>0} \log \text{p.d.f.}(e_n|x_n)$$

Results

Tobit is rejected for alcohol, durables and clothing:

(i) overpredicts zeros

(ii) H_0: P=1 is rejected by L–M test.

Rejected for tobacco by comparison with probit coefficients

Generalised P–Tobit

Clothing only: 10% subsample

Elasticity = 1.723

Prediction of zeros = 15.5%

E(c) = 0.0817

P(c) = 0.896 at E(c)

Measurement and Modelling in Economics
G.D. Myles (Editor)

ECONOMETRIC APPROACHES TO THE SPECIFICATION AND ESTIMATION OF INTERTEMPORAL CONSUMER BEHAVIOUR*

by

Richard Blundell†

1 Introduction

This paper is concerned with the construction and properties of empirical models for individual household life–cycle labour supply and commodity demand behaviour. It is not directly concerned with either aggregate savings or aggregate intertemporal substitution in the labour market. These two areas have been recently surveyed as part of the excellent papers by Deaton (1985) and Nickell (1988). However, in so far as aggregate models of intertemporal behaviour are generated from conditions for individual life–cycle optimising behaviour the properties of the various approaches discussed in this paper will have a bearing on such studies. Similarly, to the extent that the conditions for the choice of appropriate estimators remain intact after aggregation, the implications for estimation resulting from the stochastic behaviour of individual optimising rules is also useful. However, estimation on micro–data has in addition to account for the various types of censoring in the data that may arise due to corner solutions or rationing. Indeed, female labour supply and disaggregate

---

† Department of Economics, University College, London, and Institute for Fiscal Studies

* This paper is based on a previous paper published in *Econometric Reviews* 1986, 5. I should like to thank Gordon Anderson, Martin Browning, Vanessa Fry, Costas Meghir, Panos Pashardes, Richard Smith, Guglielmo Weber and Ian Walker for allowing me to draw on ideas developed in jointly authored papers. Terence Gorman, John Ham, Costas Meghir, Dale Poirier, Ian Walker and participants at the conference have provided very useful comments on earlier drafts. All errors remain mine. Finance for this research, provided by the ESRC under project B002 32207, is gratefully acknowledged.

consumption will provide the main focus of attention. Although not dealing directly with empirical tests of the life–cycle model itself, the results will hopefully be relevant for the framework chosen to represent the life–cycle model in such tests. Inadequate as certain aspects of the life-cycle model may appear, empirical tests based on an inappropriate or restrictive representation of that model provide little information on its validity. The type of econometric models for life–cycle behaviour to which we shall turn initially are those that assume intertemporal separability. These have been made popular through the important studies of Heckman and MaCurdy(1980), MaCurdy(1980) and subsequently extended in the recent work of Browning, Deaton and Irish (1985). In these models, as is the case with all models that assume intertemporal separability, unobservable life–cycle variables can be captured through a single summary statistic. Labour supply or commodity demand behaviour that is consistent with life–cycle optimisation can therefore be expressed in terms of current observable variables and a single conditioning variable summarising all intra–period allocations. In the studies referred to above the marginal utility of wealth is used for this purpose. Although the marginal utility of wealth is generally unobservable and will differ across individuals with different characteristics and expectations, for any given individual with perfect foresight it is (after suitable discounting) constant across time. Relaxing this perfect foresight assumption to allow replanning and uncertainty simply introduces an innovation error representing new information so that discounted marginal utility follows a random walk. Provided the unobservable marginal utility enters the model through an individual specific intercept term it can be eliminated, even under uncertainty, by first differencing panel data observations.

In order that marginal utility enters the behavioural model for labour supply and commodity demand in a way that allows the simple first differencing solution, life–cycle preferences will naturally be subject to restrictions. These restrictions are non–trivial in the sense that many of the important parameters reflecting the degree of flexibility in intertemporal and intratemporal substitution are restricted prior to estimation.

In Section 3 of this paper the implications of these results are considered in detail and alternative econometric approaches to recovering parameters of the same broad intertemporally separable model are suggested following the developments in MaCurdy (1983).

These alternative econometric approaches to the estimation of life–cycle models identify within period preferences using alternative choices for the conditioning variable summarising past decisions and future anticipations. Provided sufficient separability restrictions can be made on intertemporal preferences, results from two–stage budgeting under uncertainty allow the generation of a number of different empirical specifications consistent with the same underlying life–cycle model. By ingenious choice of conditioning variable, many available micro–data sets traditionally used for the estimation of static models, can be used for the estimation of life–cycle consistent models. The duality between these alternative representations of behaviour may then be exploited to retrieve many important parameters of within period and between period preferences under the least restrictive assumptions. In each approach the stochastic properties of the conditioning variable plays an important role in the choice of appropriate estimation technique.

The intertemporal separability assumption which lies behind all of the above specifications appears to be one of the least acceptable assumptions in these life–cycle models. That individual behaviour is to an extent dominated by habits and adjustment costs is difficult to dispute. However, the intertemporally separable model rules out such explicit dynamic behaviour. The introduction of such dynamics can severely complicate the empirical specification of life–cycle models but nevertheless towards the end of Section 3 various tractable empirical specifications are briefly considered. If the prime concern is to measure (consistently) short-run effects then it is possible that, where individual data is used, inclusion of sufficient conditioning variables relating to past decisions will suffice to produce consistent estimates of short run behaviour. For example, the inclusion of the detailed structure of family composition may largely replicate the important determinants of past female labour market behaviour. Inclusion of such observable characteristics may then eliminate the need for explicit dynamics entering through past consumption or labour supply behaviour.

The layout of the paper is as follows. In Section 2 various approaches to the specification and estimation of life–cycle behaviour are reviewed while in Section 3 the restrictions underlying each of these approaches is considered in detail. Section 4 provides some new results which relate to a cohort panel constructed from individual cross–section data from the UK. Finally, Section 5 concludes.

2 Intertemporal Separability, Frisch Demands and Two–Stage Budgeting

2.1 The Additively Separable Life–Cycle Framework

The principle objective of this section is to review the various alternative parameterisations of the life-cycle decision making model under intertemporal additive separability. In order to do so a common optimising framework is employed and initially we shall assume that households have perfect foresight. In each period labour supplies and commodity demands are chosen so as to maximise discounted lifetime utility subject to time and asset constraints with perfectly predicted future market wages, prices, transfer income and demographic characteristics.

Within this additively separable intertemporal framework, relaxing the perfect foresight assumption so as to allow uncertainty and replanning does not significantly alter the overall structure of the models. It does however require careful treatment in choosing an appropriate estimator. The separability assumption allows the direct application of two–stage budgeting theory and ensures that life–cycle consistent current demands can be written in terms of a single variable capturing both past decisions and future anticipations.

Defining x_s to be the choice vector in period s, lifetime utility in any period 't' may be written as the following discounted sum of concave twice differentiable period by period utility indices $U_s(x_s)$

$$V_t = \Sigma_s \phi^{s-t}\, U_s(x_s) \quad t < L \tag{1}$$

the summation over s runs from t up to L where L is the number of periods in the lifetime of the household decision maker and ϕ represents the subjective time discount factor (see Ghez and Becker (1975)). The choice vector x_s contains both the "leisure" time components (l_s) of household utility and commodity demands (q_s), while the direct dependence of the period by period utility index on "s" reflects the influence of predetermined taste shifter variables such as family size and composition variables. In Section 3 this separability assumption will be relaxed to enable U to depend explicitly on past values for x_s although we shall argue that in cross–section micro data sets where past values of x_s may not be measured directly, the "taste shifter" variables in (1) will often capture these factors adequately.

Corresponding to x_s is an associated "price" vector π_s whose elements contain both market wages (w_s) and commodity prices (p_s). This can be used to define a "full" income budget identity

$$\pi_s{}' x_s = y_s \tag{2}$$

where full income y_s is the sum of unearned income, asset decumulation and the value of the household's time endowment. By defining B_s to be the level of transfer income in period s, A_s to be the level of household assets at the end of period s and r_s to be the rate of interest earned on A_{s-1} during period s, the sum of the unearned income and asset decumulation components of y_s may be expressed as

$$\mu_s = B_s + r_s\, A_{s-1} - \Delta A_s \tag{3}$$

where Δ is the first difference operator and ΔA_s is the change in assets over period $s-1$. Letting T_s be the vector of time endowments available to individuals in the household, full income is the familiar expression

$$y_s = w_s{}'T_s + \mu_s \tag{4}$$

In this general model with borrowing and saving the 'other income' variable μ_s will differ critically from its usual definition in the static framework which omits the third term on the left hand side of (3). In the Blundell and Walker (1986) study, for example, it is negative for the majority of families with working wives. However, in developing this life-cycle framework we have added to the perfect foresight and separability assumptions the crucial perfect capital markets condition. Under this assumption the interest rate r_s is fixed and independent of A_s so that given perfect foresight any amount of future labour or nonlabour income can be discounted into current period income. In this case a sequence of asset levels A_s or savings decisions for s = 1,...,L can be freely chosen so as to maximise life–cycle utility (see King (1985)).

To complete this outline of the life–cycle framework it is useful to eliminate A_s from (3) for s=t,..,L using (2) and (4) to define the following life-time wealth constraint

$$\Sigma_s \pi_s{}'x_s = (1 + r_t)\, A_{t-1} + \Sigma_s w_s{}'T_s + \Sigma_s B_s = w_t \tag{5}$$

where all prices, wage rates and transfer incomes are discounted back to period t. The arguments of (5) now exactly correspond to those in the life-time utility function V_t. Alternatively, identifying the separate labour supply (hours of work) and commodity demand components in π_s and x_s, the wealth constraint can be rewritten

$$\Sigma_s p_s{}'q_s = (1 + r_t)\, A_{t-1} + \Sigma_s w_s{}'h_s + \Sigma_s B_s \tag{6}$$

where the vector of hours of work $h_s = T_s - l_s$ is clearly a choice variable in this framework. As each element of l_s is bounded from above by the corresponding element of T_s, each element of h_s is required to be non-negative in every period. Corner solutions for l_s at zero hours of work will then be a crucial aspect of the following analysis. As an example, for women, the corresponding element of T_s can be reduced significantly with the presence of dependent children making corner solutions a frequent occurence in such periods. Moreover, where zero hours is a choice, the ability to borrow in the life–cycle framework is likely to increase the probability of such an occurence in comparison to the static framework.

2.2 Two–Stage Budgeting and Price Aggregation

The form of (1) and (5) is ideal for the application of two-stage budgeting results (see Blackorby, Primont and Russell (1978), Gorman (1968) and Pollak (1971)). At a first stage y_t is chosen so as to equalise the discounted marginal utility of money in each period and at the second stage x_t is chosen conditional on y_t. Under two–stage budgeting the allocation of y_t is given by

$$y_t = g_t(\pi_t, \pi_{t+1}, \ldots, \pi_L, W_t) \tag{7}$$

where g_t is homogeneous of degree zero in discounted prices π_s for s = t,..,L and wealth W_t. It is clear from (7) that the outlay variable y_t summarizes the influence of all future expectations concerning economic and demographic variables on current period as well as the influence of past decisions through A_{t-1} in W_t. The second stage allocation then determines within period demands conditional on y_t according to

$$x_t = d_t\,(\pi_t, y_t) \tag{8}$$

where d_t is a vector of demand equations homogeneous of degree zero in the price vector π_t and the conditioning variable y_t.

The forms of (7) and (8) are only correct if there are no binding constraints on the elements of x_t in period t. Should, for example, the upper bound on female time bind the form of d_t for all remaining choice variables will generally change. Similarly, if there are demand side or institutional constraints on hours of work a switch will occur at the point of rationing as discussed for males in the study of Blundell and Walker (1982). The effect of such a corner in the budget constraint is precisely the same as that observed in the standard static labour supply model so long as the conditioning variable y_t is correctly measured in each of the switching regimes. Binding constraints on x_s where $s > t$ will simply alter the form of (7) and will have no direct impact on (8). In each of these cases the measurement of y_t is crucial and provided expenditure data are available, y_t can be recovered through the within period budget constraint without direct knowledge of the form of (7). Indeed, where the emphasis is on estimating life–cycle consistent within period preferences and where we are able to measure y_t directly (and treat it as exogenous for the determination of x_t), the precise form for $g_t(\cdot)$ in (7) across regimes is unimportant.

To illustrate these various points consider representing within period preferences by the following general Gorman Polar Form (Gorman (1959)) indirect utility function

$$I_t = F_t\{(y_t - a_t(\pi_t))/b_t(\pi_t)\} \tag{9}$$

where F_t is some concave monotonic transformation and the functions $a_t(\pi_t)$ and $b_t(\pi_t)$ are concave linear homogeneous functions of π_t. Assuming there are no binding constraints the allocation of life–cycle wealth described by (7) may now be written as

$$y_t = a_t\,(\pi_t) + \theta_t[(b_t(\pi_t),\, b_{t+1}\,(\pi_{t+1}),..,b_L(\pi_L),W_t - \Sigma_s a_s\,(\pi_s))] \tag{10}$$

where $a_t(\pi_t)$ and $b_t(\pi_t)$ act as price aggregators for across period allocations. From the application of Roy's identity to (9), within period demands will have the form

$$x_{it} = a_{it}(\pi_t) + b_{it}(\pi_t)(y_t - a_t(\pi_t))/b_t(\pi_t) \text{ for all i} \tag{11}$$

where a_{it} and b_{it} are the c_{it} derivatives of $a_t(\cdot)$ and $b_t(\cdot)$ respectively. The within period demands (11) are clearly independent of the choice of monotonic transformation F_t unlike the intertemporal allocation of y given by (10) where $\theta_t(\cdot)$ will depend crucially on F_t.

Particular forms of $a_t(\cdot)$ and $b_t(\cdot)$ in (9) are given in Blundell and Walker (1986) where emphasis is placed on their flexibility and the way in which they depend on demographic variables. It is worth pointing out that by choosing $a_t(\cdot)$ to contain linear wage terms the time endowment parameters T can be subsumed into estimated parameters of the labour supply equations avoiding the need to choose them arbitrarily. Generalising (11) to allow demands to be nonlinear in the income variable y_t could well be important in empirical application and will be discussed below. However, in such cases choice of the T parameters will no longer be arbitrary.

An interesting feature of (10) is the way in which the allocation of resources between periods, that is the choice of y , depends in general on the two price "aggregator" indices $a(\cdot)$ and $b(\cdot)$. This has direct implications for the specification of aggregate intertemporal models for saving, consumption and labour supply behaviour which often rely on a single price index. These issues will be considered in more detail in Section 3.2, here we shall simply concentrate on the relationship between the functional form for it and the choice of transformation F_t.

The general question concerning conditions under which a single price aggregator index is sufficient for inter–period allocations has been raised previously by Gorman (1959) and Anderson (1979). Rather surprisingly, unless homothetic preferences (unitary income elasticities) are assumed, linear Engel curves and linear earnings functions are generally ruled out. To consider this we may generalise (9) to write

$$I_t = F_t\left\{G_t\left[\frac{y_t}{b_t(\pi_t)}\right] + c_t(\pi_t)\right\} \tag{12}$$

where G_t is some monotonic increasing function and $c_t(\pi_t) = a_t(\pi_t)/b_t(\pi_t)$. When F_t is a linear transformation then the optimal level of y_t in each period of the life–cycle is independent of $c_t(\pi_t)$ and depends on the single index $b_t(\pi_t)$ alone. However, for any nonlinear F_t both price indices $b_t(\cdot)$ and $c_t(\cdot)$ enter the intertemporal allocation unless, of course, $c_t(\pi_t) = 0$ for all t which is equivalent to the homothetic preference assumption. In order for preferences to exhibit demands linear in y (as in (11)) $G_t(\cdot)$ must also be linear in order that the Gorman form of preferences result. However, if both the G_t and F_t transformations are linear then the only solution to the life–cycle optimisation problem is one where all supernumary wealth is consumed in a single period since both optimising function and constraint are linear in y. For this reason, if we are to assume a continuous form for the allocation of y (or savings) across time then this allocation cannot depend in general on a single price index unless either we rule out demand equations linear in y (for within period) behaviour or we assume homothetic within period preferences. As mentioned above the further implications of this for modelling consumption and savings will be discussed in later sections. Here we simply note the restrictiveness of the single aggregate price index model.

2.3 Constant Marginal Utility Models (λ-constant) and Frisch Demands

As an alternative parameterization consider the following familiar first order conditions (see Heckman and MaCurdy (1980)) for the solution to the intertemporal maximization of (1) subject to (5)

$$\partial U_t / \partial x_t = \lambda_t \pi_t \tag{13}$$

and

$$\lambda_{s+1} = (1/\phi(1 + r_{s+1}))\lambda_s \tag{14}$$

where the lagrange multiplier λ_s represents the marginal utility of income in period s. In the relationship (14), the marginal utility of income in each period is seen to provide the link between current and other period decisions showing its importance in the determinion of y_t and the precise form for (7). Indeed, λ_t acts as a summary of between period allocations and therefore is a suitable conditioning variable for "λ–constant" or Frisch demands much in the same way y_t enters the y–conditional demands. These Frisch demands can be viewed as a rearrangement of (13) generating

$$x_t = f_t(\pi_t, \lambda_t) \tag{15}$$

which are homogenous of degree zero in π_t and λ_t^{-1}. The general properties of demand equations (15) are described in detail in Browning, Deaton and Irish (1985) and provide a useful interpretation of life–cycle behaviour. For example, the wage elasticities conditional on λ_t identify the effect of (fully anticipated) movements along the household's lifetime wage profile. One advantage of using (15) directly is that the process (14) can be usefully exploited to eliminate λ_t in empirical implementation. However, unless care is taken at this stage in the construction of life–cycle models using (14) and (15), strong implicit restrictions can be imposed on the type of within period and intertemporal preferences that are permitted.

An equivalent derivation of the λ–constant model is given by Browning, Deaton and Irish (1985) using the individual's profit function defined by

$$\Pi(\pi_t, \lambda_t^{-1}) = \max\{U_t/\lambda_t - C(\pi_t, U_t)\} \tag{16}$$

where $C(\pi_t, U_t)$ is the consumer's expenditure or cost function (see Deaton and Muellbauer (1980)). The profit function $\Pi(\pi_t, \lambda_t^{-1})$ is linear homogenous in π_t and λ_t^{-1}, decreasing in π_t and increasing in λ_t^{-1}. They show that the λ–constant demands correspond to the negative of the price derivatives of Π, i.e.

$$x_{it} = -\partial\Pi(\pi_t,\lambda_t^{-1})/\partial\pi_{it} \tag{17}$$

Equations (17) are equivalent to demands derived from (15) when the latter can be solved explicitly.

In comparison to the y–conditional models of Section 2.2 the λ-constant demands directly measure within period as well as intertemporal preferences and therefore require an explicit choice for the monotonic transformation corresponding to F_t in (9). That is any system of λ-constant demands rests on the choice of a particular monotonic transformation. For example, if F_t in (9) is chosen to be log linear so that the marginal utility of income is simply

$$\lambda_t = \frac{1}{y_t - a_t(\pi_t)}\,. \tag{18}$$

then the λ–constant demands corresponding to (9) take the form

$$x_{it} = a_{it}\,(\pi_t) + \frac{b_{it}\,(\pi_t)}{b_t\,(\pi_t)}\,\lambda_t^{-1} \tag{19}$$

If panel data were available λ_t could be written using (14) in terms of initial period marginal utility λ_0 which could then be treated as a fixed household specific effect over the panel. Where panel data are not available but consumption data are, the alternative parameterisations like (8) become invaluable. However, to what extent these alternative parameterisations, which avoid the explicit need for panel data, can fully identify the parameters of intertemporal and intratemporal substitution that are retrievable from certain of the λ–constant models remains an important question. To answer this a suitable measure of intertemporal substitution is required.

The parameter that perhaps best captures the underlying intertemporal substitution possibilities in the life–cycle allocation problem is the intertemporal elasticity of substitution defined by Browning (1985) and given by

$$\Phi = \frac{I_y}{y\ I_{yy}} \tag{20}$$

where I_y is the y–derivative of indirect utility. As Browning notes Φ is related to measures of risk aversion since it reflects the concavity of the transformation F_t. As an example consider a loglinear choice for F_t in (9), the intertemporal elasticity is then given by

$$\Phi = \frac{y_t - a_t(\pi_t)}{y_t} \tag{21}$$

Rather surprisingly, the unknown parameters of Φ in $a_t(\pi_t)$ are equally well identified from either y–conditional demands (11) or λ–constant demands (19).

It is true, however, that for general choices of F_t (the monotonic transformation of within period preferences), y–conditional or related approaches cannot identify all of the parameters needed to construct Φ_t. Nevertheless, they will be shown to identify all the parameters in some of the popular parameterisations of the λ–constant model, thus providing an indication of the restrictiveness underlying certain of these models that arise from (14) and (15). The panel data required for the λ–constant approach should allow for less and certainly not more restricted models than a single cross–section or a time series of cross–sections that the alternative approaches require. It should therefore allow estimation of more general models with individual fixed effects not equally or more restricted models. These issues will form the theme for much of the discussion in Section 3 but before turning to them we shall briefly consider the relaxation of the perfect foresight assumption and its various implications for estimation.

2.4 Replanning and Uncertainty

When the perfect foresight assumption is relaxed to allow the more empirically realistic assumption of individual replanning following the arrival of new unanticipated information, the constancy result for the "expected" discounted marginal utility of money remains intact. To briefly illustrate this consider the optimal set of future decisions as seen from period t given a particular drawing of future uncertain variables. These decisions can be summarised by the following partial indirect utility function:

$$\Omega_{t+1} = \max \sum_{s=t+1}^{L} \phi^{s-t}\, U_s\,(x_s)$$

$$\text{subject to } \sum_{s=t+1}^{L} \pi_s{}'\, x_s = W_t - y_t \tag{22}$$

defined conditionally on current allocation y_t (see Epstein (1975) and Meghir (1985)). Any period's optimal decision (as seen from period t) can then be expressed as a sequential planning problem with Lagrangean

$$L_t = U_t(x_t) + E_t\Omega_{t+1} + \lambda_t(y_t - \pi_t{}'x_t) \tag{23}$$

where E_t is an expectation conditional on information up to and including t. Provided each U_t is strictly concave we may generate the following first order stochastic Euler conditions for the global saddle point of (23)

$$\frac{\partial L_t}{\partial x_{it}} = \frac{\partial U_t}{\partial x_{it}} - \lambda_t \pi_{it} = 0 \text{ for all i} \tag{24}$$

and

$$\frac{\partial L_t}{\partial y_t} = \frac{\partial E_t \Omega_{t+1}}{\partial y_t} + \lambda_t = 0. \tag{25}$$

The constancy result of Section (2.3) may then be generalised to

$$E_t\left[\phi(1 + r_{t+1})\lambda_{t+1}\right] = \lambda_t. \tag{26}$$

If a random variable e_{t+1} is defined to represent new information or innovations arising in period t + 1, the consequent revision of λ_{t+1} may be approximated by

$$\lambda_{t+1} = \left[\frac{1}{\phi(1+r_{t+1})}\right] \lambda_t \exp(e_{t+1}) \tag{27}$$

the particular form for e_{t+1} having been chosen to ensure $\lambda_t>0$. The stochastic process for λ_t is then written more conveniently as

$$\ln\lambda_{t+1} = \ln\left[\frac{1}{\phi(1+r_{t+1})}\right] + \ln\lambda_t + e_{t+1} \tag{28}$$

where e_{t+1} is independent of all information dated t or earlier from (26). Since λ_t is some function of the full income or expenditure variable y_t, (28) defines a dynamic process for y_t and therefore a dynamic process for life-cycle savings.

It is the simplicity of (28) and the properties of the innovation error e_t that have made the λ–constant approach so attractive for estimation. In Section 3 we shall argue that an appropriate combination of the λ–constant and y–conditional approaches may provide the best method for exploiting information in micro–data sets. However, before that discussion we turn to some of the implications for estimation in each of these approaches separately.

2.5 Implications for Estimation

Although in estimating within period preferences from the y-conditional models y_t is treated as an appropriate conditioning variable, there is no reason why it should be assumed exogenous. Any exogeneity assumption of this type should be checked against the data at the outset. Two difficulties arise at this point. How do we test this (or any other) exogeneity assumption across a system of equations (especially when there are limited dependent variables such as female labour supply in the model) and how should we estimate if exogeneity is rejected? For estimation, we shall write the system of y–conditional demands as

$$x_t = d_t\,(\pi_t, y_t) + v_{1t} \tag{29}$$

where x_t may represent a vector of earnings and expenditure shares (out of full income) and v_{1t} is a vector of stochastic error terms. If the y-conditional model is estimated from panel data then $d_t(\cdot)$ may include an individual fixed effect for each commodity. For the purposes of testing the exogeneity assumption we shall approximate the equation for y_t by

$$y_t = \gamma' z_t + v_{2t} \tag{30}$$

where z_t are a vector of observed exogenous variables with γ as their unknown constant coefficients and v_{2t} is a random error term. The question of focus here relates to the exogeneity of y_t for the estimation of the parameters of the y–conditional model (29).

Where x_t is continuously observed for all observations, so that there are no limited dependent variables, the procedure for testing and estimation is straightforward and follows the analysis of Hausman (1978) and Wu (1973). The vector of error terms v_{it} is partitioned in the following way

$$v_{1t} = \alpha\, v_{2t} + e_t \tag{31}$$

here α is a vector of correlation parameters such that $E(v_{2t}\, e_t) = 0$ and therefore each element of e_t is independent of v_{2t}. Inclusion of the reduced form residual v_{2t} in each equation in (29) and estimation of the extended system provides consistent estimates for the parameters of the y-conditional demands. In addition, an F–test for the inclusion of the vector of estimated coefficients on v_t provides an asymptotically efficient exogeneity test (see Blundell and Smith (1984)). For forms of (29) that are nonlinear in y_t, multivariate normality for v_{1t} and v_{2t} is required for asymptotic efficiency of the test.

Where limited dependent variables arise a normality assumption on v_{1t} would usually be made. Moreover in this limited dependent variable case with random effects only, Smith and Blundell (1986) show that little of the standard analysis for exogeneity testing discussed above changes. Consistency of the estimates in the extended system under the alternative remains and the exogeneity test is again asymptotically efficient. The only problem to note under this limited dependent variable case is that the form for y_t switches when any decision variable is rationed or at a corner solution. In this case the coherency conditions of Gourieroux, Laffont and Monfort (1980) need to be checked before proceeding with estimation and testing.

For the estimation of λ–constant models an instrumental variable estimator is suggested directly from the stochastic properties of e_{t+1} in the stochastic Euler equation (28). Provided the demand equations resulting from (24) can be expressed with a term linear in $\ln\lambda_t$, the unobservable marginal utility of wealth can be eliminated by first differencing panel data observations using (28). The resulting differenced model now contains the innovation error e_{t+1} as part of its disturbance term. Although e_{t+1} will not be independent of variables dated $t + 1$ appearing on the right hand side of the differenced model that are not fully anticipated, it will be independent of variables dated t or earlier. A simple consistent estimator can be derived exploiting this independence condition underlying the first order Euler conditions to suggest appropriate instruments. Where $\ln\lambda$ has this linear property, estimation of the differenced model follows from a direct application of standard instrumental variable theory (see Sargan (1958)). All explanatory variables dated $t + 1$ which cannot be perfectly predicted by the individual are treated as endogenous. Attfield and Browning (1985) rather neatly exploit the symmetry and homogeneity restrictions on a system of λ-constant demands to derive consistent estimators of this type. Given the conditions on the error term e_{t+1} (and provided no other error process entering the specification is correlated over time) all variables in the model dated t or earlier are eligible instruments and there is no reason to assume e_{t+1} has a homoscedastic distribution over time. Particular forms for this linear differenced specification will be considered in the next section where some of the restrictions underlying such a specification of the λ–constant model are outlined. Where $\ln\lambda$ (from (24) for example) is nonlinear in unknown parameters the properties of the innovation error e_{t+1} still determine the optimal choice of instruments, however (28) cannot in this case be expressed as a linear differenced model. Hansen and Singleton (1982) develops sufficient conditions on the nonlinear model for the extension of the Sargan IV technique to this application.

Two important issues have been ignored in the above discussion. The first, and possibly most important omission from this paper, is the importance of measurement error especially for models nonlinear in wage or income variables. Where measurement error is serious it is difficult to argue for nonlinear models however restrictive the linear specification might appear. This is perhaps even more crucial for the estimation of λ-constant models where differencing can exacerbate the importance of measurement error even in the linear differenced case. However, it cannot be ignored when considering general y–conditional models of the type developed in Section 2.2. It is likely that the full income variable y, although possibly endogenous, is measured more accurately in y-conditional rather than static models where unearned or asset income variables are required. Nevertheless, significant measurement error in wages or prices will largely rule out the estimation of models incorporating complex non-additive within–period preferences. Where lack of accurate information on certain variables occurs as a result of a corner solution, for example wages in an hours of work equation for women, there is no reason why consistent within period preference parameters should not be recovered from truncated estimation of the y–conditional model. The exogeneity test described above is valid in the truncated as well as other limited dependent variable models.

The second issue which seems relevant is the distinction between the estimated parameters in a cross–section (or time–series of cross-sections) random effects model compared with those from a panel data model. MaCurdy (1983) rightfully draws attention to this point noting that many included variables are likely to be correlated with unobservable individual random effects. To the extent that this persists, even after having included individual characteristics in the y–conditional random effects model, provides a crucial constraint on the use of (randomly sampled) cross–sections or time series of cross–sections. An important drawback of short panels has been recognised by Chamberlain (1984) and Hayashi (1987). They point out that the orthogonality condition on past variables used to define the "optimal" instrumental variable estimators referred to above holds for a particular individual across time and may not hold across individuals in a panel with few time series observations.

3 Restrictions and Properties of the Intertemporally Separable Models

3.1 Restrictions Underlying λ–constant Models

For direct estimation of the λ–constant models under uncertainty using (28) empirical applications have generally required they take the form:

$$g_{1i}(x_{it}) = g_{2i}(\pi_t) + \tau_i \ln\lambda_t \tag{32}$$

where $g_{1i}(\cdot)$ is some known (monotonic) transformation and $g_{2i}(\cdot)$ an appropriately homogenous function of the vector of prices and wages π_t. Moreover, $g_{2i}(\pi_t)$ is usually expressed as the outer product $d_i'f(\pi_t)$, where d_i is a vector of constant unknown parameters for the ith equation and $f(\pi_t)$ is a known function of π_t. Using (28), and assuming a discount rate equal to the interest rate, the model may then be written as the following linear differenced specification

$$\Delta g_{1i}(x_{it}) = d_i' \Delta f_t(\pi_t) + \tau_i e_{t+1} \tag{33}$$

This equation represents a reasonably general form for empirical λ-constant models under uncertainty and apart from MaCurdy (1983) (whose approach will be discussed later) the majority of models that have been estimated in the literature can be expressed in this form.

Working explicitly from the direct utility model, the form of (33) generally requires the utility derivative in (24) to be a function of x_{it} alone. That is, U_t must be explicitly additive, strongly restricting both within-period preferences (see Deaton (1974)) and intertemporal preferences. In particular, life–cycle utility in (1) takes the following additive form across time and goods

$$\Sigma_s \phi^{s-t} U_s = \Sigma_s \phi^{s-t} \Sigma_i U_{is}(x_{is}). \tag{34}$$

Heckman and MaCurdy (1980), for example, employ a utility function of this type with each

$$U_{is}(x_{is}) = \alpha_i(x_i^{\beta_i} - 1)/\beta_i \tag{35}$$

yielding the following loglinear form for (32)

$$\ln x_{it} = \sigma_i \ln\alpha_i - \sigma_i \ln\pi_{it} - \sigma_i \ln\lambda_t \tag{36}$$

with $\sigma_i = 1/(1 - \beta_i)$. The profit function corresponding to (35) may be written

$$\Pi(\pi, \lambda_t^{-1}) = \Sigma_i \alpha_i^{\sigma_i} \pi_{it}^{\beta_i \sigma_i} \lambda_t^{-\sigma_i} + \beta_0 \lambda_t^{-1} \tag{37}$$

where $\beta_i < 1$ and where we have used the linear homogeneity of $\Pi(\pi_t, 1/\lambda_t)$ in derivation. In Blundell, Fry and Meghir (1985) it is shown that this profit function may be generalised to relax additive separability only at the expense of assuming homothetic preferences. That is, relaxing additivity in (36) requires unitary within period full income elasticities. Notice that the λ–constant elasticities can be recovered through the elimination of $\ln\lambda_t$ across commodities (see Altonji (1982, 1986)) avoiding the explicit need for panel data.

When modelling labour supply decisions it is often more convenient to work with demand equations linear in x_{it}. Browning, Deaton and Irish (1985) working from the profit function derive such an alternative class of models. Moreover, their class of models implies neither explicit additivity nor homothetic preferences. It does, however, impose some strong restrictions on within–period and intertemporal preferences.

The general form for their profit function is given by

$$\Pi(\pi_t, \lambda_t^{-1}) = k_0\lambda_t^{-1} - n(\pi_t) + \Sigma_i k_i \, \pi_{it} \, (\ln(\pi_{it}\lambda_t) \tag{38}$$

where each k_i for $i = 0,..,n$ are unknown parameters and $n(\pi_t)$ is some general concave linear homogenous function of π_t. The corresponding λ-constant demands are of the form

$$x_{it} = n_i \, (\pi_t) - k_i \ln\pi_{it} - k_i - k_i \ln\lambda_t \tag{39}$$

where $n_i(\pi_t)$ is the ith price (or wage) derivative of $n(\pi_t)$. Since n_t can be a function of all prices and wages, additivity is relaxed. Within–period preferences are quasi–homothetic with the familiar Gorman Polar form (9) where

$$a(\pi_t) = n(\pi_t) - \Sigma_i k_i \pi_{it} \ln(\pi_{it}/(\Sigma_i k_i \pi_{it}) - \Sigma_i k_i \pi_{it} \tag{40}$$

and

$$b(\pi_t) = \Sigma_i k_i \pi_{it} \, . \tag{41}$$

As $y_t - a(\pi_t)$ is large relative to $a(\pi_t)$ for 'richer' individuals, $b(\pi_t)$ represents the substitution possibilities for such individuals. The Leontief form for $b(\pi_t)$ therefore restricts these substitution effects to zero.

Intertemporal preferences are also restricted through the implicit form for F which is required in order to generate (39). A useful way to consider this point is once again to use the intertemporal elasticity of substitution defined above. In terms of the indirect utility function this measure indicates precisely whether a particular choice of the monotonic transformation F_t contains additional parameters determining the intertemporal elasticity that cannot be recovered from the Marshallian or y–conditional demands.

Unlike the log transformation for F_t (whose rather surprising implications for λ–constant and y–conditional demands were outlined in Section 2.3), the Browning, Deaton and Irish model allows an additional parameter in the determination of Φ over and above those recovered from the y–conditional demands. To see this consider the corresponding indirect utility function which has exponential form

$$I_t = -\exp\left[-(y_t - a(\pi_t))/b(\pi_t)\right] \tag{42}$$

implying $\Phi_t = -b_t(\pi_t)/y_t$. From (42) we see the parameters of $b(\cdot)$ rather than $a(\cdot)$ enter Φ. Since without panel data $b(\cdot)$ can only be identified up to a normalising scalar, the intertemporal elasticity is only determined from y–conditional demands up to this scaler. The normalising scalar can therefore be recovered from the differenced λ–constant model (39) but not from the y–conditional demands. However, although the intertemporal elasticity Φ in this exponential case is always negative it approaches zero as the income allocation y increases. For an arbitrarily rich household therefore this intertemporal elasticity of substitution is zero. This contrasts rather dramatically with the loglinear transformation whose corresponding intertemporal elasticity is given in (21) and which is bounded between zero and –1 approaching –1 as the income allocation y_t rises. The loglinear transformation allows the elasticity to grow away from zero with increasing income whereas the exponential transfomation assumes the opposite. The concavity properties of Φ in the exponential case are therefore equivalent to a constant absolute risk aversion assumption.

As y–conditional demands are only designed to provide a description of within period preferences the identification result for the exponential case described above is hardly surprising. However, given the parameters of a y–conditional demand system, any remaining intertemporal parameters could perhaps be better recovered using the update rule for marginal utility λ_t at a second stage. These points provide the motivating thrust for the alternative approaches to which we turn in Section 3.2. To summarise this sub–section: the penalty for the relaxation of explicit within-period additive separability in the λ–constant approach that uses (33) is therefore the linearity of "Engel curves" in terms of full income and the restrictions on substitution possibilities imposed by the adoption of a linear Leontief form for $b(\pi)$ and an exponential form for $F_t(\cdot)$.

3.2 Alternative Representations of the Intertemporally Separable Model

The alternative representations we consider here are those that recover life–cycle consistent within period preferences by substituting out the unobservable marginal utility of wealth across two or more contemporaneous decisions (see Altonji (1982, 1986)) or eliminating it using the within-period budget constraint (see Betancourt (1971), Blundell and Walker (1986) and MaCurdy (1983)). As we saw in Section 3.1 the parameters estimated from such alternative representations may be sufficient to identify intertemporal substitution elasticities in some of the more restrictive models. Clearly, if a single household decision variable is all that is available in the data then neither of these alternative approaches are possible and the first differenced λ–constant specifications discussed in Sections 2 and 3.1 are the only available. However, many data sets record some other decisions. As examples, male and female labour supply or labour supply and food consumption are collected in the Michigan PSID, while male labour supply, female labour supply and a large number of expenditures are included in the UK Family Expenditure Survey.

Using these alternative representations the parameters of life–cycle consistent within period preferences can be recovered using only a single cross-section of data, or at least a cross–section of time–series, provided that there is sufficient price and wage variation across the sample. Moreover, these estimates of within period preferences are invariant to the choice of monotonic transformation $F_t(\cdot)$ which we have shown to be rather critical in some of the specifications of the λ–constant model. Since the monotonic transformation underlying some of the λ–constant specifications is assumed and not estimated, the intertemporal parameters of interest can often be identified from a suitable cross–section (rather than a panel) using the alternative representations. However, provided attrition is not too high, the use of panel data for estimation of the alternative models is always to be preferred.

The attraction of the alternative representations is in the way they break the link between restrictions on the form for within–period preferences and that for intertemporal preferences. MaCurdy (1983), for example, exploits this property of the alternative approaches to separately identify (and estimate) the form of the monotonic transformation and therefore the intertemporal elasticity from the update rule for λ_t conditional on the estimated within–period preferences. The idea here is to use the y–conditional model to estimate all the parameters of indirect utility up to this monotonic transformation without restricting intertemporal elasticities. In the case of the quasi–homothetic model, (9) described in Section 2.2, this would mean estimating the y–conditional model and constructing $(y_t - a_t(\pi_t))/b_t(\pi_t)$ for each time period of panel data available. If we define this variable as real supernumary outlay (income) y_t^r then, introducing new intertemporal parameters α_t and β, we could define a suitable monotonic transformation as

$$I_t - \alpha_t(y_t^r)^{\beta}/\beta \tag{43}$$

similar to the transformation of within period preferences adopted by MaCurdy (1983). In this case the intertemporal elasticity of substitution Φ equals $(1 - a_t/y_t)/\beta - 1$ and marginal utility is given by

$$\lambda_t = \alpha_t (y_t^r)^{\beta-1} / b_t(\pi_t). \tag{44}$$

Using the update rule for λ_t described by (28) we can then identify β and α_t from a first difference model of the form

$$\Delta \ln y_t^r = \sigma \Delta \ln \alpha_t - \sigma \Delta \ln b_t(\pi_t) + \sigma \ln(1 + r_t)\phi + e_t \tag{45}$$

where $\sigma = 1/(1 - \beta)$ which can be estimated using panel data conditional on the y–conditional estimates of $b_t(\cdot)$ and y_t^r.

Although MaCurdy (1983) uses additive within–period preferences in his application of this approach such a restriction is unnecessary and in Blundell and Walker (1986) and Meghir (1985) general quasihomothetic forms for within period preferences are adopted. In both studies the data strongly reject simple forms for $b(\pi)$ like those adopted by Browning, Deaton and Irish (1985) and suggests that a translog form with parameters depending on demographics might be superior. All such generalisations of preferences would be ruled out in the linear λ–constant first differenced models discussed earlier in Section 3.1.

Interestingly, the log differenced model (45) relates directly to popular consumption/savings models. For y_t to be defined over consumption expenditure alone, consumption and labour supply have to be assumed additively separable. Then setting $a(\pi)$ to zero and approximating $b(\pi)$ by the retail price index, y_t^r becomes real consumers expenditure and (45) may be written

$$\Delta \ln(y_t/b_t) = \sigma \Delta \ln \alpha_t - \sigma(\Delta \ln b_t - \ln(1 + r_t)\phi)) + e_t \tag{46}$$

The second term on the right hand side of (46) is equivalent to a real interest term introduced by Muellbauer(1983). Current dated variables in

(46) may still be correlated with the error term and as a result instrumental variable estimation is required to recover the intertemporal substitution parameters consistently.

There seems no particular reason to require a single price index for the allocation of y and, in general, equations of the form (45) can be used to recover an estimate of β conditional on the within period preference parameter estimates derived from the y–conditional model. This procedure seems attractive since it exploits all aspects of panel data (or constructed panel data, see Browning, Deaton and Irish (1985)) to their full without imposing unnecessary restrictions on within or intertemporal preferences.

Where labour supply is included in estimation (or where additive separability between time and goods is not assumed) rationing in the labour market from one period to the next will be reflected in the determination of savings behaviour through the price aggregators $a(\cdot)$ and $b(\cdot)$. In periods where constraints occur, unconstrained elements of x_t will depend on the level (and not the price or wage) of the rationed goods. As described in Section 2.2, $b_t(\cdot)$ and $a_t(\cdot)$ will then depend on the virtual price or wage (see Neary and Roberts (1980)) which is a function of the level of the ration providing a simple way of introducing involuntary unemployment or more generally nonparticipation in an optimal savings function as has been suggested by King (1985) in his review of savings models.

3.3 Some Empirical Comparisons from Micro–Data

Empirical studies using micro–data have generally confined their attention to the labour supply decisions of prime–aged males. The resulting λ–constant wage elasticity estimates are usually small, often of the correct sign but barely significant. Perhaps the most influencial study in this respect is MaCurdy (1980) who, using a model of the first differenced additively separable type described in Section 2.4, reports estimates in the range .1(.125) to .23(.095) where the parentheses contain the corresponding standard errors. The lower estimates are derived from a model that included time dummies to capture the interest rate term in the update for λ. Altonji (1986) recovers broadly similar estimates using as instruments for the log wage term both those suggested by human capital theory (see MaCurdy (1980)) and an alternative wage measure. The robustness of these results to changes in preferences is explored in Ham (1986) where, using preferences of the type adopted by Browning, Deaton and Irish (1985), had little effect on the general size and significance of this intertemporal elasticity. The corresponding elasticity from the Browning, Deaton and Irish (1985) study using a constructed panel from the UK Family Expenditure Survey is of a similarly small order.

More worrying, perhaps, than the small elasticities is the reliability of these labour supply models for prime–age males. In the study by Ham (1986) there appears to be strong evidence in the US data for rejecting this labour supply specification. Prime–aged male unemployment does not appear to be explained by substitution in the intertemporally separable life–cycle model. Married women may well fit these behavioural models rather better since their decisions with regard to transitions in the labour market are taken more regularly and often appear more responsive to economic considerations. Certainly the studies by Heckman and MaCurdy (1980,1982), Altonji (1986) and Blundell and Walker (1986) would suggest this. In Heckman and MaCurdy (1982) the λ–constant wage elasticity for leisure was estimated reasonably precisely at –.406. Given that the average hours worked by working women in any year was equivalent to an average weekly hours of 26 while the maximum available in any week was assumed to be 168, the corresponding labour supply elasticity in excess of 2.0 is relatively large. Although lower, the elasticities reported by Altonji (1986) corresponding to similar average weekly hours are still of the order .867 and .746. Given hours are generally lower for women, some increase in the elasticity over prime age males is to be expected. However, these estimates do appear significantly larger.

3.4. Relaxing Intertemporal Separability

Although easy to criticise, the underlying dynamics of the intertemporally separable models discussed are not easy to extend while retaining empirical tractability. Anderson and Blundell (1982, 1983) for example, start with the intertemporally separable specification of the y-conditional model of Section 2.2 and generalise that structure by allowing flexible interrelated dynamics. This assumes that the y–conditional model represents steady state behaviour but does not adequately capture short-run adjustment. These "ad hoc" dynamic models clearly act as a useful test of intertemporal separability and can be partially rationalised in a life–cycle framework. However, their main attraction must be the computational ease of estimation and their ability to be subjected to a barrage of data coherency tests. Where intertemporal separability is rejected their use in policy analysis is limited to the extent that they do not identify the separate sources of dynamic adjustment. For this, some sort of theoretical model of short–run adjustment under uncertainty is required that generalises the intertemporal separability assumption of the previous sections.

Deriving theoretical models that are fully consistent with life–cycle utility maximisation under costs of adjustment or habit persistence, for example, is in principle a straightforward extension of the discussion in the previous section. It can also be shown that such an extension will allow the identification of the separate sources of dynamic process. Here we shall concentrate on relaxing the critical separability assumptions in a rather specific way. In particular, (expected) life–cycle utility (1) is replaced by

$$V_t = U_t(x_t, x_{t-1}) + E_t(\Sigma_s \phi^{s-t} U_s(x_s, x_{s-1})) \tag{47}$$

where the presence of x_{t-1} may reflect adjustment costs in consumption (Weissenberger (1986)) or habit persistence (Boyer (1983), Philips and Spinnewyn (1981), Spinnewyn (1981) and Muellbauer and Pashardes (1982)).

The analysis of the optimal path for maximising (47) subject to the budget constraint (5) yields the following first–order conditions similar to those examined in the literature on rational partial adjustment models, for example Kennan (1979), Hansen and Sargent (1981) and Nickell (1988)

$$E_t\, \partial V_t/\partial x_{it} = \lambda_t\, \pi_{it} \tag{48}$$

with $E_t \partial V_t/\partial x_{it} = \partial U_t/\partial x_{it} + E_t \partial U_{t+1}/\partial x_{it}$. These are most conveniently seen as adding an additional term $E_t(\partial U_{t+1}/\partial x_{it})$ to the first order conditions (24). Homogeneity in terms of $1/\lambda_t$ and prices π_{it} is clearly retained in (48). Euler equations based on this framework have been estimated in the factor demand literature (see, for example, Pindyck and Rotemberg (1983) and Kennan (1979)). The distinguishing characteristic for demands by households is that the 'target' variable λ_t is unobserved. However, from the stochastic update rule (27) we know the process by which λ_t evolves and we shall argue that the estimation techniques described in Sections 2 and 3 can suggest suitable empirical approaches in this explicitly dynamic environment. When the term $\partial U_{t+1}/\partial x_{it}$ is non zero the "simple" dynamic process for y_t, for example (45), in the previous section will not result. Indeed, λ_t cannot now be written as a direct function of y_t via the indirect utility function as was the case for the intertemporally separable models. To generate empirical models from (48) it is necessary to assume some structure for past behaviour in U_t and we turn first to the adjustment cost specification.

For the two–good case Weissenberger (1986) generates an extremely useful approximate expression for λ_t in terms of π_t and expected wealth enabling an estimable form even though highly nonlinear. However, with static relative price expectations all uncertainty enters through expected wealth which depends only on the expected income streams. As these expectations are the same in each equation and should be derived rationally, based on all information available in period t, there are in general over-identifying restrictions which can be tested in similar ways to the tests developed by, for example, Epstein and Yatchew (1985), Muellbauer (1983) and Sargent (1978). For large systems with more than two choice variables and with interrelated adjustment costs no such expression for λ_t can be derived even under static expectations.

Following the discussion of previous sections, alternative approaches to the estimation of the preference and adjustment costs parameters suggest themselves. The unobservable λ_t could, for example, be substituted out across commodities in (48) producing dynamic equivalents to the models developed by Altonji (1982,1986) and described earlier. Eliminating λ_t in this manner across pairs of commodities generates a system of simultaneous dynamic equations. Replacement of the conditional expectations that arise in (48) by their realisations then allows the parameters of (47) to be recovered consistently (up to a normalising scalar) by instrumental variable estimation. Similarly, (48) could be used in conjunction with the stochastic update rule for λ to produce a dynamic demand system corresponding to the differential demand systems described earlier. However, these will not in general reduce to a simple linear dynamic model.

Equations of this general form probably represent the most useful empirical environment in which to generalise the intertemporal separability assumption. Indeed, Murphy (1986) has used this method to test (and reject) intertemporal separability assumption in a labour supply and commodity demand model. These methods correspond closely to the alternative approaches described earlier and clearly deserve further empirical investigation. They will certainly allow the extension to the many goods case with interrelated adjustment costs thus generalising (47).

Under certain circumstances the explicitly dynamic form of lifecycle expected utility considered in the adjustment costs model can be reduced by a suitable transformation of variables into simple forms that correspond to those considered in Section 2. For example, in the habit persistence model, utility in each period can be defined over some supernumary consumption level after removing habitual consumption (see Pollak (1970)). In this case each x_{it} in U_s may be redefined as:

$$\theta x_{it} = x_{it} - \theta_i x_{it-1} . \tag{49}$$

Here θ_i reflects the degree of habit persistence for each good i, single order dynamics having been chosen here for convenience. Naturally, when all θ_i are zero, the model will reduce to the y-conditional system, whereas when at least one θ_i differs from zero, we have a partial adjustment type model with a distributed lead target. Model (49) forms the basis of the useful empirical studies of Spinnewyn (1981) and Muellbauer and Pashardes (1983).

It may often be the case with micro–data that appropriate conditioning variables can be chosen to avoid the use of explicit dynamics in modelling short–run responses. This observation will be particularly relevant for applications of the life–cycle model to cross–section or short panel data sets. In these data sets many household/individual characteristics relating to past decisions may be observed although measures of *actual* past behaviour over *many* periods may not.

To illustrate this point within a more conventional framework consider the use of demographic variables, most especially the size and composition of the household, in labour supply and commodity demand studies. At the time choices are made composition, labour supply and commodity demand may be jointly determined. At any future time in the life-cycle of the household current composition will then reflect past consumption and labour supply behaviour. Current decisions can therefore either be written in terms of the *past* levels of consumption/labour supply or current composition. In this framework, composition acts as an appropriate conditioning variable that eliminates, in this case, the need for explicit dynamics. If long–run responses are required the dynamics of the composition choice should be considered. However, for short–run life-cycle consistent responses this is unnecessary. Aggregate data that does not measure these conditioning variables accurately cannot, therefore, condition appropriately. For this reason we might well expect to observe explicit dynamic behaviour in aggregate studies where it may be necessary to model these dynamics even for the estimation of short–run responses.

In general, past behaviour x_{t-1} is included in current utility along with current composition or demographic structure z_t. However, x_{t-1} will be directly related to other choices made in that past period – choices over composition, durable stocks, accommodation, education etc. Although x_{t-1} may not be observed the outcomes of these other choices from past periods may well be recorded. In this case the short–run or current period utility index $U(x_t,x_{t-1},z_t)$ in (50) for example may be rewritten as $U(x_t,\overset{*}{z}_t,z_t)$ where $\overset{*}{z}_t$ represents the currently observed outcomes of past decisions. Redefining z_t to include these $\overset{*}{z}_t$ variables current period utility may be rewritten $U(x_t,z_t)$ precisely as in the separable model (1). Optimal short–run behaviour within the life–cycle model can then be recovered from y–conditional demands which include z_t. The exogeneity of z_t and y_t for the parameters of $U(\cdot)$ remaining an empirical issue.

4 Estimation from a Cohort Panel

4.1 Methodology

For the purposes of illustrating the arguments outlined above we provide an illustration using a cohort panel data set to estimate the parameters of a Euler equation specified over consumption and conditional on labour supply behaviour. Our illustration is drawn from the recent results described in Blundell, Browning and Meghir (1989) and we begin by noting that our data is taken from fifteen annual surveys for the UK stretching over the period 1970–1984. Rather than being a panel following the same individuals across time, the data set is a pooled time series of cross sections. As a result we construct a pseudo panel using cohort averages in order to estimate the model. To relate these cohort averages to our discussion of within period preferences consider the Euler equation for marginal utilty 'λ' for each individual. Then take expectations conditional on the person belonging to cohort c. With sufficiently large cohort groups we may replace the expectation terms by their corresponding cohort averages.

We first take the following functional form for period t utility

$$I_t = V_t^{\{1+\rho\}}$$

where { } represents a Box–Cox transformation and V_t describes the indirect utility function identified from Marshallian demand analysis. Taking cohort averages, log marginal utility is given by

$$(\ln\lambda_t)_c = \rho(\ln V_t)_c + (\ln V_t'))_c \tag{50}$$

where subscript c refers to an average across cohort members. Providing entry into the cohort (through immigration) and exit (through death and emigration) are uncorrelated with the marginal utility of wealth, we can use (28) to write

$$\Delta(\ln\lambda_t)_c - \ln(\phi(1 + r_{t+1})) = (e_t)_c, \text{ with } E_{t-1}(e_t) = 0 \tag{51}$$

where again Δ is the first difference operator. In our empirical application we allow $\rho(\ )$ to be written

$$\rho(\cdot) = \rho_0 + \Sigma_k \rho_k z_{tk} \tag{52}$$

where the z_{tk} contain labour supply variables. This implies that (50) involves the cohort means of $\ln(V')_t$, of $\ln V_t$ and the cross–moments of $\ln V_t$ with the characteristics z_t.

Given this consistent aggregation procedure, the ρ parameters in (52) can be estimated using a generalised method of moments estimator applied to

$$\Delta(\ln V'_t)_c + \rho_0 \Delta(\ln V_t)_c + \Sigma_k \rho_k \Delta(z_{tk} \ln V_t)_c - \ln(\phi(1+r_{t=1})) = (\epsilon_t)_c \quad (53)$$

where the subscript c on the differenced terms points to the fact that we are differencing cohort averages rather than individual observations (see Blundell, Browning & Meghir (1989) for details). Before this Euler equation can be estimated we need to estimate the parameters of within period preferences which enter through indirect utility V and its derivatives. For this we turn to the household data itself and recover V from a standard application of Marshallian demand analysis as described in Section 3.2.

4.2 Estimation of the Demand System

Since our pooled cross–section data set contains in excess of 60,000 observations our approach is to estimate the parameters from each unrestricted demand equation separately, cross–equation restrictions being imposed at a later stage using Amemiya's principle otherwise known as Minimum Chi Square (MCS). The attraction of the MCS for microeconomic analysis of consumer behaviour relates to the separate stages of unrestricted and restricted estimation. At the first stage estimates of the parameters of the unrestricted model are recovered. For a standard demand system (PIGL in this case) this would involve estimating unrestricted share equations. These estimates together with their covariance matrix summarise all information available in the data concerning preference parameter estimation. In effect they act as sufficient statistics for the purposes of demand system estimation on the vast quality of micro level data.

Denoting these unrestricted estimates as $\hat{\theta}$, theoretical restrictions (symmetry, homogeneity, for example) on these may usually be expressed as

$$\theta = S\theta^*$$

To impose these restrictions the Minimum Chi Square (MCS) method chooses an estimator of θ^* so as to minimise the quadratic form

$$m = (\hat{\theta} - S\theta^*)' \, \Sigma_{\theta}^{-1}(\hat{\theta} - S\theta^*)$$

where Σ_θ is the estimated variance–covariance matrix for $\hat{\theta}$. Indeed, consistency of the resulting MCS estimator simply requires that the restrictions are correct and that $\hat{\theta}$ is a consistent estimator. Any positive definite weight matrix can be used to replace Σ_{θ}^{-1}. However, where the correct weight matrix is used the MCS estimator is asymptotically equal to the maximum likelihood estimator and the minimized value of m is an optimal chi-squared test of the restrictions (see Ferguson [1958]). The MCS estimator itself is given by

$$\theta^* = (S^{+\prime} S^+)^{-1} S^{+\prime} \theta^+$$

where $S^+ = \Sigma_{\theta}^{-1/2} S$ and $\theta^+ = \Sigma_{\theta}^{-1/2}\hat{\theta}$. For large samples, this method proceeds as follows. Firstly, the maximum likelihood (or an asymptotically equivalent) estimator is obtained by estimating each unrestricted demand equation efficiently, as described above. Then the efficient <u>restricted</u> estimators are recovered by fitting the restrictions in the dimension of θ which can be significantly less than the number of observations. The application of the MCS method therefore makes symmetry constrained estimation and testing available in very large samples. To implement constrained maximum likelihood estimation all information in the unrestricted model is exploited and summarized by simply defined sufficient statistics in the first stage of estimation.

In our first attempt to obtain an estimated demand system we have concentrated on the following broad commodities groups: food, alcohol, fuel, clothing, transport services and other. In terms of sample selection, the results reported here refer to a sample of 66,650 GB households from 1970–1984 inclusive whose head is not retired or self employed.

We turn first to the symmetry restricted estimates for the detailed expenditure allocations from a 'standard' Almost Ideal model (see Deaton and Muellbauer (1980) of within period preferences. In Table 1 the price and income coefficients that correspond to γ_{ij} and β_i parameters of the

Almost Ideal share equations are presented. In all equations, we consistently find that both own and cross price variables are significant. We should note the interpretation of the β coefficient on the logarithm of real expenditure lnC. This coefficient is found to display seasonal and demographic variation. For food it is negative for all households which implies that food is a strong necessity. For example, the implied budget elasticity (at the means) of 0.647 for households with children, .589 for others. The difference between households with and without children arises because of the interaction term DlnC between a child dummy and the logarithm of real expenditure: the presence of similar terms to account for quarterly variation has been ignored, as the seasonal dummies multiplying the real expenditure term sum up to zero, so that we can interpret the coefficient on lnC itself as an annual mean. In all regressions lnC was treated as endogenous.

This research, reported in more detail in Blundell, Pashardes and Weber (1988), indicates the large number of characteristics which were allowed to influence the α_i parameter in each share equation. Despite the large number of other characteristics prices have a significant impact. These results look reasonably plausible and persuasive. The price parameters are precisely estimated and we are able not to reject the homogeneity restriction implied by the theory, which proves a major stumbling block for other demand studies. This contrasts markedly with results on aggregate data – see, for example, Deaton and Muellbauer (1980). Moreover, both in the Deaton and Muellbauer study and in many that follow – Anderson and Blundell (1983), for example – dynamic mis–specification is suggested as the root cause of homogeneity rejections. As was noted earlier, the omitted characteristics in aggregate models implied from this study may evolve in a way that is captured by the introduction of dynamic adjustment or trend like terms.

Figures in parentheses are asymptotic standard errors. Quarterly seasonal dummies, regional dummies and a linear time trend were included throughout. A detailed data description and list of other included variables is provided in the Blundell, Pashardes and Weber (1988) paper.

Table 1: The Estimated Almost Ideal Demand System

	Food	Alcohol	Fuel	Clothing	Transport
PFOOD	0.09549	0.008900	–0.015619	0.002953	–0.040691
	(.00986)	(.00670)	(.00548)	(.00801)	(.01110)
PALC		–0.058948	0.059719	–0.005651	0.041823
		(.00875)	(.00589)	(.00617)	(.01052)
PFUEL			0.007386	–0.000340	–0.048787
			(.00666)	(.00531)	(.00872)
PCLOTH				0.015460	–0.004968
				(.00938)	(.01069)
PTRPT					.049782
					(.02384)
DlnC	0.002042	–0.002260	0.000856	0.001940	–0.004648
	(.00032)	(.00026)	(.00018)	(.00035)	(.00045)
S1LnC	–0.006488	–0.005083	–.005069	–0.004163	0.015229
	(.00238)	(.00192)	(.00137)	(.00256)	(.00330)
S2LnC	–0.012108	–0.004799	0.004974	–0.006389	0.009867
	(.00238)	(.00192)	(.00137)	(.00256)	(.00330)
S3LnC	–0.015181	–0.008093	0.012065	–0.006251	0.005501
	(.00233)	(.00188)	(.00135)	(.00251)	(.00323)
LnC	–0.131479	0.052971	–0.059122	0.035100	0.032784
	(.00209)	(.00169)	(.00122)	(.00225)	(.00290)
Homog(1)	0.5	0.4	0.4	0.9	0.2

Source: Blundell, Pashardes and Weber (1988).
Notes: Columns referring to Services and Other Goods categories have been excluded for space reasons.

4.3 The Euler Equation Estimates

Given these estimates of within period preferences and therefore the within period indirect utility indices V_t we turn to the Euler equation estimates. The instruments used in the method of moments estimator are lagged cohort averages and do not involve any estimated parameters. The stochastic process underlying the Euler condition does require careful attention in applications to micro–data as has been noted above. For example, it is quite possible that innovations which are uncorrelated across time for any individual may be correlated across individuals in any time period. For this reason "N" asymptotics may not be sufficient to guarantee the consistency of generalised Method of Moments parameter estimates using lagged instruments but few time series observations. Our solution to this potentially important problem is to exploit our comparatively long time series of cohort observations and to carefully assess the validity of chosen instrumental variables at each step in estimation.

Thus to estimate our Euler equations we construct cohorts using the same data as the one used in the conditional demand system estimation. We have ten cohorts each covering a five year band. This choice leads to cohorts with approximately 300–400 members each (per year) a number hopefully sufficient to make the error in measurement problem mentioned by Deaton (1985) of second order significance. Moreover, in removing any household whose head is over 60 years of age we hope to minimise the effects of non random attrition due to death. After allowing for the different periods over which each cohort is observed and the loss of observations due to differencing and lagging the instrument set, the resulting data set comprises 100 observations For the interest rate we use the simple bank lending rate available at the end of period. Since most households hold bank deposits this seems an appropriate choice.

In Table 2 we present the estimated coefficients for alternative estimators of the cohort Euler equation using our estimated demand system described above . All standard errors in Table 2 are heteroscedasticity adjusted and the GMM estimator exploits this adjustment to improve upon the efficiency of the simple IV estimator. Moreover, the estimator variation in the parameters underlying the estimation of the indirect utility parameter of V is allowed for. All instruments were dated t–2. This timing of instruments was used to ensure consistency and under MA(1) errors. The r_1 diagnostic in Table 2 is a one degree of freedom of this hypothesis and is distributed asymptotically as N(0,1) in the null. It is quite clear from Table 2 that the use of OLS on such an Euler equation generates considerable bias. Our discussion of results therefore centers on the GMM estimates. The Sargan criteria at the foot of the Table suggests that the assumed properties on the instruments is acceptable.

The first coefficient ρ_0 is self explanatory and feeds in directly to the intertemporal substitution elasticity which may be written as:

$$\Phi_h = \frac{\ln C_t^h}{\rho_0 + \Sigma_k \rho_k z_{tk}^h - \ln C_t^h} \tag{54}$$

for household type h. The remaining coefficients refer to the ρ_k parameters in (53) and only switch on when the household type possesses one of these characteristics. The base level household represents a childless couple living in rented accommodation with the head of household in employment. For this type of household the value of $\ln C_t^h$, the log of real expenditure, is approximately –2.5 indicating a very large and negative intertemporal elasticity using the OLS estimates but a value insignificantly different from –1 for estimates from the GMM estimators.

Table 2 suggests that it is the labour market status variables that dominate the determination of Φ once $\ln C_t^h$ is allowed for. The presence of a head unemployed in the household significantly reduces ρ_t. Conversely, households with a working wife (WWORK) have a significant increase in ρ_t as do households with multiple adults (MULTADLT). In each column of Table 2 we present the mean value of Φ, the intertemporal elasticity of substitution, implied by these estimates.

In the empirical results reported in Blundell, Browning and Meghir (1989) it was suggested that the curvature of the Engel curves from the Almost Ideal Model provided a better description of within period preferences than either a corresponding quasihomothetic model or a quadratic model. That paper also attempted to establish whether such mis-specification would lead to severe mis–specification in the intertemporal model.

While the results for the AI model and for a quasihomothetic one were similar – the intertemporal elasticity of substitution remaining smallest

Table 2: The Cohort Model Estimates

	OLS	GMM
ρ_0	2.22963	–1.9915
	(0.4868)	(1.9915)
HUNEMP	–0.1203	–1.3816
	(0.1469)	(0.4413)
OWNER	0.1987	0.5702
	(0.1579)	(0.4988)
LA	0.3997	0.4165
	(0.1491)	(0.3854)
SGLAT	0.0449	1.7230
	(0.1744)	(0.6563)
MULTAD T	0.2669	0.8776
	(0.1286)	(0.3801)
CHO2	–0.008	0.7989
	(0.2073)	(0.6563)
CH34	–0.1145	0.1845
	(0.1918)	(0.3971)
CH510	0.1162	0.1077
	(0.1057)	(0.2131)
CH118	0.1162	0.2049
	(0.1051)	(0.2162)
WWORK	0.1450	1.0047
	(0.0999)	(0.3078)
Sargan	–	22.6180
Φ	–2.9700	–0.7842
r_1		–2.3124
r_2		1.0975

Source: Blundell, Browning and Meghir (1989)

Notes: Standard errors in parentheses. r_1 and r_2 are first and second order serial correlation test statistics respectively (N(0,1)) under the null. The Sargan test is a 24 degree of freedom test of instrument validity.

for the group where both adults are unemployed, a quadratic model was quite different in this respect, showing that mis–specification in the Engel curves can indeed lead to misleading results on intertemporal substitution.

5 Conclusions

For many aspects of household or individual decision making, the life–cycle framework provides an attractive modelling environment. The ability of such models to distinguish between responses to evolutionary and unanticipated changes in exogenous economic and demographic factors represents a clear advantage over static and ad–hoc dynamic models. To what extent this advantage exists for modelling aggregate responses is more open to question and certainly requires careful treatment when aggregating over individuals some of whom may face constraints on behaviour or corner solutions in the goods and labour market. This paper has concentrated on the construction and properties of models for individual behaviour and individual data. For certain sections of the population (married women in the labour market, for example) the usefulness of the life–cycle framework would seem indisputable. A formal intertemporal optimising framework has been used throughout this paper in order to pinpoint some of the more unreasonable a priori assumptions on inter- and intra–temporal substitution underlying many of the previously estimated life–cycle models. Furthermore, within this framework the interpretation of static and ad–hoc dynamic models is more easily analysed and 'constraints' on behaviour (involuntary unemployment or more generally labour market non participation, for example) can be captured easily through the standard application of rationing theory.

A major part of this paper has been concerned with the representations of life–cycle decision making under uncertainty and intertemporal separability. Our attention has focused on the implicit preference restrictions behind each empirical representation, in particular the popular first differenced models. Although to a large extent the representation adopted in any empirical application is dictated by the accuracy and type of available data, it is still of interest in interpreting estimated models to determine the degree of underlying restrictions. For example, the linear marginal utility constant models (at least under uncertainty) require within-period preferences to lie in two restrictive classes. As neither of these two classes seem reasonable as maintained hypotheses in empirical application, the resulting estimates are very likely to be biased and misleading. The alternative econometric approaches which have been the motivation for this paper attempt to break the link between intertemporal and within-period preferences and allow more general preferences to be estimated.

REFERENCES

Altonji, J G (1982): 'The Intertemporal Substitution Model of Labour Market Fluctuations: An Empirical Analysis', *Review of Economic Studies*, 49, 783–824

——— (1986): 'Intertemporal Substitution in Labour Supply: Evidence from Micro-Data'; *Journal of Political Economy*, 94, (3.2), S176-S215.

Anderson, R W (1979): 'Perfect Price Aggregation and Empirical Demand Analysis'; *Econometrica*, 47, 1209-1230.

Anderson, G J & R W Blundell (1982): 'Estimation and Hypothesis Testing in Dynamic Singular Equation Systems'; *Econometrica*, 50, 1559-1571

——— & ——— (1983): 'Testing Restrictions in a Flexible Dynamic Demand System: An Application to Consumers Expenditure in Canada'; *Review of Economic Studies*, 50, 397-410.

Attfield, C L & M J Browning (1985): 'A Differential Demand System, Rational Expectations and the Life-Cycle Hypothesis', *Econometrica*, 53, 31-48.

Betancourt, R.R. (1971), Intertemporal Allocation under Additive P : Implications for Cross–Section Data, *Southern Economic Journal*, 37, 458-468

Blackorby, C, D Primont,& R R Russell (1978): *Duality, Separability and Functional Structure* (New York, North–Holland).

Blundell, R W (1986): 'Econometric Approaches to the Specification of Life-Cycle Labour Supply and Commodity Demand Behaviour'; *Econometric Reviews*, 1.

———, M J Browning & C Meghir(1989): 'A Microeconometric Model of Intertemporal Substitution and Consumer Demand', UCL Economics Discussion Paper, 89–11.

———, V Fry & C Meghir (1985): 'λ–Constant and Alternative Empirical Models of Life–Cycle Behaviour under Uncertainty', forthcoming in Laffont et al (eds.), *Advances in Microeconometrics*, Blackwells, Oxford, 1989.

———, P Pashardes & G Weber (1988): 'What do we Learn About Consumer Demand Patterns From Micro Data' Institute For Fiscal Studies, Working Paper 88–10.

——— & R J Smith (1984): 'Separability Exogeneity and Conditional Demand Models', University of Manchester, Department of Econometrics, Mimeo.

——— & I Walker (1982): 'Modelling the Joint Determination of Household Labour Supplies and Commodity Demands', *Economic Journal*, 92, 351–364.

——— & ——— (1986): 'A Life–Cycle Consistent Empirical Model of Family Labour Supply using Cross–Section Data', *Review of Economic Studies*, 53, 539–558.

Boyer, M (1983): 'Rational Demand and Expectations Patterns under Habit Function', *Journal of Economic Theory*, 58, 99–122.

Browning, M J (1985): 'Which Demand Elasticities Do we Know and Which Do we Need to Know for Policy Analysis', McMaster Economics Discussion Paper 85–13.

———, A S Deaton & M Irish (1985): 'A Profitable Approach to Labour Supply and Commodity Demands over the Life–Cycle', *Econometrica*, 53, 503–544.

Chamberlain, G (1984): 'Panel Data'; in Z. Griliches ed, *Handbook of Econometrics*, North Holland.

Deaton, A S (1974): 'Reconsideration of the Empirical Implications of Additive P', *Economic Journal*, 84, 338–348.

——— (1985a): 'Panel Data from Time Series of Cross Sections', *Journal of Econometrics*, 30, 109–126

——— (1985b): 'Life–cycle Models of Consumption: Is the Evidence Consistent with the Theory?'; in T Bewley (ed) *Advances in Econometrics*, Vol II, Cambridge.

——— & J Muellbauer (1980): *Economics and Consumer Behaviour*; Cambridge University Press.

Epstein, L G (1975): 'A Disaggregate Analysis of Consumer Demand Under Uncertainty'; *Econometrica*, 43, 877–895.

Epstein, L G & A Yatchew (1985): 'The Empirical Determination of Technology and Expectations: A Simplified Procedure'; *Journal of Econometrics*, 27(2), 235–258.

Ferguson, T (1958): 'A Method of Generating Best Asymptotically Normal Estimates with Application to the Estimation of Bacterial Densities'; *Annals of Mathematical Statistics*, 29, 1046–1061.

Ghez, G & G S Becker (1975): *The Allocation of Time and Goods over the Life Cycle*; New York, Columbia University Press.

Gorman, W M (1959): 'Separable Utility and Aggregation'; *Econometrica*, 27, 469–481.

——— (1968): 'Conditions for Additive Separability'; *Econometrica*, 36, 605–609.

Gourieroux, C, J J Laffont & A Monfort (1980): 'Coherency Conditions in Simultaneous Linear Equation Models with Endogenous Switching Regimes'; *Econometrica*, 675–95.

Ham, J C (1986): 'Testing Whether Unemployment Represents Intertemporal Labour Supply'; *Review of Economic Studies*, Vol LIII (4), No 175, 559–578.

Hansen, L P & T Sargent (1981): 'Linear Rational Expectations Models for Dynamically Interrelated Variables', in R Lucas and T Sargent (eds), *Rational Expectations and Econometric Practice*, University of Minnesota Press, Minneapolis.

——— & K J Singleton (1982): 'Generalised Instrumental Variables Estimation of Non–Linear Rational Expectations Models', *Econometrica*, 50, 1269–1286.

Hausman, J A (1978)" 'Specification Tests in Econometrics', *Econometrica*, 48, 697–720.

Hayashi, F (1987): 'Tests for Liquidity Constraints: A Critical Survey and Some New Results', in T Bewley (ed), *Advances in Econometrics: Fifth World Congress*, Vol II, Cambridge University Press.

Heckman, J J & T MaCurdy (1980): 'A Life Cycle Model of Female Labour Supply'; *Review of Economic Studies*, 47, 47–74.

——— & ——— (1982): 'Corrigendum on a Life–Cycle Model of Female Labour Supply', *Review of Economic Studies*, 49, 659–660.

Kennan, J (1979): 'The Estimation of Partial Adjustment Models with Rational Expectations', *Econometrica*, 47, 1141–1165.

King, M A (1985): 'The Economics of Saving: A Survey of Recent Contributions', in K Arrow & S Honkapojha (eds), *Frontiers in Economics*, Basil Blackwell (Oxford).

MaCurdy, T E (1981): 'An Empirical Model of Labour Supply in a Life Cycle Setting', *Journal of Political Economy*, 89, 1059–1085.

——— (1983): 'A Simple Scheme for Estimating an Intertemporal Model of Labour Supply and Consumption in the Presence of Taxes and Uncertainty', *International Economic Review*, 24, 265–289.

Meghir, C (1985): 'The Comparative Statics of Consumer Demand under Uncertainty', UCL Economic Discussion Paper 85–21.

Muellbauer, J (1983): 'Surprises in the Consumption Function', *Economic Journal* (conference papers),34–50.

——— & P Pashardes (1982): 'Tests of Dynamic Specification and Homogeneity in Demand Systems', Birkbeck College Discussion Paper, No.125.

Murphy, A (1986): 'Intertemporal Substitution and Consumption', mimeo, Maynooth College, Dublin.

Neary, J P & K Roberts (1980): 'The Theory of Household Behaviour under Rationing', *European Economic Review*, 13, 25–42.

Nickell, S J (1988): 'The Short–Run Behaviour of Labour Supply'; in T F Bewley (ed), *Advances in Econometrics: Fifth World Congress* Vol II, Cambridge University Press.

Philips, L & F Spinnewyn (1981): 'Rational and Myopic Demand System'; in R Bassman & J Rhodes (eds) *Advances in Econometrics*, JAI Press.

Pindyck, R S & J J Rotemberg (1983): 'Dynamic Factor Demands under Rational Expectations', *Scandinavian Journal of Economics*, 85.

Pollak, R A (1970): 'Habit Formation and Dynamic Demand Function', *Journal of Political Economy*, 78, 77–78.

——— (1971): 'Conditional Demand Functions and the Implications of Separability', *Southern Economic Journal*, 37, 423–433.

Sargan, J D (1958): 'The Estimation of Econometric Relationships Using Instrumental Variables', *Econometrica*, 26, 393–415.

Smith, R J & R Blundell (1986): 'An Exogeneity Test for the Simultaneous Equation Tobit Model with an Application to Labour Supply', *Econometrica* ,54, 679–686.

Spinnewyn, F (1981): 'Rational Habit Formation', *European Economic Review*, 15, 91–109.

Weissenberger, E (1986): 'An Intertemporal System of Dynamic Consumer Demand Functions', *European Economic Review*, 30, 859-892.

Wu, D M (1973): 'Alternative Tests of Independence between Stochastic Regressors and Disturbances', *Econometrica*, 41, 733–750.

Conference Discussion

Note: If we imagine a consumer maximising discounted utility (1) subject to the life–cycle wealth constraint (5) then one way of viewing the first order conditions is to write them as in (13) and (14).

Equations (13) are the usual conditions for within period utility maximisation where π_{it} is the price of good i in period t while (14) is the update rule or 'Euler' equation describing the optimal evolution of the marginal utility of wealth λ_t.

i Intertemporal Separability

TG If you've additive utility, you've normalised up to an affine transformation, therefore the marginal utility of income being constant has a constant meaning: while it no longer has a constant meaning if you haven't got such a normalisation. So the curves you get depend on your normalisation and your views about utility.

RB I think that will come out very clearly and the normalisation point is crucial to all of this. This (equation 13) is the set of standard conditions that give Marshallian demands within period and these are invariant to normalisations of the standard kind; whereas this update rule (14) isn't and it is essentially going from (13) to (14) ie, going from standard Marshallian demands to this update rule that many of the restrictions that have been imposed in estimated models enter in an unwitting way.

Note: There are various ways of incorporating uncertainty into these models. So far we have assumed perfect foresight. The simplest generalisation is to perfect certainty where point expectations are allowed to be revised at each time period, but once revised the assumed outcomes are considered to hold with certainty. Full uncertainty would allow a distribution of outcomes for any uncertain variable and one would have to take expectations over the whole distribution.

Under uncertainty, maximising life–cycle utility conditional on expectations formed in period t simply involves rewriting (14) as

$$E_t(\lambda_{t+1}) = (1/\phi(1 + r_{t+1}))\lambda_t$$

as is given by equation (25).

ii JM Is the interest rate r stochastic or non–stochastic ?

RB I'm assuming it's non–stochastic. Only things in λ_t are stochastic, but clearly by re–arranging the equation above you could write the update rule with a stochastic r by using the conditional expectation (ie $E_t(\phi(1 + r_{t+1})\lambda_{t+1}) = \lambda_t$).

iii TG I think it's an awful shame we don't use the discount factor rather than the interest rate, because here it would be the expectation of the discount rate that would come in and be simpler.

RB If the interest rate equals the discount rate, then $\phi(1 + r) \equiv 1$.

TG No, not the discount rate, the discount factor.

RB Okay, I've kind of set it up like that. Parameter ϕ is the discount factor and I could take the expectations on the RHS and work directly with the discount factor.

Note: If λ_t is distributed log–normally then (26) can be rewritten

$$E_t \ln\lambda_{t+1} = \ln(1/\phi(1+r_{t+1})) + \ln\lambda_t + (1/2)\sigma_t$$

where σ represents the conditional variance of λ_{t+1}. After removing the expectation, this generates an equation as described by (28). If the conditional variance can be assumed constant it simply adds an additional intercept term to the stochastic update rule (Euler equation) for the marginal utility of wealth λ_t.

iv TG If you use a discount factor and make it log normal as well, in my belief you get no variance term. You'd just get the log normal means, basically, and you'd get the excess bit which was another normal residual.

RB That's right. The main point to note is that if you did that, and this was an uncertain variable, putting the actual values in here would be correlated with the innovations because.....

TG If you make it jointly log–normal you'd get an additive as an extra factor.

v SP What's the objection to sticking with the original equation rather than switching to logs?

RB Nothing really. Nothing whatsoever. You really want λ to be positive everywhere, and it's rather nice to choose a stochastic rule which doesn't involve adding on a truncated distribution.

Note: An example of the type of Euler equations typically estimated is given by assuming an explicitly additive form for within and across period preferences as in (37). The Euler equation then takes the form

$$\Delta \ln x_{it} = \Delta \sigma_i \ln \alpha_{it} - \sigma_i \, \Delta \ln \pi_{it} + e_{it} \qquad (*)$$

where Δ is the first difference operator. This is indeed the form used by Heckman and MaCurdy (1980).

vi JM You're also assuming no stochastic element in marginal utility?

RB Yes, but you get round that by saying well let's just say α_i was stochastic across individuals, a fixed effect say. So if there were any individual specific fixed effects and if you were using panel data, you could difference those out. So it does allow a very wide distribution of unobservables across individual agents in their determination of this model. You could have individual or household specific α_{it}'s provided they were constant across the panel. They would just come into the fixed effect and could be differenced out.

vii CB It's not just the additivity within periods and it's not just that it's additivity across time, it's not an ordinal function. You couldn't proceed if you took any monotonic transformations.

RB This model [(*) above] fixes the monotonic transformation in a very peculiar way and in a way that seems completely unjustifiable.

CB The story I would try to tell myself is that there's more than consumer out there and that the whole world generates through essentially a very additive structure like this.

TG Well, it's a sort of aggregate existing for every individual good isn't it really? Over time as well which is really not very acceptable. But John's point: If you make the π, α and λ all jointly log–normal, you'll get an error term which is of expectation zero added on to the first equation. Admittedly when you take the Δ's you'll have autocorrelation problems.

RB The nice thing about that earlier equation is that it gives properties of e_{it+1} after having taken differences. If you thought there were measurement errors in the π_{it}, this would clearly lead to an MA type error structure and indeed, in estimation, you'd want to allow for that kind of thing by choosing instruments carefully as we do in our illustration.

Note: The log linear form for the Euler equation in (*) above imposes very strong restrictions on both within and across period preferences. Essentially it is fixing substitution possibilities by assuming explicit additivity across goods and across time. Browning, Deaton and Irish (1985) try to relax the restrictions in this model but only partially succeed. To their model indirect utility has the quasi–homothetic form given in (42) clearly relaxing within period additivity. However, they restrict the intertemporal elasticity ϕ_t to tend toward zero for richer households. Moreover, within period p display zero substitution as the household becomes richer. Here we interpret a household as getting 'richer' if their current period allocation of wealth y_t increases.

viii TG I think that there is another serious point about that. Consider it as a function of y_t forgetting about the other bits. It is the constant absolute risk aversion function. In a savings set–up that says after you've got what you absolutely need your extra expenditure will be the same in every period if you get more money. Of course you have to adjust for a and b but that's quite....

RB Yes, I think I agree with that. The way I've looked at those problems is to work out the corresponding intertemporal elasticity. If you think of the elasticity in terms of the movement of total prices in the period and think of the effect on consumption in that period. I use this Φ definition in (20) which is one of Martin Browning's definitions.

TG But in my case it was unexpected income. If you got a little bit of income you'd plan to spend the same amount extra in each period and that's strong.

RB It's certainly strong. You're absolutely right.

ix TG You said you thought it was quite sensible to use the linear expenditure system $a_t = \Sigma_i a_i \pi_{it}$ in (42). I do not believe that to be true because additive utility gives fantastically low substitution in the case of the Linear Expenditure System that frequently means the contractual expenditure has to be a high proportion of average total expenditure. So for a lot of people it's above total expenditure so they're consuming where the indifference curves go the wrong way round.

RB I wouldn't support it but in some cases I quite like the idea of fixed bundles of goods. What I'm really driving at is that if you're going to generalise L.E.S. preferences along the lines Browning, Deaton and Irish were suggesting, I wouldn't choose to take Leontief preferences for the rich and general substitution preferences for the poor.

Note: It is possible to relax the restrictions underlying these models without extra data. We need to choose a method that estimates general within period preferences that are invariant to the monotonic transformation on each period's preferences. Then by specifying a general monotonic transformation F which may be time dependent, we can recover an intertemporal elasticity that is not bound by the restrictions in the above models. For example, we could use a general transformation of quasi–homothetic preferences.

$$F = \frac{\alpha_t((y_t - a_t)/b_t)^\beta}{\beta}$$

as underlies equation (43). The forms for a and b can be recovered by 'best practice' Marshallian demand estimation without restricting within period substitution. Intertemporal preferences could then be recovered from the more general Euler equation (45). The intertemporal elasticity ϕ_t is given by equation (47). Indeed, a more general form for within period preferences, for example John Muellbauer's PIGL generalisation of the quasi-homothetic model, could be used to represent within period preferences. The intertemporal model can be made to nest the Hall type consumption models as can be seen from the form of the Euler equation given in equation (46).

As a start in this direction we present some results (described in Blundell, Browning and Meghir (1989) see Section 4) using pooled UK Family Expenditure Survey data for the years 1970–1984. There are about 65,000 observations in such a pooled sample and care needs to be taken in estimation and specification. The Almost Ideal version of John Muellbauer's PIGL model was used to represent within period preferences.

x SP What about regional price data?

RB We have regional dummies so those will cope with that to some extent. There are a lot of dummies in this model.

xi SP Do you have one of these Stone indices for each household ?

RB Yes.

SP But the budget shares you use for weighting are going to be awfully random.

RB Yes we've done various things and each time we do something someone criticises us for it. There is no doubt that it's not insensitive to the choice of the price index and strictly speaking we should go back and use the correct price index. However, we do instrument the real expenditure variable that includes these weights.

xii JM There's also the smoker/non–smoker problem. The food equation is going to have different parameters for smokers and non–smokers.

RB We have a dummy for smoking in each expenditure share equation.We don't estimate a tobacco smoking equation.

xiii TG You get dominance on the whole almost 2:1 of gross complementarity, which is quite interesting. If you look at the numbers, the negatives tend to be bigger than the positives, so if you added them up the average is probably more dominant.

RB I haven't really had time to look carefully at those.

xiv SP There are a lot of big problems with dynamic adjustment, particularly for fuel.

RB Oh yes, we condition on as many stock variables as we could, durables etc.

SP Did you notice any big problems in 1973/74 ?

SP You have the three–day working week dummy in as well.

RB We have dummies for most of these things. I'm not quite sure if they're in here. You're right, I'm sure we should worry more about dynamic adjustment for fuels but it's difficult to know how to do that with this data. The route we took was to condition on durables including fuel using appliances.

Note: Given these first stage estimates of within period preferences we then write down the expression for marginal utility λ_t for each individual household in each cohort in each year in terms of the estimated within period preference parameters and the additional unknown intertemporal parameters α and β. These are then estimated (see Section 4) using the stochastic Euler equation for $\ln\lambda_t$ using cohort means to replace actual variables as the Family Expenditure Survey is not a true panel. Of course, here we are taking cohort means of functions of variables. For example, in the quasi-homothetic model described on page 37 of the paper this would involve constructing cohort means of $\ln y_t^{[}$ and $\ln b_t\,(\pi)$ from the first stage estimations. The intertemporal parameters α and β parameters would then be estimated from (45) at the second stage.

xv AS This is effectively an assumption saying that within any cohort there is a distribution of unobserved household parameters.

RB That's right.

AS And that distribution remains the same.

RB Conditional on all these bits and pieces we've estimated so people in the cohort could be unemployed or not, they could have central heating or not and that all affects their preferences via the observed conditioning variables.

AS I'm just thinking of an unobserved taste parameter which determines their consumption pattern on top of all these observed characteristics. It's the distribution of tastes, effectively the unobserved parameter, that has the same distribution.

RB Yes, I think that's right. They way I like to look at it is through the Euler equation. This is true for any household, so under what circumstances is it true for a cohort? Well, really it says that you're conditioning on information from period t but it's a different group sampled from the same cohort. What you've got to say is that all bits of information that were relevant for the past cohort in period t are as relevant for the new sample from that same cohort.

AS The assumption here is that tastes vary across age because you're taking age cohorts. What would be the difference between taking household size cohorts ?

RB You've really got to justify it in terms of whether the information relevant for the past cohort is relevant now.

AS This is saying that it's their historical background.

RB Yes.

JM But you do allow the cohorts to vary by unemployment and tenure type?

RB Oh yes, their internal preferences are very general.

TG If the intertemporal elasticity measure is between two time periods you'd expect the variables for each time period to come in, not just one of them.

RB That's right. What you're saying is, given all the structure of preferences in the current period, what happens if I increase prices in the period without changing anything else? It's that kind of elasticity.

Measurement and Modelling in Economics
G.D. Myles (Editor)

MORE MEASURES FOR FIXED FACTORS

by W M Gorman*

Abstract

Existing 'perfect' aggregates can be considered as quantities, $v = (v_m)$ of intermediate goods $m \in M$: intermediate, eg, between fixed inputs $u = (u_f)$, say, and current goods measured as outputs.

$$x_i = \Sigma_f x_{fi} = \Sigma_f \pi_i^f(p,u_f) = \Sigma_f \partial\pi^f/\partial p_i = \psi^i(p,v), \text{ say,} \quad (1)$$

where $v = \phi(p,u)$, in an obvious notation. Here I take (1) and its Muellbauer style analogue $x_i/x_1 = \theta^i(p,v)$ as definitions, and explore their implications informally. These are that corresponding subaggregates $v_f = (v_{fm})_{m \in M}$ exist for each firm f such that $v = \Sigma_f v_f$ in an appropriate normalisation: that its gross profit function $\pi^f(p,u_f) = \lambda_m^f(a(p),\ u_f) + \mu^f(p)$, say, where $a(p) = (a^m)(p))_{m \in M}$ is a vector of shadow prices for the aggregates so normalised; and that $v_{fm} = \lambda_m^f(a)p)$, $u_f = \partial\lambda^f/\partial a^m$ as one would like. These results clearly apply to aggregates for Labour, or Food, as well as for Capital, and to groups of households as well as firms.

Various related topics are discussed briefly in the text and in appendices.

1 Introduction

The discussion is cast in terms of fixed factors. With appropriate changes it can be applied to vector social welfare, vector income measures, equivalent adult constructions,... as could the earlier results.[2]

* Nuffield College, Oxford.

[1] To the Quantative Economic Workshop there in January 1978 and to Jim Mirrlees in Oxford later that year. I thank the participants, especially Jim, for their comments.

[2] In particular they can be taken as vectors of aggregates for current goods. See page 387, between equations (38) and (39).

Last year I generalised the discussion in Gorman (1967) to the case where there might be a vector $\phi(y)$ of aggregates for capital, y being the extended vector of fixed inputs Y into the firms t = 1,2,...,T in the community. I required that

$$g(p,y) = \Sigma g^t(p,y_t) = G(p,\phi(y)), \tag{2}$$

be the gross profit function for the economy as a whole. This implied the existence of subaggregates

$$\phi^t(y_t) = (\phi^{1t}(y_t),..,\phi^{Rt}(y_t)), \tag{3}$$

such that

$$g^t(p,y_t) = \Sigma_r a^r(p)\phi^{rt}(y_t) + b^t(p), \tag{4}$$

$$g(p,y) = \Sigma_r a^r(p)\phi^r(y) + b(p), \tag{5}$$

$$g = \Sigma_t g^t,\ \phi^r = \Sigma_t \phi^{rt},\ b = \Sigma_t b^t, \tag{6}$$

so that the apparent generalisation of Gorman (1967) gave no extra mileage.

In this paper I will seek capital aggregates of the form

$$\phi(p,y) = (\phi^1(p,y),..,\phi^R(p,y)), \tag{7}$$

into which the prices p of the variable goods may enter. Since g(p,y) is already of this type, we cannot merely require that such aggregates enter into the gross profit function. Instead I explore two related problems:

$$g_i(p,y) = \partial g/\partial p_i = G^i(p,\phi(p,y)), \tag{8}$$

$$g_i(p,y)/g_1(p,y) = H^i(p,\phi(p,y)), \tag{9}$$

each good i. That is the net supplies of the variable goods, or the relative net supplies, depend on y only through these aggregates.

2 Net Supply Functions

We require

$$g_i(p,y) = \Sigma_t g_i^t(p,y_t) = G^i(p,\phi(p,y)), \text{ each } i, \tag{10}$$

and "show" that this is so if and only if

$$g^t(p,y) = K^t(a(p),y_t) + b^t(p), \text{ say}, \tag{11}$$

$$g(p,y) = K(a(p),y) + b(p), \text{ say} \tag{12}$$

$$\Sigma g^t = g,\ \Sigma K^t = K,\ \Sigma b^t = b;\ a(p)$$

$$= (a^1(p),\ldots,a^R(p)), \tag{13}$$

each $a^r(\cdot)$ a marginal profit function, in which case we may take

$$\phi^{rt}(p,y_t) = K_r^t(a(p),y_t) := \partial K^t/\partial a^r,\ [=: \theta^{rt}(a(p),y_t)] \tag{14}$$

$$\phi^r(p,y) = K_r(a(p),y) = \Sigma_t \theta^{rt}(a(p),y_t)\ [=: \theta(a(p),y)]. \tag{15}$$

Now

$$K^t = \Sigma_r K_r^t a^r;\ K = \Sigma_r K_r a^r, \tag{16}$$

by homogeneity. Hence

$$g^t(p,y_t) = \Sigma_r a^r(p)\phi^{rt}(p,y_t) + b^t(p), \tag{17}$$

$$g(p,y) = \Sigma_r a^r(p)\phi^r(p,y) + b(p), \tag{18}$$

$$g = \Sigma_t g^t,\ \phi^r = \Sigma_t \phi^{rt},\ b = \Sigma_t b^t, \tag{19}$$

which is remarkably similar to (1), (4)–(6) and the corresponding result in Gorman (1967).

Note that the net supply functions

$$x_{ti} = g_i^t = \Sigma_r K_r^t a_i^r + b_i^t = \Sigma_r a_i^r \phi^{rt} + b_i^t, \tag{20}$$

$$x_i = g_i = \Sigma_r K_r a_i^r + b_i = \Sigma_r a_i^r \phi^r + b_i, \tag{21}$$

may be derived by differentiating (17) and (18) and neglecting the dependence of the aggregates on p. That is, we can treat the aggregates as pure measures of capital.

To to all this we proceed much as in Gorman (1967) and (1976). I eschew rigour to a remarkable degree.

Assume $R < n$ = number of goods. There is no problem otherwise. Solve (10) for $\phi(p,y)$ in terms of $g_1(p,y),\ldots,g_R(p,y)$[3] and normalise to take those as the new $\phi^1(p,y),\ldots,\phi^R(p,y)$. Then

$$\phi^r(p,y) = \Sigma_t \phi^{rt}(p,y_t);\ \phi^{rt}(p,y_t) = g_r^t(p,y_t), \tag{22}$$

and (10) becomes

$$\Sigma_t g_i^t(p,y_t) = G^i(p,\Sigma_t \phi^t(p,y_t)), \text{ say, each } i, \tag{23}$$

$$\phi^t = (\phi^{1t},\ldots,\phi^{Rt}). \tag{24}$$

3 So that the appropriate Jacobian is of rank $R \geqq 1$ in the neighbourhood in question, we then number the appropriate rows 1,2,...,R.

This is just the affinity[4] equation of Gorman (1967) and (1976) and, as shown there, in a wide class of cases its solution is (17) as required.

I will sketch the argument again here. First

$$g^t_i(p,y_t) = G^{ti}(p,\phi^t(p,y_t))), \text{ say.} \tag{25}$$

To show this hold p,y_s, $s \neq t$ constant, and vary y_t holding $\phi^t(p,y_t)$ constant in (23). Now choose a reference vector $\bar{y}$, define

$$b^t_i(p) = G^{ti}(p,\phi^t(p,\bar{y}_t)) = g^t_i(p,\bar{y}_t),$$

$$b_i(p) = G^i(p,\phi(p,\bar{y})) = g_i(p,\bar{y}), \tag{26}$$

$$K^{ti}(p,\phi^t(p,y_t)) = G^{ti}(p,\phi^t(p,y_t))$$

$$- b^t_i(p)[= g^t_i(p,y_t) - g^t_i(p,\bar{y}_t) =: R^t_i(p,y_t),] \tag{27}$$

$$K^i(p,\phi) = G^i(p,\phi(p,y)) - b_i(p), [= g_i(p,y) - g_i(p,\bar{y}) =: h_i(p,y),] \tag{28}$$

to get

$$K^{ti}(p,z) = K^i(p,z), [z = \phi(p,y)], \tag{29}$$

$$\Sigma K^i(p,z_t) = K^i(p,z_t), [z_t = \phi^t(p,y_t)], \tag{30}$$

in (25). This is the linearity equation yielding

$$K^i(p,z) = \Sigma_r a^{ri}(p) z_r, \tag{31}$$

---

4 [By analogy with Cauchy's "linearity equation".]

under very general conditions. (25)–(27) then yield[5]

$$g_i^t(p,y_t) = \Sigma_r a_i^r(p)\phi^{rt}(p,y_t) + b_i^t(p), \tag{32}$$

and (10), (28)

$$g_i(p,y) = \Sigma_r a_i^r \phi^r(p,y) + b_i(p). \tag{33}$$

Now $g^t = \Sigma_i p_i g_i^t$, $g = \Sigma_i p_i g_i$, etc. Hence

$$g^t(p,y_t) = \Sigma_r a^r(p)\phi^{rt}(p,y_t) + b^t(p), \tag{34}$$

$$g(p,y) = \Sigma_r a^r(p)\ \phi^r(p,y) + b(p). \tag{35}$$

Alternatively, integrating (32), (33) we get

$$g^t(p,y_t) =: K^t(a(p),y_t) + b^t(p), \tag{36}$$

$$g(p,y) =: K(a(p),y) + b(p), \tag{37}$$

and differentiating with p_i and identifying with (32)–(33),

$$K_r^t(a(p),y_t) = \phi^{rt}(p,y_t);\ K_r(a(p),y) = \phi^r(p,y) \tag{38}$$

as required.

This "proves" (2)–(21).

---

[5] The essential condition is that we should be able to manipulate enough y's while holding the others constant. To do so we may need to group firms as in Section 6. If so, the appropriate Jacobian for the economy as a whole, and for each subeconomy formed by throwing out just one of the groups, must be of the same rank R. If you like, none of the 'factors' must be specific in a particular group, when the group has been appropriately assembled – for this it seems to be enough that none be specific to a single firm. The replacement of a^{ri} in (31) by $a^r{}_i$ in (32) is justified on page 399 below.

It might not seem that this further weakening of our requirements has got us very far either. On the other hand (36)–(37) is reasonably wide.

Some interpretation is called for.

Suppose for the moment that $a^1(\cdot),\dots,a^R(\cdot)$ are actual profit functions, not merely marginal profit functions. $b^1(\cdot),..,b^T(\cdot),b(\cdot)$ certainly are by (26). Then we can treat them as unit profit functions. As such they will have dual constant returns 'transformation functions' corresponding to $a(\cdot)$, $b(\cdot)$, $f(\cdot) = (f^r(\cdot))$, $h(\cdot) = (h^t(\cdot))$. These are quantities of fictitious variable goods – ie, variable goods aggregates – and

$$z_r := f^r(x_r) = \partial K/\partial a_r = \phi^r(p,y), \tag{39}$$

$$z_{tr} := f^r(x_{tr}) = \partial K^t/\partial a_r = \phi^{rt}(p,y_t), \tag{40}$$

so that each unit of fixed aggregate r yields a unit of variable aggregate r without other inputs! – either in the economy at large or the individual firm.

$$w_t = h^t(x_{t_0}) = \partial g^t/\partial b^t = 1, \text{ each } t, \tag{41}$$

is, if you like, the overhead variable good in firm t, $w = h(x_0) = 1$ for the economy as a whole.

Where the $a^r(\cdot)$ are not profit functions, it is better to interpret them as processes, and the $b^t(\cdot)$, $b(\cdot)$ still as common overheads. Then one unit of $\phi^r(p,y)$ supports process r at level one where it is available.

When process r is running at intensity z_r it yields $z_r a_i^r(p)$ units of variable good i, at prices p. When I say that $\phi^r(p,y)$ supports it, I mean that $z_r = \phi^r(p,y)$, $z_{tr} = \phi^{rt}(p,y_t)$, as in (39)–(40).

Note that the shadow price of the rth capital factor is $a^r(p)$ everywhere, without there being a market for it. This is as in Gorman (1967) and (1976).

3 Relative Supplies

I require

$$g_i(p,y)/g_1(p,y) = \Sigma_t g_i^t(p,y)/\Sigma_t g_1^t(p,y),$$

$$= H^i(p,\phi(p,y)), \text{ say,} \tag{42}$$

as a further relaxation.

Solve the first[6] R equations and normalise to get

$$\phi^r(p,y) = g_r(p,y)/g_1(p,y) = \psi^r(p,y)/\psi^1(p,y), \text{ say} \tag{43}$$

where

$$\psi^r(p,y) := g_r(p,y) = \Sigma g_r^t(p,y_t) =: \Sigma \psi^{rt}(p,y_t), \text{ say,} \tag{44}$$

$r = 1,\ldots,R + 1.$

(42) now reads,

$$\Sigma_t g_i^t(p,y) = [H^i(p,\psi/\psi^1)\cdot \psi^1 =]\ G^i(p,\psi(p,y));\ G^i(p,\cdot) \text{ conical,}$$

$$= G^i(p,\Sigma\psi^t(p,y_t));\ \psi^t = (\psi^{1t},\ldots,\psi^{R+1,t}). \tag{45}$$

Treating this affinity[7] equation just like (23), we get

---

6 So that the appropriate Jacobian is of rank $R \geqq 1$ in the neighbourhood in question, we then number the appropriate rows 1,2,....,R.

7 By analogy with Cauchy's "linearity equation".

$$g_i^t(p,y_t) = \Sigma_r a_i^r(p)\psi^{rt}(p,y_t) + b_i^t(p), \tag{46}$$

$$g_i(p,y) = \Sigma_r a_i^r(\pi)\psi^r(p,y), \tag{47}$$

since

$$b(p) = \Sigma_t b^t(p) = 0, \tag{48}$$

because $G^i(p,\cdot)$ is conical in ψ we also get

$$g^t(p,y_t) =: K^t(a(p),y_t) + b^t(p), \text{ say}, \tag{49}$$

$$g(p,y) =: K(a(p),y), \tag{50}$$

$$\psi^{rt}(p,y_t) = K_r^t(a(p),y_t);\ \psi^r(p,y) = K_r(a(p),y), \tag{51}$$

$$g^t(p,y_t) = \Sigma_r a^r(p)\psi^{rt}(p,y_t) + b^t(p), \tag{52}$$

again as in Section 2.

(49) may be thought rather a coincidence. In particular it could not hold if we required firms to be born and die. If so we would have

$$\text{each } b^t(p) = 0, \tag{53}$$

in (48) and (51).

It is easily seen that (42) is satisfied with

$$\phi^r(p,y) = K_{r+1}(a(p),y)/K_1(a(p),y) = \psi_{r+1}/\psi_1. \tag{54}$$

Compare with the results in Section 2 and notice how little extra mileage the further relaxation (42) of our requirement on the aggregates yields.

4 Aggregates Independent of Prices

If, in Section 2,

$$g_i(p,y) = G^i(p,\phi(y)), \text{ each } i, \tag{55}$$

$$g(p,y) = \Sigma p_i G^i(p,\phi(y)) = G(p,\phi(y)), \text{ say}, \tag{56}$$

and vice versa, the case discussed in Gorman (1967) and (1976). We are left with the case discussed in Section 3,

$$g_i(p,y)/g_1(p,y) = H^i(p,\phi(y)). \tag{57}$$

Using (48), (49)[8]

---

[8] (59) is not obvious. Here is one way of deriving it. (58) and (49)-(50) imply

i $\qquad \Sigma_{mt}K^t_m(a(p),y_t)a^m_i(p) = H^i(p,\phi(y))\Sigma_{mt}K^t_m(a(p),y_t)a^m_1(\hat{p}).$

Set $p = \overline{p}$, $a(\overline{p}) = \overline{a}$ and define

ii $\qquad \theta^{tr}(y_t) = \Sigma K^t_m(\tilde{a},y_t)a^m_r(\overline{p}), \ r = 1,2 \ldots R+1$

appropriately chosen, to get

iii $\qquad \Sigma_{mt}K^t_m(a(p),y_t)a^m_i(p) = H^i(p,\Sigma\theta^t(y_t))\Sigma_{mt}K^t_m(a(p),y_t)a^m_1(p)$

in a slightly different notation, where $H^i(\cdot)$ is positively homogeneous of degree zero.

Now set $\tilde{y}_t = y_r, t \neq s$, $\theta^s(\tilde{y}_s) = \theta^s(y_s)$, write down (iii) for y too, and difference, to get

iv $\qquad \Sigma_m\{K^{bs}_m(a(p),\tilde{y}_s)a^m_i = K^s_m(a(p),y_s)\}a^m_i(p)$

$\qquad\qquad = H^i(p,\Sigma\theta^t(y_t))\Sigma_m\{K^s_m(a(p),\tilde{y}_s) - K^s_m(a(p),y_s)\}a^m_i(p)$ each i,

so that <u>either</u>

$$K_m(a,y)/K_1(a,y) = h^m(a,\phi(y)), \text{ say}. \tag{58}$$

Writing this down for $a = \bar{a}$, $m = 2,3,...R + 1$, say, solving for $\phi(y)$ and substituting back, we get

$$\Sigma_t K_m^t(a,y_t) = L^m(a,\Sigma_t \psi^t(y_t))\Sigma_t K_1^t(a,y_t), \text{ say}, \tag{59}$$

where $\psi^t = (\psi^{1t},..,\psi^{R+1t})$, and

$$\psi^{rt}(y_t) = G_r^t(\bar{a},y_t),\ r = 1,2,..,R + 1, \tag{60}$$

$$L^m(a,\cdot) \tag{61}$$

is positively homogeneous of degree zero.

Hold a and each y_s constant, $s \neq t$ and vary y_t to $\tilde{y}_t$, where $\psi^t(\tilde{y}_t) = \psi_t(y_t)$. (61) then yields

$$K_m^t(a,\tilde{y}_t) - K_m^t(a,y_t) = L^m(a,\Sigma\psi^t(y_t))(K_1^t(a,\tilde{y}_t) - K_1^t(a,y_t)). \tag{62}$$

Unless

$$K_1^t(a,\tilde{y}_t) - K_1^t(a,y_t) = 0, \tag{63}$$

---

v $\quad \Sigma_m\{K_m^s(a(p),\tilde{y}_s - K_m^s(a(p),y_s\}a_i^m(p) = 0,$

or

vi $\quad H^i(p,\Sigma\theta^t(y_t)) = h^i(p)$, say,

each i, since it is independent of y_t, each $t \neq s$, each d.

Now (vi) certainly holds for $i = 1$, since $H = 1$. By footnote 2 it does not hold for nay of $i = 2...R + 1$. Hence (80) holds for all such i, and hence for $i = 1$, too, by multiplying (80) by p_i and adding, we get

vii $\quad K^s(a(p),\tilde{y}_s) = K^s(a(p),y_s) =: K^r(a(p),O^s(y_s)).$

The proof of (59) is now left to you.

we can divide across by it, to find

$$L^m(a,\Sigma\psi^t(y_t)) = l^m(a), \text{ say.} \tag{64}$$

Rejecting this degenerate case, we have (63), which of course implies

$$\overset{t}{K}_m(a,\tilde{y}_t) = \overset{t}{K}_m(a,y_t) = \overset{t}{M}_m(a,\psi^t(y_t)), \text{ say,} \tag{65}$$

in (62), so that (59) becomes

$$\Sigma_t\overset{t}{M}_m(a,\psi^t) = L^m(a,\Sigma_t\psi^t)\Sigma_t\overset{t}{M}_1(a,\psi^t). \tag{66}$$

Set now

$$\tilde{\psi}^t = \psi^t + \mu;\ \tilde{\psi}^\theta = \psi^\theta - \mu,\ \tilde{\psi}^s = \psi^s, \text{ otherwise,} \tag{67}$$

to get

$$\begin{aligned}&(\overset{t}{M}_m(a,\psi^t + \mu) - \overset{t}{M}_m(a,\psi^t)) - (\overset{\theta}{M}_m(a,\tilde{\psi}^\theta + \mu) - \overset{\theta}{M}_m(a,\tilde{\psi}^\theta)),\\ &= p^m(a,\Sigma\psi^t)\{(\overset{t}{M}_1(a,\psi^t + \mu) - \overset{t}{M}_1(a,\psi^t))\\ &- (\overset{\theta}{M}_1(a,\tilde{\psi}^\theta + \mu) - \overset{\theta}{M}_1(a,\tilde{\psi}^\theta))\},\end{aligned} \tag{68}$$

which once again implies (64), unless $\{\cdot\} = 0$, implying

$$\begin{aligned}&\overset{t}{M}_m(a,\psi^t + \mu) - \overset{t}{M}_m(a,\psi^t) = \overset{\theta}{M}_m(a,\tilde{\psi}^\theta + \mu) - \overset{\theta}{M}_m(a,\tilde{\psi}^\theta),\\ &= \lambda_m(a,\mu), \text{ say,}\end{aligned} \tag{69}$$

and, in turn

$$\lambda_m(a,\mu + \nu) = \lambda_m(a,\mu) + \lambda_m(a,\nu), \tag{70}$$

so that

$$\lambda_m(a,\mu) = \Sigma c^r_m(a)\mu^r, \text{ say}, \tag{71}$$

and

$$K^t_m(a,y_t) = M^t_m(a,\psi^t) = d^t_m(a) + \Sigma_r c^r_m(a)\ \psi^{rt}, \text{ say}, \tag{72}$$

where

$$d^t_m(a) = M^t_m(a,0). \tag{73}$$

This is the same as

$$K^t(a,\psi^t) = d^t(a) + \Sigma_r c^r(a)\psi^{rt}, \tag{74}$$

yielding

$$K(a,\psi) = \Sigma_r c^r(a)\ \psi^r, \tag{75}$$

where

$$\Sigma_t d_t(a) = 0, \tag{76}$$

by (7).

Clearly (75)–(76) also implies (58) with

$$\phi^r(y) = \psi^{r+1}(y)/\psi^1(y),\ r = 1,2,\ldots,R. \tag{77}$$

(77) may seem rather a coincidence. If we require aggregation to remain possible when firms die, it implies

$$\text{each } d^t(a) = 0. \tag{78}$$

Look now at (49). Merge $d^t(a)$ into $b^t(p)$ there, and renormalise to call c^r, a^r.

(75) then becomes

$$g^t(p,y_t) = b^t(p) + \Sigma_r a^r(p)\psi^{rt}(y_t), \tag{79}$$

$$g(p,y) = \sum_{r=1}^{R+1} a^r(p)\psi^r(y);\ \psi^r(y) = \Sigma_t\psi^{rt}(y_t), \tag{80}$$

with

$$\Sigma b^t(p) = 0, \tag{81}$$

always, and

$$\text{each } b^t(p) = 0, \tag{82}$$

if firms may die.

The conditions (79)–(80) are clearly sufficient for (58) as well as necessary.

5 Additive Homogeneity

Consider the solution

$$g(p,y) = K(a,y) + b(p),$$

$$= N(a + be,b,y),\ \text{say},\ e = (1,..,1), \tag{83}$$

in Section 2. Clearly

$$N(a,0,y) = K(a,y), \tag{84}$$

$$N(c,y) = N(c - c_R e,y) + c_R, \text{ say}, \tag{85}$$

$$N(c + de,y) = N(c - c_R e,y) + c_R + d = N(c,y) + d, \tag{86}$$

which I call additive homogeneity, for an obvious reason. More precisely, it is of degree one.

I came upon it first in this context a couple of years ago when I was discussing polynomial Engel curves, and again in a problem in social choice theory. The general condition for it is (83) with $b(\cdot)$ functionally independent of a. Note that, were it not, it could be subsumed into $H(\cdot,y)$.

Now look at (83) with this functional independence.

Write

$$h(p,y) = g(p,y) - b(p) = \tilde{G}(p,\psi) - \tilde{G}(p,0), \text{ say}, \tag{87}$$

a marginal cost function and take

$$\tilde{a}^r(p) = h(p,y^r), \tag{88}$$

where y^r is a fixed y vector. Then solve the equations

$$\tilde{a}^r = \Sigma_s a^s \psi^s(a,y^r) = \Sigma_s a^s \psi^{sr}(a), \text{ say}, \tag{89}$$

for the a's in terms of the $\tilde{a}$'s, substitute back to get

$$\Sigma a^r(p)\psi^r(a,y) = \Sigma \tilde{a}^r(p)\tilde{\psi}^r(\tilde{a},y), \tag{90}$$

in general, and drop the $\sim$'s. The new $\tilde{a}$'s are now marginal profit functions and

$$c^r(p) = g(p,y^r) = b(p) + a^r(p),$$

$$r = 1,2,..,R + 1, \; b(p) = \tilde{G}(p,0), \tag{91}$$

are ordinary profit functions. Finally

$$g(p,y) = \sum_1^R a^r(p)\psi^r(a,y) + b(p),$$

$$= \sum_1^{R+1} c^r(p)\psi^r(c,y); \; \Sigma\psi^r(c,y) = 1,$$

$$= N(c,y), \text{ say.} \tag{92}$$

Differentiating with regard to p_i, we get

$$g_i = \sum_1^{R+1} N_r(c,y)c_i^r. \tag{93}$$

Multiply this by p_i, add, and note the fact that $g(\cdot,y)$, and each $c^r(\cdot)$, is a profit function, to get

$$g(p,y) = \sum_1^{R+1} N_r(c,y)c^r(p); \tag{94}$$

so that we take

$$\psi^r(c,y) - N_r(c,y). \tag{95}$$

Using the solution of (88) again and substituting into (83), we see that it holds in the new normalisation too, with (91). Hence (86) holds, or differentiating with regard to d, and setting d = 0,

$$\Sigma N_r(c,y) = 1, \tag{96}$$

which is what we would expect from (92) and (95).

One advantage of this form is that the $c^r(\cdot)$ are profit functions – though there seems no necessity for $N(\cdot,y)$ to be one. It is conical of course, but does not seem to need to be closed convex. If $f^r(\cdot)$ is the dual of $c^r(\cdot)$, however

$$z_r = f^r(x_r), \text{ say; } x = \Sigma_r x_r,$$

$$= N_r(c,y); \tag{97}$$

rather as at the end of Section 2.

A simple example is the affine cost function

$$g(p,u) = a(p)u + b(p) = a(p) + b(p))u + b(p)(1-u),$$

$$= c^1(p)u + c^2(p)(1-u), \tag{98}$$

where

$$c^1(p) = g(p,1); \; c^2(p) = (p,0). \tag{99}$$

6 A Hint of Rigour

The critical assumption underlying the arguments from linearity above, is that the components [of the ϕ^{rt}] can be varied independently.

There are important cases where this is not so. For example y_t might be a scalar and R > 1 in Section 2, or R > 0 in Sections 3 and 4.

It is quite easy to rearrange[9] the argument to take account of this. Group the "firms" into "industries", = 1,2,.., each sufficiently large for

$$\phi^{r\theta}(p,y_\theta) = \sum_{t\in\theta} \phi^{rt}(p,y_t), \tag{100}$$

to be independently variable. Then (23) becomes

$$\Sigma_\theta \overset{\theta}{g}_i(p,y) = G^i(p,\Sigma_\theta \phi^{r\theta}(p,y_\theta)), \text{ each } i, \tag{101}$$

where

$$g^\theta(p,y_\theta) = \sum_{t\in\theta} g^t(p,y_t), \tag{102}$$

so that the argument leading up to (31) implies

$$\overset{\theta}{g}_i(p,y_\theta) = \Sigma_r \overset{r}{a}_i(p)\phi^{r\theta}(p,y_\theta) + \overset{\theta}{b}_i(p), \tag{103}$$

and thus

$$\sum_{t\in\theta} \overset{t}{g}_i(p,y_t) = \sum_{t\in\theta} \{\Sigma_r \overset{r}{a}_i(p)\phi^{rt}(p,y_t)\} + \overset{\theta}{b}_i(p) \tag{104}$$

by (101) and (102). Hence

$$\overset{t}{g}_i(p,y_t) = \Sigma_r \overset{r}{a}_i(p)\phi^{rt}(p,y_t) + \overset{t}{b}_i(p), \text{ say,} \tag{105}$$

where

---

9 Read in conjunction with footnotes 2 and 5.

$$\sum_{t\epsilon\theta} b^t(p) = b^\theta(p). \tag{106}$$

(105) is the result from which everything else in Section 2 flows.

A precisely similar argument works in Section 3, and, with only a little, rather obvious, adaptation in Section 4.

The other difficulty lies in the normalisations used. They have functionally independent components at national level, and the equations determining them have to be invertible. For this, I presume, one might have to use different bases in different sub–domains. I have not gone into this.[10]

A further point worth taking up is this: how does

$$\overset{t}{g}_i(p,y_t) = \overset{t}{b}_i(p) + \Sigma_r a^{ri}(p)\phi^{rt}(p,y_t), \tag{107}$$

implied by (25), become

$$\overset{t}{g}_i(p,y_t) = \overset{t}{b}_i(p) + \Sigma \overset{r}{a}_i(p)\phi^{rt}(p,y_t), \tag{108}$$

in (31). In (107), i is only an index, in (108) it denotes differentiation.

Here is the construction.

First define

$$h^t(p,y_t) = g^t(p,y_t) - g^t(p,\bar{y}_t) = g^t(p,y_t) \tag{109}$$

$$- b^t(p),\ h(p,y) = \Sigma h^t(p,y_t) = g(p,y) - g(p,\bar{y}) = g(p,y) - b(p), (109)$$

where $\bar{y}$ is taken from (26). Then take fixed vectors $y^1,..,y^R$ and set

$$\tilde{a}^r(p) = h(p,y^r) = g(p,y^r) - b(p). \tag{110}$$

The natural version of (107) then yields

10 See Nuffield Discussion Paper No 2 for a treatment.

$$\tilde{a}_i^s(p) = \Sigma_r a^{ri}(p)\phi^r(p,y^s), \tag{111}$$

$$= \Sigma_r a^{ri}(p)\psi^{rs}(p), \text{ say.}$$

Solve this for $a^{ri}(p)$ as

$$a^{ri}(p) = \Sigma_b \theta^{rs}(p)\tilde{a}_i^s(p), \text{ say,} \tag{112}$$

and define

$$\tilde{\phi}^{st}(p,y_t) = \Sigma_r \theta^{rs}(p)\theta^{rt}(p,y_t), \tag{113}$$

$$\tilde{\phi}^s(p,y) = \Sigma\theta^{rs}(p)\phi^r(p,y),$$

$$= \Sigma_t \tilde{\phi}^{st}(p,y_t). \tag{114}$$

Substitute from (112)–(114) into (107), and drop the ~'s, to get (108).

APPENDICES

APPENDIX A

Equations like (A1)–(A4) below often turn up in the theories of aggregation and decentralisation. In the past I had dealt with them on an ad hoc basis, with the help of whatever else I knew of the structure. Here I guess at what may be the general solution or something like it. The discussion follows a revised version written in 1980, which is slightly less opaque than the version presented at the LSE in 1978.

Consider the differential equations

$$g_i(x) = \lambda^r(x)f_i(x), \text{ each } i \in r, \text{ each } r \in R, \tag{A1}$$

where

$$x = (x_i)_{i \in I}, \tag{A2}$$

$$R = \{r \in R\} \text{ is a partition of I}, \tag{A3}$$

$$\text{the } \lambda(\cdot)\text{'s are distinct and } \lambda^r(x) > 0, \text{ each } r \in R. \tag{A4}$$

Given appropriate smoothness and connectivity which I will assume throughout,

$$g(x) = \phi^r(y_r, f(x)), \text{ say, each } \in R, \tag{A5}$$

where

$$x_r = (x_i)_{i \in r},\ y_r = (x_i)_{i \notin r}, \text{ each } r \in R, \tag{A6}$$

and

$$\overset{r}{\phi}_0(y_r,f(x)) := \partial\phi^r/\partial f = \lambda^r(x) > 0, \text{ each } r \in R. \tag{A7}$$

When $\#R = 1$

$$g(x) = \phi(f(x)), \text{ say,} \tag{A8}$$

of course, given appropriate connectivity.

When $\#R = 2$, (A5) may be written

$$g(x,y) = \phi(x,f(x,y)) = \psi(y,f(x,y)), \tag{A9}$$

so that

$$f(x,y) = \psi^{-1}(y,\phi(x,f(x,y))) = \phi^{-1}(x,\psi(y,f(x,y))) \tag{A10}$$

since $\psi(y,\cdot)$ and $\phi(x,\cdot)$ are strongly increasing. With luck this may be solved iteratively. Similarly

$$g(x,y) = \phi(x,\psi^{-1}(y,g(x,y))) = \psi(y,\phi^{-1}(x,g(x,y))), \tag{A11}$$

and as we will see later

$$h(x,y) = \tilde{\phi}(y,f(x,y)) = \tilde{\psi}(x,f(x,y)), \tag{A12}$$

some $\tilde{\phi}(\cdot)$, $\tilde{\psi}(\cdot)$ iff

$$h(x,y) = H(f(x,y),g(x,y)), \text{ some } H(\cdot). \tag{A13}$$

This suggests the following construction for any finite $R=\{1,2,...n\}$, say,

$$a^r(x) = \alpha^r(x_r,\alpha^{r+1}(x_{r+1},\alpha^{r+2}(\ldots,\alpha^{r-1}(x_{r-1},a^r(x))..))), \tag{A14}$$

each $r \in R$, where

$$\alpha^r(x_r,\cdot) \text{ is strongly increasing, each } r \in R. \tag{A15}$$

Assuming that (A14) converges for each $r \in R$,

$$a^r(x) = \alpha^r(x_r,\alpha^{r+1}(\ldots,\alpha^{s-1}(x_{s-1},a^s(x))..)), \tag{A16}$$

each $r \in R$,

where we identify α^{n+p} with α^p throughout, that is work modulo n. Hence

$$a^r(x) = A^{rs}(x_r,..,x_{s-1},a^s(x)), \text{ say, each } r,s \in R, \tag{A17}$$

$$= A^r(x_r,..,x_n,\ a^1(x)), \text{ say, each } r \in R, \tag{A18}$$

and equally, (A16) yields

$$a^r(x) = B^r(x_1,x_2,..,x_{r-1},a^1(x)), \text{ say, each } r \in R, \tag{A19}$$

when we put $r = 1$, $s = r$ in (A17) and invert.

Thus $f = a^r(x)$, $g = a^s(x)$ satisfy equations like (A5), each $r,s \in R$. Now consider any well behaved

$$f(x) = F(a^r(x)\,|\,r \in R) \tag{A20}$$

$$[:= F(a(x)),\ a(x) = (a^r(x)_{r\in R})].$$

It yields

$$f_i = \sum_{r\in R} F_r a_i^r. \tag{A21}$$

Now (A18) and (A19) imply

$$\overset{r}{a}_i(x) = \mu^{rt}\overset{1}{a}_i(x), \text{ say, each } i \in t, \ r,t \in R, \tag{A22}$$

so that

$$f_i(x) = \Sigma_{r \in R} F^r \mu^{rt} \overset{1}{a}_i = u^t(x)\overset{1}{a}_i(x), \text{ say, each } i \in t \in T, \tag{A23}$$

and hence

$$g_i(x) = \lambda^t(x) f_i(x) \text{ each } i \in t, \text{ by (A1)}, \tag{A24}$$

$$= \sigma^t(x)\overset{1}{a}_i(x), \text{ say, each } i \in t \in R. \tag{A25}$$

I will now show that this implies

$$g(x) = G(a^r(x) \,|\, r \in R), \text{ say.} \tag{A26}$$

This will be so iff

$$g_i(x) = \Sigma_{r \in R} \pi^r(x)\overset{r}{a}_i(x), \text{ say,} \tag{A27}$$

so we want to throw (A25) into that form.

To do so, introduce

$$\rho^{rs}(x), \text{ each } r,s \in R, \tag{A28}$$

such that $\Sigma_{s \in R} \rho^{rs}(x) = 1$, each $r \in R$,

to get

$$g_i(x) = \sigma^r(x) \sum_{s \in R} \{\rho^{rs}(x) a_i^s(x) / \mu^{sr}(x)\}, \ i \in r \in R, \tag{A29}$$

by (A22) and (A25), which implies (A27) if

$$\sigma^r \rho^{rs} / \mu^{sr} = \pi^s, \text{ each } r,s \in R, \tag{A30}$$

or, equivalently

$$\rho^{rs} = \pi^s \mu^{sr} / \sigma^r = \pi^s \theta^{rs}, \text{ say}, \tag{A31}$$

so that

$$\sum_{s \in R} \theta^{rs}(x) \pi^s(x) = 1, \tag{A32}$$

by (A28), which is always possible unless

$$\sum_{s \in R} \theta^{rs}(x) q^s(x) = 0, \text{ some } q \neq 0, \tag{A33}$$

or, equivalently,

$$\sum_{s \in R} \mu^{sr}(x) q^s(x) = 0, \tag{A34}$$

itself equivalent to the functional dependence of the $a^r(\cdot)$. Thus (A1) and (A20) imply

$$g(x) = G(a(x)), \text{say}, \ [a(x) = (a^r(x))_{r \in R}] \tag{A35}$$

when the $a^r(\cdot)$ are functionally independent. It is trivial that (A20) and (A33) imply (A1).

One could begin this argument with (A14)-(A15) instead of (A1)-(A4). Because of (A18) and (A19), f depends only on $a^r(\cdot)$, given y_r. Because of the strict monotonicity, a knowledge of f,y_r yields a^r and hence each a^s,.. hence any $G(a^r(x)|r \in R)$. That is at least one side of the argument.

I am now in a position to state a conjecture.

Conjecture: The pairs $f(\cdot)$, $g(\cdot)$ satisfying equations like (A1) can be constructed in the following manner:

Define a <u>structure</u> Ω on R = (1,2,..n). It is a group of subsets $S \subseteq R$ closed under intersection. To be precise,

$$R, \theta \in \Omega; \text{ and if } S, T \in \Omega, S \subset T \in \Omega. \tag{A36}$$

Call R, θ the <u>trivial</u> elements of Ω, the others its <u>proper elements</u>, and the minimal proper elements, bottom elements. For each bottom element $S \in \Omega$ construct a vector.

$$a^S(x_S) = (a^r(x_S))_{r \in S}, \; x_S = (x_r)_{r \in S}, \tag{A37}$$

of independent functions of x_S, as in (A14).

The next step is to construct similar functions on the higher elements of Ω. If a bottom element $S \subset T \in \Omega$, $a^S(x_S)$ will be among the arguments of $a^T(x^T)$. That is how the x_r, $r \in S$, enter into $a^T(\cdot)$. The simplest way to deal with this is to regard the $a^S(x_S)$ which have been constructed at any stage as parameters entering into each $a^T(\cdot)$, $S \subset T \in \Omega$, and remove the corresponding sectors $r \in S$ from the listing of each $T \in \Omega$ with $S \subset T$, including R. Call the correlates of R, Ω, $S \in \Omega$:R^*, Ω^*, $S^* \in \Omega^*$. Ω^* is clearly a structure on R^*. Take its bottom elements. Let T^* be one of them, and T its correlate in Ω. Then

$$a^T(x_T) = a^{T^*}(x_{T^*}, a^S(x_S) \mid S \subseteq T \text{ and } S \in \Omega), \tag{A38}$$

and so on up the graph of Ω.

This construction will yield just n a's, and each sector $r \in R$ will have a unique vector $a^S(\cdot)$, $S \in \Omega$, into which it first enters, and through which it enters into each $a^T(\cdot)$, $S \subset T \in \Omega$. It is this which ensures that (1) holds and

$$f(x) = F(a(x)),\ g(x) = G(a(x)). \tag{A39}$$

Remember that (A23) and (A25) hold by construction for any element of a(x) into which x_t first enters.

Given sufficiently good behaviour, then, this procedure does generate solutions of our problem.

What remains to be proved is that they are the only solutions.

"Proved" is far too strong a word for the arguments in this appendix. "Argued" or "suggested" might be better.

Note in particular that I have done nothing at all about:

i) The possible functional dependence of the $a^r(\cdot)$ generated in this way, either on paper or in my own mind. My thoughtless guess is that this would be avoided were we to insist that each $\alpha^r(\cdot)$ depends on both its arguments – except for single member bottom elements. For them $\alpha^r = \alpha^r(x_r)$, or, for an $S^* \in \Omega^*$, $\alpha^r(x^r$; lower α's). If so, its dependence on x_r is probably all that is required;

ii) The Condition that the $\lambda^r(\cdot)$ be distinct; or even the precise meaning to be given to that term;

iii) My arbitrary ordering of the elements of R as 1,2,...,n.

APPENDIX B – PROCESSES

In the text we found profit functions

$$g(p,y) = b(p) + G(a^1(p),..,a^R(p);y), \tag{B1}$$

where the $a(\cdot)$'s, though conical, were not necessarily convex. I suggested that they might be thought of as representing processes. They were in fact marginal profit functions but, if not convex, could not be profit functions, or more particularly unit profit functions. I will call them quasi profit functions.

Suppose one of these were not convex. If actually concave it is just a loss function, and the observed behaviour loss minimising. If not, the situation is a little peculiar.

Let there be only two goods for simplicity x, y; take the latter as numeraire and let the price of the latter be p. I will consider the locus of the net supplies x,y from this process as p varies.

Suppose for definiteness, that $a''(p) > 0$ for $p < p^*$, $a''(p^*) = 0$ and $a''(p) < 0$ for $p > p^*$. The quasi profit function is then convex – or 'normal' – for $p < p^*$, concave for $p > p^*$ and has inflection at p^*.

Now

$$x = a'(p),\ y = a(p) - pa'(p), \tag{B2}$$

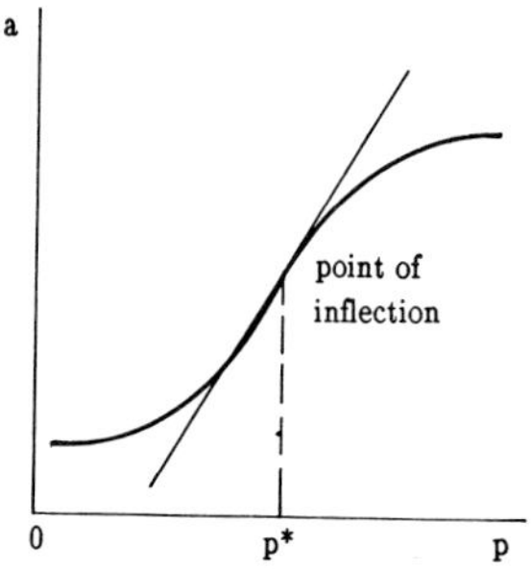

Figure 1

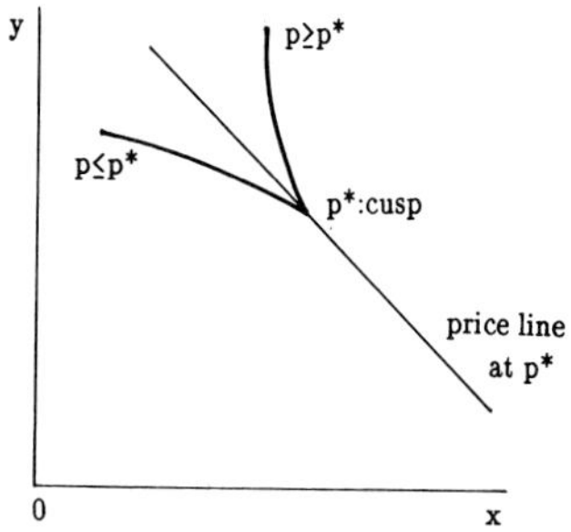

Figure 2

here. Incidentally I have taken both x and y to be outputs in the illustration in Figure 1. Now let the price p increase from zero towards p^*. In this convex section of the quasi profit function, (x,y) trace out a normal transformation locus in Figure 2, x increasing and y decreasing in particular. At p^*, however, $dx/dp = dy/dp = 0$, so that it comes to a halt, momentarily, and then reverses direction with x increasing, y decreasing. Of course it still lies on the price line at p^* to a first approximation in x – or second in p since $a''(p^*) = 0$. Hence it goes off in a cusp as in Figure 2.

Had there been a higher $p^{**} > p^*$, such that $a''(p)$ became positive again after p^{**}, we would have something like that depicted in Figures 3 and 4.

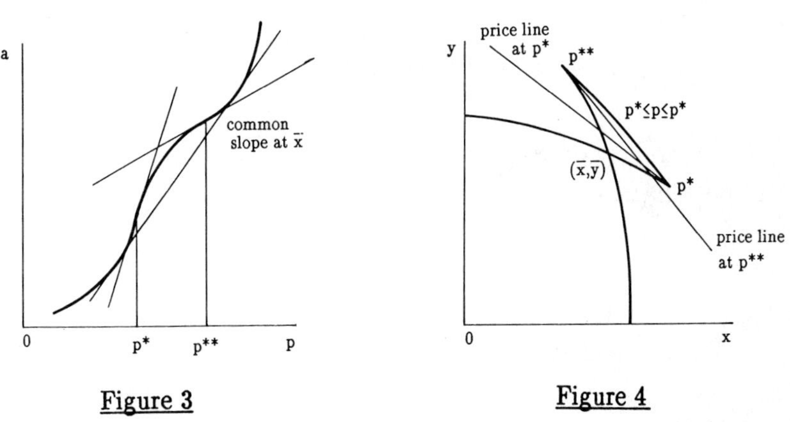

Figure 3 Figure 4

You must decide for yourself whether Figure 4 is acceptable as a picture of process production.

APPENDIX C – Market Production

As I mentioned in discussing Polynomial Engel Curves, that system is not very realistic. Nevertheless it may be worthwhile tying the present discussion up with it.

Set then

$$\Sigma f^{ti}(p,y_t) = F^i(p,\phi(y)), \tag{C1}$$

where y_t is now income. Arguments like those in Sections 2 and 6 quickly yield

$$f^{ti}(p,y_t) = b^{ti}(p) + \Sigma_r a^{ri}(p)\phi^{rt}(y_t) \tag{C2}$$

which was the starting point for Polynomial Engel Curves.

Similarly, in the market shares version

$$\frac{\Sigma_t f^{ti}(p,y_t)}{\Sigma_t f^{t1}(p,y_t)} = F^i(p,\phi(y_t)) \tag{C3}$$

we first set $p = \bar{p}$ to get

$$\Sigma f^{ti}(p,y_t) = H^i(p,\Sigma\psi^t(y_t))\Sigma f^{t1}(p,y_t) \tag{C4}$$

as in Section 3, where

$$H^i(p,\cdot) \text{ is now positively homogeneous of degree } 0, \tag{C5}$$

and then arguments like those in that Section 6 quickly yield

$$f^{ti}(p,y_t) = b^{ti}(p,y_t) + \sum_1^{R+1} a^{ri}(p)\psi^{rt}(y_t), \tag{C6}$$

$$\Sigma f^{ti}(p,y_t) = f^i(p,y) = \sum_{1}^{R+1} a^{ri}(p)\psi_r(y_t), \tag{C7}$$

where

$$\psi^r(y) = \Sigma\psi^{rt}(y_t);\ \Sigma b^{ti}(p) = 0. \tag{C8}$$

Homogeneity gives that

$$\phi^{rt}(y_t) = c_{rt}y_t(\log y_t)^{\beta_{rt}} \tag{C9}$$

without loss of generality and then in (C8)

$$\alpha_{rt} = \alpha_r, \beta_{rt} = \beta_r. \tag{C10}$$

The argument for (C9) is given in Polynomial Engel Curves. As I mentioned, I think one would deflate y_t in defining Engel Curves. In doing so one would lose (C9) and (C10).

REFERENCES

Gorman, W M (1967): "Measuring the quality of fixed factors", in J Wolfe (ed), *Essays in Honour of Sir John Hicks.*

Gorman, W M (1967): "More About Fixed Factors": unpublished paper presented to the Conference of the Mathematical Economics Study Group of the SSRC.

Conference Discussion

Handout 1 Variations on a theme of Cauchy.

$$s(u + v) = s(u) + s(v) \tag{1}$$

$\Rightarrow$

$$s(v) = av,\ a.v,\ Av, \tag{2}$$

according to the dimensions of s,v under weak maintained conditions.

$$s(\Sigma v_f) = \Sigma s^f(v_f), \tag{3}$$

$\Rightarrow$ (2) if each $s^f(0) = 0$.

$$g(\Sigma u_f) = \Sigma g^f(u_f) \tag{4}$$

$\Rightarrow$

$$g^f(u_f) = s^f(v_f) + c_f = s(v_f) + c_f = s(u_f) + b_f, \text{ say,} \tag{5a}$$

where

$$g(\Sigma u_f) = s(\Sigma u_f) + b,\ b = \Sigma b_f, \tag{5b}$$

by changing the origins, where s(.) satisfies (2).

Example: existence of a representative consumer with cost function:

$$g(p,\Phi(u)) = \Sigma g^f(p,u_f). \tag{6}$$

Set $p = \bar{p}$, $u^* = g(\bar{p}, \Phi(u))$, $u_f{}^* = g(\bar{p},u_f)$, to get

$$g^*(p,\Sigma u_f{}^*) = \Sigma g^{*f}(p,u_f{}^*), \tag{7}$$

$$g^{*f}(p,u_f^*) = a(p)u_f^* + b_f,\ g^f(p,0) = b_f. \tag{8}$$

2 Demand independent of distribution of income

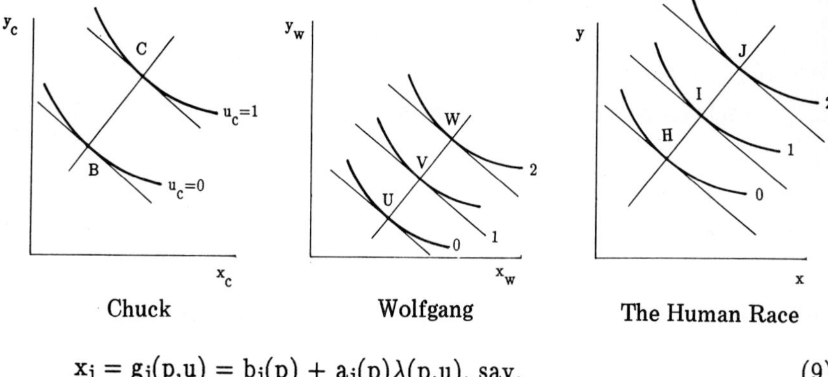

Chuck Wolfgang The Human Race

$$x_i = g_i(p,u) = b_i(p) + a_i(p)\lambda(p,u), \text{ say,} \tag{9}$$

where

$$b(p) = g(p,0),\ a(p) = g(p,1) - g(p,0) \tag{10}$$

⇒

$$g(p,u) = \Sigma p_i g_i = b(p) + a(p)\lambda(p,u) \tag{11}$$

$$= b(p) + a(p)u,$$

after differentiating across and normalising.

Note the affine function in u: this recurs.

i CJB This isn't so clear, we haven't been given a range for λ.

TG λ goes from $+\infty$ to $-\infty$ to give the whole line.

CJB So we can hit axes?

TG But actually, no, sorry; in principle it could be a straight line, but I'm going to start it at $u = 0$, the bottom curve on Figure 1. Should the Engel curve cut the axis it would have to bend, which is not allowed. Should you require market demand to be independent of the distribution of income without such a bend, the utility function would have to be homothetic. So we're starting at a given utility level.

Handout 3 Labour, Plastic Goods, etc.
Technology without externalities

$$T(u) = \Sigma T^f(u_f) \tag{12}$$

ii AS Does (12) exclude anything being a fixed factor for more than one firm?

TG Oh, no, any number of firms may use lathes. However if smoke nuisance is produced by two firms, and you regard the licence to produce so much smoke as a fixed input, that would be an externality. They could be grouped together as a single unit; and so could other groups elsewhere. If they do not overlap we could use these groups as the units and the analysis would go through. You really need very little independence.

Handout Partitioned vector of current net products

$$x = \Sigma x_f = (x_m)_{m \in M}, \tag{13}$$

Aggregates as intermediate goods

$$T(u)\colon 1 \geq f(x,u) \equiv F(\Phi(x),u), \tag{14}$$

where

$$\Phi^m(x_m) = v_m,\ \Phi(x) = (\Phi_m(x_m))_{m \in M} = v. \tag{15}$$

iii CB Terence, why've you switched here from a set notation to a production function? I would have thought ...

TG I do agree, I was tempted to keep sets, but I wanted to make certain the $\Phi(x)$ came in naturally.

Handout Turn back to figure on page 2.

Gross profit functions

$$h^m(p_m,v_m) = \max\{p_m.x_m \mid \Phi^m(x_m) \leq v_m\}, \tag{16}$$

$$= a^m(p_m)v_m + b^m(p_m), \text{ say,}$$

iv CB This is a strong assumption and I think it's worth telling people what it is you're buying. It's really saying that the Hessian of f is continuous in u and x so that you can manipulate the demands for every final good.

TG Oh no, for every current good, it turns out,

CB For every good, not just final goods?

TG Except that changing u_m might not change some component of x_m over a considerable range.

CB So what you're avoiding by this is an economy that decomposes itself into little bits.

TG No, no. They'd have to be very little bits. There have to be two firms f,g whose demand for labour I can affect by changing their endowments u_f, u_g Two is enough to give me linearity. The fact that the others don't come on at all means they have no Engel curve to be linear.

v WE $\Phi(x)$ in (15) is a scalar valued function?

TG $\Phi^m(\cdot)$ is scalar valued but $\phi(\cdot)$ is a vector of the $\Phi^m(\cdot)$'s.

WE $\Phi(x)$ is?

TG $\Phi(x)$ is a vector, but $\Phi^m(x_m)$ is a scalar in (16).

WE What is $\Phi(x)$, a production function?

TG Look at (16). Think of v_m as the quantity of an intermediate good produced from the fixed inputs u and itself producing current goods x_m.

AS $\Phi(x)$ is the vector of aggregates.

CB But you can think, Wolfgang, of each $\Phi^m(\cdot)$ a production function, if you prefer, merely measuring the x's as net inputs.

WE I am against vector valued production functions because the mapping is a correspondence.

TG These aren't vector valued. Let's look at the case of labour; $-x_m$ would then be a vector of labour inputs used to produce an aggregate amount v_m of labour, to be used with the fixed inputs and aggregates for other classes of current inputs to produce, eg Plastic Goods, and hence individual plastic goods.

vi IJ The whole thing must be convex presumably.

TG Oh, indeed. $g(\cdot,u)$ is convex. It's just the individual $a^m(\cdot)$'s that need not be.

IJ Suppose $b^m(\cdot)$s linear, $a^m(\cdot)$ would need to be convex.

TG I believe so.

Handout

$$g(p,u) = \max \{\Sigma a^m(p_m)v_m \,|\, F(v,u) \leq 1\} + \Sigma b^m(p_m) \tag{17}$$

$$= G(a(p),u) + \Sigma b^m(p_m), \text{ say, } a(p) = (a^m(p_m)),$$

vii TS In (5) why aren't there homogeneity properties from these sub-profit functions for $a^m(\cdot)$?

TG $a^m(\cdot)$is homogenous but not necessarily convex. It is homogenous and under my assumptions probably continuous, but it doesn't necessarily have convexity properties. You can force convexity on it by requiring that I can so vary the u's that I can persuade the people not only to employ all the people in the world at the moment but all the people that in an infinite time might be bought. The fact the different Engel growth curves can't cut then forces $a^m(\cdot)$ to be convex.

viii AS What is the interpretation of $h^m(\cdot)$ in (17)?

TG I broke this thing down into stages, u produces v, a vector of v_m's; and from those v_m's I produce x_m. Look at m as a profit centre producing x_m from v_m, $h^m(p_m,v_m)$ is the profit it makes when given v_m and told to get to it.

AS So this is basically viewing the aggregates as production functions?

TG It does seem to be a natural interpretation of them. Economists have certain sorts of intuition so it seems simple to treat them as intermediate goods.

Handout (6) ⇒

$$g^f(p,u_f) = G^f(a(p),u) + c^f(p), \text{ say, each } f, \tag{18}$$

$$v_m = G_m(a(p),u) := \partial G/\partial a^m, \tag{19}$$

$$= \Psi^m(p,u), \text{ say,}$$

4 Aggregates as intermediate goods

$$u \Rightarrow v \Rightarrow x \text{ via } v \in R(u),\ x \in S(v), \tag{20}$$

Gross profit functions

$$g(p,u) = \max\ \{p,x \,|\, x \in T(u)\} \tag{21}$$

$$= \Sigma g^f(p,u_f)$$

$$h(p,v) = \max\{p.x \,|\, x \in S(v)\}, \tag{22}$$

ix GM Are additive technology and additive gross profit functions equivalent?

TG In my belief yes. Oh no, not quite equivalent as technologies need not be convex.

Handout

$$g(p,u) = \max\ \{h(p,v) \,|\, v \in R(u)\} \tag{23}$$

$$= h(p,\psi(p,u)) \text{ see (ix), say,}$$

where h(p,v) attains its maximum at

x CB That (non convexity of $a^m(\cdot)$) is only really a problem if you wanted to go back to the primal.

TG That's right.

CB So $h(\cdot)$ is the maximum of x in S?

TG No, h(p,v) is the profit function corresponding to the technology $x \in S(v)$. Now that is equal to the maximum of p.x given that x belongs to S(v).

CB And this $\psi(u)$?

TG Oh, how right you are. It should be $\psi(p,u)$ of course as in (5) below. That's absolutely critical, thank God you said it.

xi AS u and u_f have the same dimension?

TG No. For different firms one may have quite different lists of inputs, there's no way the fifth component of u_1 corresponds to the fifth component of u_2, for instance.

AS Why can't we have a large enough specification so that each firm just has a lot of zeros?

TG There's no advantage.

Handout

$$v_m = \Psi^m(p,u), \text{ say, each m, cf (16)} \tag{24}$$

where $v = (v_m)_{m \in M}$, #M less than each dim u_f, and less than dim x; and work with the excess profit or quasirent functions,

$$s(p,v) = \Sigma r^f(p,v_f), \tag{25}$$

in the spirit of (4)–(5) (main text).

xii AS What is to stop u_f containing some element that doesn't actually affect firms f's. production?

TG Well, it's quite likely. The firm has some wretched thing lying around the place which a scrap metal firm would charge to take away, so it just leaves it. It doesn't get in the way, it's in some out–of–the–way shed. Entirely possible.

AS So in equation (10) one could be saying "let's change ..." in this r^f function: this is conceivably asking the question of what is the effect of a machine in British Leyland on some other firm.

TG Oh, no, definitely not. These are firm specific.

AS What is making the v_f firm specific?

TG Because I, in my role as God, called it that.

CB They've been artificially constructed if you go back to (25) you're forcing dummy v_f's on the economy and move u^f's.

TG Oh, no, there's a big difference between v_f and v_f^*.

AS Is there an implicit assumption that there is no interdependence between these firms?

TG That is the critical assumption. Equation (22) defines a technology without externalities.

Handout And the extra supply functions

$$s_i(p,v) = \Sigma r_i^f(p,v_f). \tag{26}$$

xiii CB I don't follow, what is an r^f?

TG Take a $\bar{u}_f$ in the domain of u_f, and measure v_f as $u_f - \bar{u}^f$, $r^f(p,v_f)$ is the extra profits or quasi–rent f get at prices p, because it is endowed with u_f rather than $\bar{u}_f$.

Handout Take an independent set M of the extra supply functions in (7). Call the corresponding current goods $m \in M$, set

$$v_m^* := s_m(p,v) = \Sigma_f r_m^f(p,v_f) =: \Sigma_f v^*_{fm}, \text{ each } m \in M, \tag{27}$$

solve these for

$$v_m = \Psi^m(p,\Sigma_f v^*_{fm}), \tag{28}$$

in the spirit of (0.6–7), and substitute back into (6) to get

$$s^*(p,\Sigma v_f^*) := s(p,v) = \Sigma_f r^f\,(p,v_f), \tag{29}$$

Now set $v_g = 0$, $g \neq f$, implying $v_g{}^* = 0$, $g \neq f$ and hence

$$r^f(p,v_f) = s^*(p,v_f^*). \tag{30}$$

substituting into (10) and using (0.1)–(0.2), (0.4)–(0.5), we get,

$$g^f(p,v_f) = a(p).v_f + b^f(p),\ g(p,v) = a(p)\ v + b(p) \tag{31}$$

xiv CB You could have proceeded directly couldn't you? From the T in (12). You know that has to be equal to the sum of the T^f's so you could have constructed the $v_m{}^*$'s directly off T by projection.

TG I think it's entirely possible.

CB It just means you won't have to rely on an inversion argument.

xv CB What Terence doesn't point out, what's ingenious, is this: he starts out asking if you look at this set of equations, you have $s_i = \Sigma f_i^f$ and he's looking for aggregates to explain them. Now you treat these as demand and supply functions and look for a subset that are linearly independent.

TG Functionally independent.

CB The v's that he started out looking for, the demand functions themselves actually become the dummies and so the reason I was suggesting you work in the primal is the assumption that you can solve this.

TG I believe I've proved all these things when second derivatives exist. If you have at least #M aggregates to explain production, there will in a certain neighbourhood always be at least #M functionally independent supply equations.

CB I want to claim that it makes no difference. Suppose you had just two of these things and you would't find two independent you would know that as the Jacobian vanishes the first is linearly dependent on all the others so you could use Euler's theorem to add up the profit functions.

TG Oh, I see.

CB You would get your result whether you inverted or not.

TG I'll give you a counter example. Ok, it won't do. Ok, at 12.53 I agree with what you said.

Handout Aggregates in supply functions

(26) may be replaced by

$$s^i(p,v) = \Sigma r^f_i(p.v_f) \tag{32}$$

where the i superscript on s is now only an index. We get the same results, just as we did when we dropped the idea of a representative consumer on page 2.

This does not imply that an aggregate appears in the primal, because of lack of convexity, which was also a danger in my previous analysis.

Each yields

$$g^f(p,u_f) = G^f(a(p),u_f) + b^f(p),\ a(p) = a^m(p)_{m \epsilon M}, \tag{33}$$

$$g(p,u_f) = G(a(p)u) + b(p),\ b(p) = \Sigma b^f(p), \tag{34}$$

$$v^*_{fm} = G^f_m = \partial G^f/\partial a^m;\ v^*_m = G_m = \Sigma G^f_m = \Sigma_f v^*_{fm}.$$

Measurement and Modelling in Economics
G.D. Myles (Editor)

USE AND MISUSE OF SINGLE–CONSUMER RESULTS IN A MULTI–CONSUMER ECONOMY: THE OPTIMALITY OF PROPORTIONAL COMMODITY TAXATION

by
Charles Blackorby*, Russell Davidson#, and William Schworm†

1 Introduction

Many economic models are first tested in the context of an economy with a single consumer or a single firm or both. In fact much of our intuition comes from analyzing these simple environments; furthermore it is a sensible modelling strategy to abstract from those issues which are not immediately germane to the question at hand. However sensible such a modelling strategy may be, it is also important to see how robust these results are to the particular abstractions which have been made and to assess their importance. In the paper which we presented at the conference we analyzed the robustness of some results which were derived by Ian (Jewitt (1981)). Ian provided necessary and sufficient conditions for the optimality of piecemeal second–best policy in a single consumer environment. He found that the preferences of that single agent had to be implicitly separable[1] with respect to a partition of the variables into those which a planner could control and those which were irrevocably distorted. Our aim at the time was to analyze the robustness of Ian's result in an economy with more than one consumer.

---

* University of British Columbia, and GREQE.

Queen's University.

† University of British Columbia and University of Western Ontario.

[1] Ian, following Gorman (1972, 1976) calls this pseudoseparability. We find this nomenclature to be slightly misleading and have adopted implicit, as Ian and Tim Besley had at the conference.

At the conference, Tim Besley and Ian Jewitt (1987), in the context of a particular model, analyzed the conditions under which proportional commodity taxation would be optimal in a single consumer economy. The Simmons (1975)–Deaton (1979,1981) sufficient conditions which they presented are known to be necessary as well if agents have endowments of some commodities. We analyze such a model and examine its robustness to the single consumer assumption. The moral of the story is that the presence of many consumers yields results which are substantially more restrictive. By itself this is perhaps not surprising, as we are asking for proportional commodity taxes to be optimal in a more general model. However, by providing necessary and sufficient conditions in this more general model we gain on two fronts. First of all, we know precisely what restrictions are required in a multi–consumer economy. Secondly, we may get a feel for the sort of thing which we miss when we analyze single consumer models.

2 Technical Assumptions

We consider a competitive economy composed of H price–taking consumers, each of whom has well–behaved preferences. The preferences of consumer h are represented by $U^h\colon \Omega \to \mathfrak{R}$ where $\Omega \subset \mathfrak{R}^N_+$ is closed, convex, and does not contain the origin. Each U^h is assumed to satisfy conditions R_1: for all $X_h \in \text{int}\ \Omega$

(i) U^h is twice continuously differentiable,

(ii) $\nabla_x U^h(X_h) >> 0_N$, and

(iii) $y \neq 0_N$ and $y^T \nabla_x U^h(X_h) = 0$ implies that $y^T \nabla_{xx} U^h(X_h) y < 0$.

An alternative representation is given by the expenditure function of each consumer, E^h, defined by

$$E^h(u_h,P) = \min_X \{P \cdot X \,|\, U^h(X) \geq u_h\}, \tag{1}$$

for u_h in the range of U^h, $X \in \text{int}\ \Omega$.

For any $X \in \text{int}\ \Omega$, at $u_h = U^h(X)$ and $P = \nabla_x U^h(X)/X^T \nabla_x U^h(X)$, E^h satisfies $\overline{R}_1$:

(i) E^h is twice continuously differentiable,

(ii) $\nabla_P E^h(u_h,P) >> 0_N$, and

(iii) $\nabla_u E^h(u_h,P) > 0$,

(iv) $y \neq 0_N$ and $y^T P = 0$ imply that $y^T \nabla_{PP} E^h(u_h,P) y < 0$, and

(v) E^h is homogeneous of degree one in P.

A proof that R_1 and $\overline{R}_1$ are equivalent can be found in Blackorby and Diewert (1979) and so we shall speak simply of regularity conditions R. U^h and E^h are said to be differentially strongly pseudoconcave.

3 Community Preferences

The economy consists of H consumers with preferences characterized by "no–worse–than" sets defined by

$$B_h(u_h) = \{X_h \mid U^h(X_h) \geq u_h\} \tag{2}$$

for $h = 1,\ldots,H$. The set $B_h(u_h)$ is the set of commodity vectors that enable consumer h to achieve at least the utility level u_h. Under our regularity conditions, R, on preferences, $B_h(u_h)$ is strictly convex and the correspondence $u_h \rightarrow B_h(u_h)$ is continuous and increasing.

The economy commodity vectors that can provide a given vector of utilities $u = (u_1,\ldots,u_H)$ for the consumers are given by the *Scitovsky set* (see Scitovsky (1942) and de Graaf (1957)):

$$B(u) = \sum_h B_h(u_h) = \{X \mid U^h(X_h) \geq u_h,\ h = 1,\ldots,H, \sum_h X_h \leq X\} \tag{3}$$

If $X \in B(u)$ then there exists an allocation, $(X_1,\ldots,X_H)$, that is feasible in the sense that $\sum_h X_h \leq X$, and is such that each consumer can achieve utility level u_h for $h = 1,\ldots,H$. Community preferences are represented by the collection of Scitovsky sets B(u) for all u in the Cartesian product of the ranges of the utility functions.

There are two equivalent ways of characterizing the efficient outcomes possible in the economy defined by the Scitovsky sets. The first is a direct representation of them which we call the *Paretian utility function.* It is given by

$$U^g(X,u_{(g)}) = \max_{\{x_h\}}\{U^g(X_g)\,|\,U_h(X_h) \geq u_h,\ h\neq g, \text{and} \sum_{h=1}^{H} X_h = X\} \quad (4)$$

where $u_{(g)}$ is the vector u without its gth component. In general, if a planner maximizes U^g subject to technological constraints, the outcome is efficient. Thus U^g characterizes the possible efficient outcomes. The Paretian utility function U^g is the natural analogue in a multi–consumer economy of a single agent's utility function.

An alternative and useful way to characterize the Scitovsky sets is by their support function which we call the *Scitovsky expenditure function*; it is defined by

$$E(u,P) = \min_{x}\{P\cdot X\,|\,X \in B(u)\}. \quad (5)$$

Notice that by use of (1), (2) and (3) it can also be written as

$$E(u,P) = \sum_{h=1}^{H} E^h(u_h,P) \quad (6)$$

where E^h is the expenditure function of consumer h given by (1). (6) follows from the additivity of the Scitovsky sets in (3). Notice that, conditioned on the *vector* of utilities u, the Scitovsky expenditure function has all the properties of the expenditure function of a single agent. However, conditioned on the price vector, the Scitovsky expenditure function is additive in utilities – a property that cannot be shared with the expenditure function of a single consumer. It is from this observation that the mathematical results of our paper flow. If one places some restriction on the preferences of a single agent, this shows up as restriction on the expenditure function of that agent, and there is nothing more to be said. If, however, there is more than one consumer, a similar restriction placed on the Scitovsky expenditure function may interact with its property of being additive in utilities to force additional stronger restrictions on the Scitovsky expenditure function and hence on the Scitovsky sets themselves.

4 Implicit Separability

The notion of implicit separability was introduced by Gorman (1976) (he called it pseudo–separability); we shall define it and then relate it to other separability concepts, in particular homothetic implicit separability and direct separability, which are shown to be special cases.

We shall say that a function $F(X^c,X^d)$ is implicitly separable with respect to the partition $\{I^c,I^d\}$ if there exist functions F^c and F^d that, whenever $y = F(X^c,X^d)$, then

$$F^c(y,X^c) = F^d(y,X^d). \tag{7}$$

The idea of implicit separability extends very naturally from functions to preferences, either individual preferences or community preferences as expressed by a collection of Scitovsky sets or a Paretian utility function. The Scitovsky sets can be implicitly represented by means of a function T as follows:

$$X \in B(u) \Leftrightarrow T(u,X) \geq 0 \tag{8}$$

The relation with the Paretian utility function is

$$T(u,X) \geq 0 \Leftrightarrow U^g(X,u_{(g)}) \geq u_g. \tag{9}$$

If the variables in X are partitioned into two sectors as before, the Paretian utility function or the corresponding collection of Scitovsky sets are said to be implicitly separable with respect to the partition $\{I^c,I^d\}$ if there is some implicit representation T with the form

$$T(u,X) \equiv \overline{T}(u,X^c,T^d(u,X^d)), \tag{10}$$

where $\overline{T}$ and T^d are continuous real–valued functions and $\overline{T}$ is increasing in its last argument. It is easy to see from (10) along with (9) that the function U^g is then implicitly separable in its X arguments in the sense of the definition (7). As discussed below, direct separability of the preferences results if the aggregator T^d in (10) is independent of u.

We find it convenient to have a dual characterization of implicit separability, and indeed need this representation to solve our problem. First define the expenditure functions which are dual to $\overline{T}$ and T^d by

$$C(u,P^c,\xi) = \min_{X^c}\{P^c\cdot X^c \,|\, \overline{T}(u,X^c,\xi) \geq 0\}, \tag{11}$$

and

$$D(u,P^d,\xi) = \min_{X^d}\{P^d\cdot X^d \,|\, T^d(u,X^d) \geq \xi\}. \tag{12}$$

The former of these is the minimum cost of getting each consumer to utility level u_h, h = 1,...,H, by means of group c commodities given that the level of group d is ξ, whereas the latter is the minimum cost of providing ξ given the utility vector u. Note that C and D are homogeneous of degree one with respect to P^c and P^d respectively. It can be shown that the overall cost of providing utility vector u, given by the Scitovsky expenditure function (5), is related to (11) and (12) as follows:

$$E(u,P) = \min_{\xi}\{C(u,P^c,\xi) + D(u,P^d,\xi)\}. \tag{13}$$

That is, total expenditure is minimized by choosing the level of the aggregate, ξ, so as to minimize the sum of the sector expenditures. A Scitovsky expenditure function is dual to a Scitovsky set with one implicitly separable group if and only if it can be written as in (13); we refer to a Scitovsky expenditure function that satisfies (13) as conditionally additive. Moreover if $\xi = \Xi(u,P)$ yields the minimum[2] in the above problem (13) then Ξ is implicitly separable in (I^c,I^d).

The structure of the expenditure function (13) shows that the economy decomposes into two sectors. As constructed, C is the conditional expenditure function dual to $\overline{T}$ in (10) and D is the conditional expenditure function dual to T^d (also in 10)). Together they are used to form the overall expenditure function E.

Expenditure in sector c is given by $C(u,P^c,\xi^*)$ where ξ^* is the value of the optimiser in (13); similarly the expenditure in sector d is given by $D(u,P^d,\xi^*)$. To see this, use Shephard's Lemma and the envelope theorem in (13) to obtain

$$\overset{*}{X}{}^c = \nabla_c E(u,P) = \nabla_c C(u,P^c,\xi^*), \tag{14}$$

and

$$\overset{*}{X}{}^d = \nabla_d E(u,P) = \nabla_d D(u,P^d,\xi^*), \tag{15}$$

where

$$\xi^* = \Xi(u,P) \tag{16}$$

is determined by the first–order condition

2 We do not in all this wish to exclude the possibility that E should actually be additive. In this case ξ, if it exists at all, is artificial and may therefore be chosen to have any property we please. In the somewhat less artificial case of an optimising ξ that is not unique, all that we claim is that there exists an optimising ξ that is implicitly separable.

$$C_\xi(u,P^c,\xi^*) + D_\xi(u,P^d,\xi^*) = 0. \tag{17}$$

Total expenditure in sector c is given by

$$P^c \cdot \overset{*}{X^c} = (P^c)^T \nabla_c C(u,P^c,\xi^*) = C(u,P^c,\xi^*) \tag{18}$$

where the last equality follows from Euler's Theorem. Similarly, total expenditure in sector d is

$$P^d \cdot \overset{*}{X^d} = (P^d)^T \nabla_d D(u,P^d,\xi^*) = D(u,P^d,\xi^*). \tag{19}$$

Expenditure in sector c(d) depends on prices in sector d(c) only through an aggregator, the optimiser ξ^*. It is again clear from examining (13) that implicit separability is a completely symmetric notion.

Depending on circumstances, there are many possible interpretations that can be given to the optimiser ξ^*. It is only through its value that changes in the consumption bundle in one sector can have any influence on preferences over consumption bundles in the other sector. It must therefore represent some kind of aggregate effect engendered by consumption in one sector that can influence preferences in the other sector. In general, any change in a consumption bundle will affect preferences over all other goods. In the present case, the intrasector influences must be captured by the variation of one single aggregate, the optimiser.

By imposing special conditions on one sector or the other we depart from the symmetry of implicit separability. For convenience we focus on sector d. Suppose that the aggregator T^d in (10) is homothetic in X^d but still dependent on the utility vector. We call this *homothetic implicit separability.*[3]

---

[3] This has been called quasi–separability by Gorman (1976) and implicit separability by Blackorby, Primont, and Russell (1978), and Deaton (1979). Besley and Jewitt (1987) suggest this name.

The expenditure function characterization of homothetic implicit separability is particularly revealing. Assuming that T^d is homothetic in X^d, it follows from the duality relation (12) that the conditional expenditure function D can be written in the form[4]

$$D(u,P^d,\xi) = \overline{D}(u,P^d)\xi. \tag{20}$$

The overall expenditure function is therefore

$$E(u,P) = \min_{\xi}\{C(u,P^c,\xi) + \overline{D}^*(u,P^d)\xi\} \tag{21}$$

$$= \overline{E}(u,P^c,\overline{D}(u,P^d)). \tag{22}$$

Hence homothetic implicit separability (of sector d) of the direct preferences implies that sector–d prices are separable from sector–c prices, conditional on u, in the expenditure function.

Homothetic implicit separability is a necessary condition in a variety of problems: see Blackorby and Russell (1976, 1978) for restrictions on within–group elasticities of substitution; Blackorby, Lovell, and Thursby (1976) for its relation to neutral technical progress. Homothetic implicit separability appears as a necessary condition in the optimal commodity tax problem to be discussed below.

Direct separability is also a special case of implicit separability. Suppose that in the implicit representation (10) there is no explicit u-dependence in the aggregator T^d. Then (9) can be solved for the utility of agent g to obtain

$$U^g(X,u_{(g)}) = \overline{U}^g(X^c,T^d(X^d),u_{(g)}). \tag{23}$$

4 See the proof of Theorem 3.6 in Blackorby, Primont, and Russell (1978).

Then we have ordinary direct separability. If on the other hand the implicit representation can be written as

$$U^g(X,u_{(g)}) = \overline{U}^g(U^c(X^c) + U^d(X^d),u_{(g)}) \tag{24}$$

then we have an additively separable representation. Implicit separability is weaker than either of these two notions simply because the aggregates themselves can depend on the distribution of wealth (or utility).

The expenditure function which is equivalent to the direct separability of the primal function can be written as

$$E(u,P) = \min_{\xi} \{C(u,P^c,\xi) + \overline{D}(P^d,\xi)\}. \tag{25}$$

Since u does not appear in the aggregator T^d, it plainly cannot appear in the conditional cost function $\overline{D}$.

Finally, note that homothetic (direct) separability is a special case of both homothetic implicit separability and direct separability. Homothetic separability means that T^d in (23) is homothetic, and that the expenditure function can be written as

$$E(u,P) = \min_{\xi}\{C(u,P^c,\xi) + \hat{D}(P^d)\xi\} \tag{26}$$

$$= \overline{E}(u,P^c,\overline{D}(P^d)) \tag{27}$$

This characterization is well known; see Blackorby, Primont, and Russell (1978), section 3.4.2 and references there.[5]

---

[5] Quasi–homothetic separability is weaker than homothetic separability and appears as a result in many aggregation problems. In this case one can write $\overline{D}(P^d,\xi) = \hat{D}(P^d)\xi + \mathring{D}(P^d)$. In this note this does not play a role as it is eliminated by maintaining homothetic implicit separability.

5 The Optimality of Proportional Commodity Taxes

It is well known that in many single consumer models homothetic implicit separability is necessary and sufficient for optimal commodity taxes to be proportional (Simmons (1974) and Deaton (1979, 1981)). The literature on this subject is extensive and results are very sensitive to small changes in the way the problem is modelled.[6] At one extreme the only reason for commodity taxes is that no lump–sum taxes are available and no lump–sum income is allowed. This is how Besley and Jewitt model the problem; this can certainly be justified from the point of view of a purist, and, in addition, it leads to technical complications which are interesting theoretically and may well be useful in other applications as well. With respect to the former, we do not see any need to be so pristine. Governments have many reasons why they may want or need to raise some money by means of commodity taxes and it does not seem necessary to model all this at the same time as one addresses the proportionality questions. As for the latter, we are interested in the multi–consumer aspects of the problem and wish to avoid as many other technical complications as possible.

We consider the consumers to have endowments of commodities in sector c, and that although lump–sum taxes are available, some revenue must be raised by commodity taxation. The problem itself can be written as

$$\text{Max } u^g \tag{28}$$

subject to

$$E(u,Q) = \omega^T Q^c + R \tag{29}$$

and

[6] See Atkinson and Stiglitz (1972), Sandmo (1974), Simmons (1974), Deaton (1979, 1981), Mirrlees (1984), and Besley and Jewitt (1987).

$$t^T \nabla_d E(u,Q) = R \tag{30}$$

by choice of t where $P^d + t = Q^d$ is the vector of consumer prices in the taxed sector, P^d is the vector of producer prices and ω is the aggregate vector of endowments. The Besley–Jewitt formulation of the problem can be obtained by setting both ω and R equal to zero. This restricts the set of prices and utilities to a manifold of smaller dimension so that the prices cannot all be moved independently. In our case, this has to hold for any set of initial endowments and hence no such restriction is implied.

It is reasonably easy to see why homothetic implicit separability is sufficient in the one consumer case. Let the expenditure function in (29) be that of a single consumer and suppose that preferences are homothetically implicitly separable. Then the expenditure function can be written as

$$E(u,Q) = \min_{\xi}\{C(u,Q^c,\xi) + D(u,Q^d)\xi\}. \tag{31}$$

Given (31), the first–order necessary conditions for the above problem are given by

$$\lambda \nabla_d^T D(u,Q^d)\overset{*}{\xi} + \mu t^T \nabla_{dd} D(u,Q^d)\overset{*}{\xi}$$

$$+ \mu t^T \nabla_d D(u,Q^d)\overset{*}{\xi}_D \nabla_d^T D(u,Q^d) + \mu \nabla_d D(u,Q^d) = 0 \tag{32}$$

Postmultiplying this by Q^d yields

$$\lambda \overset{*}{\xi} + \mu t^T \nabla_d D(u,Q^d)\overset{*}{\xi}_D + \mu = 0 \tag{33}$$

which we substitute back into (32) to obtain

$$t^T \nabla_{dd} D(u,Q^d) = 0. \tag{34}$$

Given our regularity conditions, the zero eigenvector of $\nabla_{dd}D(u,Q^d)$ is unique up to a constant of proportionality. By the homogeneity of the expenditure function it must therefore be proportional to Q^d itself. Hence the tax vector t must be proportional to $Q^d(=P^d+t)$ which demonstrates sufficiency.

Although homothetic implicit separability remains necessary in the multi-consumer problem it is not consistent with the fact that the Scitovsky expenditure function is, by definition, additive in the vector of utilities, given prices. The necessary and sufficient conditions for the optimality of proportional commodity taxation in a multi–consumer economy are given in the following

Theorem: Assume **R**, and that $H \geq 3$. Then proportional commodity taxes are optimal in (28)–(30) if and only if sector d is homothetically implicitly separable,

$$E(u,Q) = \min_{\xi}\{C(u,Q^c,\xi) + \overline{D}(u,Q^d)\xi\} \tag{35}$$

and either

$$\text{I:} \qquad C(u,Q^c,\xi) = \overline{C}(Q^c)\phi(\xi) + \hat{C}(u,Q^c) \tag{36}$$

with no further restrictions on $\overline{D}$, or

$$\text{II:} \qquad \overline{D}(u,Q^d) = \breve{\overline{D}}(Q^d) \tag{37}$$

with no further restrictions on C, or

$$\text{III:} \qquad C(u,Q^c,\xi) = \Phi(u)\overline{C}(Q^c,\xi) + \hat{C}(u,Q^c) \tag{38}$$

and

$$\overline{D}(u,Q^d) = \Phi(u)\breve{D}(Q^d). \tag{39}$$

where Φ is additive.

Proof: (35) follows from the single consumer argument of Simmons (1975) and Deaton (1979) while the rest follows from the conjunction of homothetic implicit separability and the results in Blackorby, Davidson, and Schworm (1988) (BDS) which are derived for the more general case of implicit separability of the Scitovsky expenditure function. There it is shown that the multi–consumer aspect of the problem forces an implicitly separable economy to satisfy three cases. Cases I and II above follow immediately from imposing homothetic implicit separability on Cases I and II of BDS. In case III of BDS it is shown that C and D as given by (11) and (12) must satisfy

$$C_\xi(u,Q^c,\xi) = \overline{C}_\xi(Q^c,\xi)\Phi(u,\xi) \tag{40}$$

and

$$D_\xi(u,Q^d,\xi) = \overline{D}_\xi(Q^d,\xi)\Phi(u,\xi) \tag{41}$$

where

$$\Phi(u,\xi) = \sum_h \Phi^h(u_h,\xi). \tag{42}$$

Homothetic implicit separability implies that

$$D_\xi(u,Q^d,\xi) = \breve{D}(u,Q^d) \tag{43}$$

which in conjunction with (41) yields

$$D_\xi(u,Q^d,\xi) = \mathring{D}(Q^d)\breve{\Phi}(u). \tag{44}$$

This in conjunction with (40) yields (38) and (39). ♦

It is obvious from inspection that there are many questions which can be asked in the single consumer case that make no sense in the multi-consumer economy. In order to highlight these additional restrictions it is useful to compare the following ratios of changes in quantities demanded in both the single and multi–consumer cases,

$$\frac{\dfrac{\partial X_i}{\partial P_m}}{\dfrac{\partial X_j}{\partial P_m}}, \quad i,j \in I^c \text{ and } m \in I^d \tag{45}$$

and

$$\frac{\dfrac{\partial X_i}{\partial P_m}}{\dfrac{\partial X_i}{\partial P_n}}, \quad i \in I^c \text{ and } m, n \in I^d. \tag{46}$$

In order to facilitate the comparison we use the image of the optimiser in each problem and compute the ratios in (45) and (46). For the single consumer problem, which implies only homothetic implicit separability, the optimiser is given by

$$\xi^* = \overline{\Xi}(u,Q^c,\overline{D}(u,Q^d)) \tag{47}$$

and using (14) and (15) we obtain

$$\frac{\dfrac{\partial X_i}{\partial P_m}}{\dfrac{\partial X_j}{\partial P_m}} = \frac{\Xi_i(u,Q^c,\overline{D}(u,Q^d))}{\Xi_j(u,Q^c,\overline{D}(u,Q^d))}, \ i, j \in I^c \text{ and } m \in I^d \tag{48}$$

and

$$\frac{\dfrac{\partial X_i}{\partial P_m}}{\dfrac{\partial X_i}{\partial P_n}} = \frac{\overline{D}_m(u,Q^d))}{\overline{D}_n(u,Q^d))}, \ i \in I^c \text{ and } m,n \in I^d. \tag{49}$$

That is, ratios of compensated demand changes in sector c with respect to a price in sector d are the same for all prices in sector d, and depend on sector d prices only through the price index $\overline{D}$; it depends on the entire vector of utilities and sector c prices directly. On the other hand (49) demonstrates that the ratio of change in the compensated demand for all sector c commodities with respect to two different sector d prices is the same, and is independent of sector c prices. As leisure is usually untaxed, this means that the change in labour supplied with respect to any two taxed commodities is independent of the wage rate. These then are some of the implications of optimal proportional commodity taxation in the single consumer model.

Let us compare these results with cases I, II, and III which arise in the multi–consumer model. In Case I the optimiser is given by

$$\xi^* = \Xi^I(\overline{C}(Q^c),\overline{D}(u,Q^d)) \tag{50}$$

and

$$\frac{\dfrac{\partial X_i}{\partial P_m}}{\dfrac{\partial X_j}{\partial P_m}} = \frac{\overline{C}_i(Q^c)}{\overline{C}_j(Q^c)}, \; i,j \in I^c \text{ and } m \in I^d \tag{51}$$

with no change in (49). In addition to the restrictions in the single consumer case given by (48), this ratio is now independent of sector d prices, is independent of the utility vector completely, and is the same for all prices in sector d.

In Case II the optimiser is given by

$$\xi^* = \Xi^{II}(u, Q^c, \overline{D}(Q^d)) \tag{52}$$

and while the ratio in (48) remains basically the same, the one in (49) becomes

$$\frac{\dfrac{\partial X_i}{\partial P_m}}{\dfrac{\partial X_i}{\partial P_n}} = \frac{\overline{D}_m(Q^d)}{\overline{D}_n(Q^d)}, \; i \in I^c \text{ and } m,n \in I^d \tag{53}$$

and is now independent of the utility vector as well as the sector c prices.

In Case III the optimiser is given by

$$\xi^* = \Xi^{III}(Q^c, \overline{D}(Q^d)); \tag{54}$$

computing the ratios (48) and (49) yields

$$\frac{\dfrac{\partial X_i}{\partial P_m}}{\dfrac{\partial X_j}{\partial P_m}} = \frac{\Xi_i^{III}(Q^c,\overline{D}(Q^d))}{\Xi_j^{III}(Q^c,\overline{D}(Q^d))},\ i,j \in I^c \text{ and } m \in I^d \tag{55}$$

and

$$\frac{\dfrac{\partial X_i}{\partial P_m}}{\dfrac{\partial X_i}{\partial P_n}} = \frac{\overline{D}_m(Q^d)}{\overline{D}_n(Q^d)},\ i \in I^c \text{ and } m,n \in I^d \tag{56}$$

which are now both independent of the vector of utilities while the latter is also independent of sector c prices.

In the single consumer case it makes sense to wonder how changes in wealth might affect the ratio of changes in the demands for goods in sector c. In the multiconsumer case the answer is known; in cases I and III, there can be no effect. Similar arguments make the comparison in sector d in Cases II and III. Modelling the multiconsumer aspect of the problem explicitly demonstrates that extensive reliance on single consumer models may be quite misleading.

To conclude, we write out the two constraints in each of the four problems given above in the belief that it is perhaps more revealing to simply examine the structure of the optimal tax problem in the single consumer case and then in the three cases which are consistent with the existence of more than one consumer.

In the homothetic implicit separability case the constraints (29) and (30) can be written in a slight abuse of notation as

$$E(u,Q^c,D(u,Q^d)) = \omega^T Q^c + R \tag{57}$$

and

$$t^T E_D(u,Q^c,D(u,Q^d))\nabla_d D(u,Q^d) = R. \tag{58}$$

By way of contrast we write out these constraints for cases I, II, and III. In Case I they become

$$E(\overline{C}(Q^c),D(u,Q^d)) + \hat{C}(u,Q^c) = \omega^T Q^c + R \tag{59}$$

and

$$t^T E_D(\overline{C}(Q^c),D(u,Q^d))\nabla_d D(u,Q^d) = R. \tag{60}$$

In the second case the constraints are written as

$$E(u,Q^c,\breve{D}(Q^d)) = \omega^T Q^c + R \tag{61}$$

and

$$t^T E_D(u,Q^c,\breve{D}(Q^d))\nabla_d \breve{D}(Q^d) = R \tag{62}$$

whereas in Case III they can be written as

$$\Phi(u)E(Q^c,\breve{D}(Q^d)) + \hat{C}(u,Q^c) = \omega^T Q^c + R \tag{63}$$

and

$$t^T \Phi(u)E_D(Q^c,\breve{D}(Q^d))\nabla_d \breve{D}(Q^d) = R. \tag{64}$$

In each of these three cases it is clear that there is substantially more structure to the problem than in the model with only one consumer. This of course tightens the model in one sense, but in another makes it very clear that relying upon results which have been derived only for a single consumer model can be misleading. That is, one could in fact be asking questions about the empty set.

References

Atkinson, A B and J E Stiglitz (1972): "The Structure of Indirect Taxation and Economic Efficiency", *Journal of Public Economics*, 10, pp 157–173.

Besley, T and I Jewitt (1987): "Optimal Uniform Taxation and The Structure of Consumer Preferences", Discussion Paper 4 in *Measurement and Modelling*, Nuffield College, Oxford.

Blackorby, C, R Davidson and W Schworm (1987): "Use and Misuse of Single Consumer Results in a Multi–Consumer Economy: an Application to the Optimality of Piecemeal Second–Best Policy"; Discussion Paper 10 in *Measurement and Modelling*, Nuffield College, Oxford.

———, ——— & ——— (1988): "The Implications of Decomposing Systems of Market Demand Functions"; Discussion Paper 88-35, UBC, December.

——— & W Diewert (1979): "Expenditure Functions, Local Duality, and Second Order Approximations"; *Econometrica*, 47, pp 579–601.

———, C Knox Lovell & M Thursby (1976); "Extended Hicks Neutral Technical Change", *The Economic Journal*, Vol 86, no 343, December pp 845–852.

———, D Primont and R Russell (1978): *Duality, Separability, and Functional Structure: Theory and Economic Applications*; North-Holland, New York.

——— & R Russell (1976): "Functional Structure and the Allen Partial Elasticities of Substitution: An Application of Duality Theory"; *Review of Economic Studies*, Vol XLIII, no 134, June, pp 285–291.

——— & ——— (1978): "Indices and Subindices of the Cost of Living and the Standard of Living"; *International Economic Review*, Vol 12, no 1 February, pp 229–240.

Deaton, A S (1979): "The Distance Function and Consumer Behaviour with Applications to Index Numbers and Optimal Taxation"; *Review of Economic Studies*, 46, pp 391–405.

——— (1981): "Optimal Taxes and the Structure of Preferences"; *Econometrics*, 49, pp 1245–1260.

Gorman, W M (1976): "Tricks with Utility Functions", *Essays in Economic Analysis: Proceedings of the 1975 AUTE Conference*, eds M J Artis and A R Nobay, CUP.

Graaf, J de V (1957): *Theoretical Welfare Economics*; London, CUP.

Jewitt, I (1981): "Preference Structure and Piecemeal Second Best Policy"; *Journal of Public Economics*, 16, pp 215–231.

Mirrlees, J (1984): "Taxing Work (Correctly)?", unpublished, Nuffield College.

Sandmo, A (1974): "A Note on the Structure of Optimal Taxation", *American Economic Review*, Vol 70, pp 1–22.

Scitovsky, T (1942): "A Reconsideration of the Theory of Tariffs"; *Review of Economic Studies*, 9, pp 89–110.

Simmons, P (1974): "A Note on Conditions for the Optimality of Proportional Taxation"; unpublished, University of York.

Conference Discussion

Handout Many macro–models use additively–separable aggregate utility functions

$$u(X) = u^*(\Sigma_t \delta^t u(X^t))$$

where t indicates time and δ the discount rate. We analyse the restrictions that assumptions such as this place on individual preferences.

i **On Introduction**

TG May I just mention how absurd quasi–homotheticity is if you've got micro goods and people with different incomes.

CB It's not just quasi–homothetic, it's the same aggregator as well.

Handout Define Paretian utility functions

$$u(X) = \max_{\{X_1,\ldots,X_N\}}\{u^1(X_1)\,|\,u^h(X_h) \geq u_h,\ \Sigma X_h \leq X\}$$

$u_h = U^h(X_h)$: Individual utility functions, household h

$C^h(u_h,Q) = \min_X \{Q.X_h\,|\,U^h(X_h) \geq u_h\}$: Individual cost function.

$B_h(u_h) = \{X_h\,|\,U^h(X_h) \geq u_h\}$: "No–worse–than" set.

$B(u) = \Sigma B_h(u_h),\ u = (u_1,\ldots,u_H)$: Scitovsky set.

$C(u,Q) = \min_X\{Q.X\,|\,X \in B(u)\}$ Scitovsky expenditure function. (1)

$$C(u,Q) = \Sigma_h C^h(u_h,Q) \qquad (2)$$

Production Possibility Frontier: $P.X = 1$ (3)

The Second Best Problem (SB) $\min_{Q^1} \{P.\nabla_Q C(u,Q^1,Q^2)\}$ (4)

Piecemeal Policy when Optimal $Q^{*1} = K(Q^2,P,u).P^1$ (5)

where Q^{*1} solves (SB).

ii IJ It's Lipsey and Lancaster's model though, isn't it?

CB I don't think that's quite right, but it's not important. The way Boadway and Harris formulated the problem is as in the optimal tax literature; I guess if they hadn't formulated the problem in that way it wouldn't have been as easy for you.

IJ It would have been easier.

iii TG I don't know any economists under forty at the time who didn't know Lipsey and Lancaster and who didn't know everything they said that was true before they said it.

CB That was 1950.

AS You were under forty at the time.

CB That was the year I graduated from high school and the only second best problem I was worried about had to do with women.

Handout Pseudo–separability and the Jewitt solution to (SB)

Let

$$T(u,X) \geq 0 \leftrightarrow X \in B(u) \tag{6}$$

T is pseudo–separable w.r.t. $I = \{I^1, I^2\}$, $I = \{1,\ldots,N\}$ if

$$T(u,X) = \overline{T}(u,X^1,T^2(u,X^2)) \tag{7}$$

Let

$$C^1(u,Q^1,\chi) = \min \{Q^1.X^1 | T^1(u,X^1,\chi) \geq 0\} \tag{8}$$

and

$$C^2(u,Q^2,\chi) = \min \{Q^2.X^2 | T^2(u,X^2) \geq \chi\} \tag{9}$$

T can be written as in (7) i.f.f. C can be written as

$$C(u,Q) = \min_{\chi} \{C^1(u,Q^1,\chi) + C^2(u,Q^2,\chi)\} \tag{10}$$

iv Pseudo–separability

IS I don't think you said that quite right, it's that if there exists an implicit representation, not a particular one you've chosen.

CB Yes, quite right.

v TG It's easiest to think of C^2 as the cost of an intermediate good Y. A firm produces Y from X^2 and then uses Y to produce X^{-2}. In which case you can think of –C as the difference between receipts and costs. So it's as if the production has a single aggregator.

Handout Let χ^* solve (10); the envelope theorem yields

$$\nabla_{Q^1}C(u,Q) = \nabla_{Q^1}C^1(u,Q^1,\chi^*)$$

and

$$\nabla_{Q^2}C(u,Q) = \nabla_{Q^2}C^2(u,Q^2,\chi^*)$$

Sufficiency of (10); rewrite the second–best problem as

$$\min_{Q^1} \{P^1\nabla_{Q^1}C^1(u,Q^1,\chi^*) + P^2\nabla_{Q^2}C^2(u,Q^2,\chi^*)\}$$

Theorem 1 Given R, robustness in aggregation, and $H \geq 4$, piecemeal policy is optimal i.f.f.

$$C(u, Q) = \min_{\chi} \{C^1(u,Q^1,\chi) + C^2(u,Q^2,\chi)\} \tag{11}$$

and either

I: $$C^1(u,Q^1,\chi) = \overline{C}^1(Q^1)\chi + \hat{C}^1(u,Q^1) \tag{12}$$

or

II: $$C^2(u,Q^2,\chi) = \overline{C}^2(Q^2)\chi + \hat{C}^2(u,Q^2) \tag{13}$$

or

III $$C^i_{\chi}(u,Q^i,\chi) = \overline{C}^i_{\chi}(Q^i,\chi) \sum_h D^h(u_h,\chi) \quad i = 1,2. \tag{14}$$

vi Theorem 1

TG Like you I've been working on long–run aggregates. The first two are self–affine and other–affine. The third one looks like a non–affine <u>aggregate</u>. In this particular peculiar case you can so normalise the aggregate that everyone consumes the same amount of it. In which case the shadow prices the households assign to it add up to that for the economy as a whole; or you can normalise in such a way that the shadow price for each household is the same, but the quantities add up to that for the economy as a whole.

So this peculiar aggregate can be represented as a private or a public good.

Handout **Theorem 2** In Theorem 1 the households' expenditure functions are of the form

$$C^h(u_h,Q) = \min_{\chi_h} \{C^{h1}(u_h,Q^1,\chi_h) + C^{h2}(u_h,Q^2,\chi_h)\} \tag{15}$$

and either

I: $$C^1(u,Q^1,\chi_h) = \overline{C}^1(Q^1)\chi_h + \hat{C}^{h1}(u_h,Q^1) \tag{16}$$

$h = 1,\ldots, H$, or

II: $$C^2(u_h,Q^2,\chi_h) = \overline{C}^2(Q^2)\chi_h + \hat{C}^{h2}(u_h,Q^2) \tag{17}$$

$h = 1,\ldots, H$, or

III: $$\overline{C}^{h1}_{\chi}(u_h,Q^1,\chi_h) = \overline{C}^1_{\chi}(Q^1,\chi_h)d^h(u_h,\chi_h)$$

and

$$\overline{C}^{h2}_{\chi}(u_h,Q^2,\chi_h) = \overline{C}^2_{\chi}(Q^2,\chi_h)d^h(u_h,\chi_h)$$

Theorem 3

Suppose that I^2 is separable in B so that $\exists\ \overline{B}$ such that $X \in B(u) \longleftrightarrow (X^1,\chi^2(X^2)) \in \overline{B}(u)$ and that

R_1: $\dfrac{\partial^2 C^h(u_h,p)}{\partial u \partial p_i} \neq 0$ for all i at some (p,u_h).